M000169400

DARKER
STARS

NEW EVIDENCE:
THE SCOPE OF OUR GROWING SOLAR SYSTEM, PLANET X,
INVISIBLE PLANETOIDS, GAS GIANTS, COMETS, PLANET NINE, AND MORE...

Andy Lloyd

Original Artwork:
Andy Lloyd

Covers Design, Editing, Formatting, Image Optimization:
Bruce Stephen Holms

Darker Stars: New Evidence: The Scope of Our Growing Solar System, Planet X, Invisible Planetoids, Gas Giants, Comets, Planet Nine, and More...

© 2018 by Timeless Voyager Press

ISBN 9781892264558

TIMELESS VOYAGER PRESS
PO Box 6678
Santa Barbara, CA 93160
TimelessVoyager.com

Editing: Bruce Stephen Holms

Table of Contents

Acknowledgments

I would like to take this opportunity to thank the following people for their help in recent years: Luis Acedo, Bob Avery, Shad Bolling, Alan Cornette, Carlos de la Fuente Marcos, Mark Keller, Eric and Jolene Harrington, Matthew Holman, Mattia Galiazzo, Jim Gilfillan, Alan Jackson, Wayne James, Dave Jewitt, Lorenzo Iorio, John Keebaugh, Mark Keller, Sasha-Alex and Janet Lessin, Renu Malhotra, Karen Meech, Richard Munisamy, Je Nolan, Jesse Socher, Barry Warmkessel.

I am grateful to my publisher, Bruce Stephen Holms, for his longstanding support, and to my editing team Lee Covino and Monika Myers for their many contributions and suggestions.

The writing of this book would not have been possible without the patience and steadfast support of my wonderful wife and sons. I love you!

This book is dedicated to the memory of Phil Whitley, an eclectic researcher and author who is sorely missed.

Timeless Voyager Press

Introduction

This book follows **Dark Star**, my previous non-fiction title, published 13 years ago. **Dark Star** presented a hypothesis about the existence of a sub-brown dwarf located far out in the depths of the outer solar system. It looked at anomalies, and it discerned patterns. It brought together contemporary scientific discoveries and more speculative notions of the epistemological power of ancient myth.

Although there is a fine tradition supporting this form of discourse, it runs the risk of alienating many scientists. Less predictably, it can also irritate many of those exploring the postmodern domain of alternative knowledge. Some readers liked it (you know who you are), perhaps because of its earnest desire to bridge this divide. It is always going to be difficult to create the right balance, but I believe that it is worth the effort to do so. Science progresses through both fact based inquiry, and also ideas. Sometimes, those ideas are necessarily challenging.

While exploring some of the same themes, this new book leans more towards the science. This is largely because of the emerging evidence which has often (but not always) supported the ideas presented in **Dark Star**. However, this is not an updated version of **Dark Star**. This is a completely new book. It contains many new ideas, as well as the most up-to-date science which has emerged over the intervening years.

I have taken full advantage of the greatly enhanced access to the scientific literature now available through the Internet. It feels to me like an ancient library has been unearthed. I have been able to 'translate' these modern scientific texts, as it were, and incorporate the remarkable information gleaned from them into this book. A skeptic may accuse me of cherry-picking the bits that work for me, and ignoring the bits that do not. Conscious of this potential bias - which is also common among a great many scientists - I have tried to provide counterpoint within this book.

Hopefully, this will help you to consider the merits of conflicting solutions. Without conducting a thorough meta-study of my own research methodology, I would admit a preference towards work that challenges preconceived notions. I like the papers that highlight uncomfortable new bits of evidence. So, I sometimes challenge myself and my own ideas, too.

Timeless Voyager Press

There is a question that I suspect scientifically-trained readers will find themselves asking as they read through this book. Why don't I just write this up in a paper, or series of papers, and get published in the literature? My answer is – I have tried, but so far I have failed. I will continue to try. I will be honest when I say that the responses I have received have sometimes been less than courteous. I realized pretty quickly that trying to go it alone in the scientific literature is fraught with difficulties.

In one of the more constructive bits of feedback, it was recommended that I seek out a published academic who might be willing to help me 'break into' the literature. I tried to interest academics (mostly astrophysicists) in aspects of this work which directly touched upon their areas of study. Again, this has proven difficult. If I asked about an aspect of their work, I might receive a reply. If I tried to interest them in a new idea that needed testing, I heard nothing back. It seems that no one wants to test this material out, even if just to rule it out. It may be because this book contains many broad-brush concepts, often questioning generally accepted assumptions. These significant challenges may not be easy to prove in a simple experiment, or computer simulation. They often involve a shift in perspective.

Without someone within the scientific establishment willing to take a risk, it will prove very difficult for the ideas in this book to become published within the scientific literature. There is a "chicken and egg" argument at work. Researchers need to establish a track record in publishing to get published as sole writers, and the gateway to that is through undertaking research degrees to become part of the establishment. Outsiders face a significant obstacle here. The guardians who stand sentinel at the gateways to academic publishing turn back contributors they do not recognize. Academic progress is a process of certified initiation.

Actually, I am no longer an outsider. Now that I work as a university lecturer, I am by definition an academic. I write papers and chapters for textbooks, conduct research and undertake post-graduate study. As a result of this work, I have refined my writing style into a more academic format, and I make use of more 'grown-up' conventions (like Harvard referencing). However, my work in Health and Social Care counts for little among astrophysicists, as the gateway to publication is subject-specific. Nonetheless, I am using the language and temperament of an academic within these pages. The credibility issue extends beyond that, however. To even offer ideas and opinions within the blogosphere invites criticism from some quarters. Some years ago, I received an email asking who I was. The correspondent was respectfully asking why I considered myself an authority on the subject I was writing about. He questioned why he should pay the slightest attention to what I had to say. I answered as best as I could, and then heard no more from my fly-by-night inquisitor.

He had clearly read enough to discern that I had some kind of idea about what I was talking about. I suspect he may have become unsettled by some of what I had to say. Perhaps it challenged some long-held assumptions. Perhaps it provided some evidence for something he had hitherto dismissed. Perhaps he just found it irritating.

Whatever the case may be, he took the time to write to ask me a question which essentially boiled down to: "Are you credible academically?"

Knowledge is an interesting concept. It can have a relationship with Truth, sometimes not. More often, it provides the best available understanding of a particular subject. Our system of education initially develops a selection of broad introductions to knowledge. Different subject areas favor different approaches to understanding. Once we have discovered what approaches to knowledge work for us as individual learners, we make choices about what we want to study further. We may simply want to learn about a certain subject, or we may study to help us in our career. A personal aptitude for a particular discipline is important, as is motivation to study.

For some reason, I have developed a strong interest in developing my knowledge in several, seemingly unrelated subjects. I don't really know why this is. Perhaps I am just naturally curious. I advanced my studies in natural science to a high level academically. I then worked in healthcare for decades, and now I teach social science. I also exercise creativity through art and music. Throughout all of this, I independently explored two further areas of study: Alternative epistemology, and astronomy. I did not undertake qualifications in these subjects, but I became well versed in them. Then I started writing books. In so doing, I drew upon a mixture of these influences and modes of thought.

Generally, people become experts in a subject by specializing. As we progress, our higher educational system requires us to focus on an increasingly narrow range of knowledge. We are required to concentrate our minds on areas of knowledge we need to master, in order to gain the qualifications we need to get on in life. While this is very helpful in many ways, it has a downside. The narrow focus, so important to post-graduate achievement, can close our eyes to other perspectives and, more importantly, to the overlapping areas of knowledge which our chosen discipline rubs up against.

Academics tend to share a common mind-set, but can easily fall into a "silo mentality". They may exclude other approaches through an unspoken pooling of shared values. Specialisms become exclusive, and their practitioners train the next generation to think like they do. This puts up barriers to those who have different perspectives. Until recently, this sat well with the prevalent modernist approach to learning. Universities were repositories of expensive, difficult to access journals containing the specialized knowledge needed to master an academic discipline. More recently, however, those old certainties are no longer quite so clear-cut. Many universities have ditched books and paper journals for Online sources of information, access to which is secured through pay-walls. There is a multiplicity of resources emerging from a marketization of information. Increasingly, these Online resources are becoming unfordable to many smaller academic institutions. Some educational institutions have to choose what their students can have access to, and what they can't.

Timeless Voyager Press

At the same time, there is a blurring of knowledge Online. It is becoming more difficult to discern what constitutes a reliable source of information. At the same time, expert opinion is questioned by a skeptical media, and by politicians anxious to justify their own agendas. Traditional sources of knowledge and expertise have been thrust into this new Wild West of information, and no one seems quite sure what is what anymore.

There are clear advantages to be gained from our Information Age. We all have information at our fingertips. Scientific data and information spills out across the Internet's floor. Academics are rightfully proud of their research work. Their scientific ethic predisposes them to share with their peers. They may even personally wish to counter the increasingly stagnating effect of information marketization. Naturally enough, scientists want their work to be read, and often do not appreciate having their work hidden away behind electronic barriers. There is an unacknowledged irony; it can sometimes be easier to access a discipline's body of knowledge through what is openly available on the Internet than it is through academic library searches. This democratization of knowledge is useful to an independent researcher. The bottom line is that you can find stuff out there, if you know where to look and whom to ask. You can engage in lifelong learning to the "nth degree", and immerse yourself in an academic topic without ever having to visit a university library.

Some of our most prolific scientists and academics are self-taught. A few who come to mind are: Elon Musk (SpaceX rocket scientist), Michael Faraday (the Faraday Cage), Gregor Mendel (theory of genetics) and George Smith (decoded Sumerian clay tablets). A different mind-set can be what is needed to make a breakthrough. Improved access to privileged information opens up new discourses. In this book, it is my mission to bring together the many strands of research that I have pursued over the last 20 years or so. I have personally explored what seems an immense amount of source material in the hope of basing this work upon a firm foundation of facts and evidence. The reference section of this book contains many links to what is readily available. This will provide you with many avenues to pursue in detail should you wish to explore any of topic further.

There is a quite separate advantage to presenting this material in a book format. Material from different disciplines of knowledge inter-meshes in these pages. Researchers sometimes get tunnel-vision, and don't always see how things fit together. Their very specialist expertise can act as a barrier to seeing the bigger picture. Also, they unconsciously work with a set of fairly immutable assumptions that result from their discipline's collective mind-set.

I am a chemist by training. Chemistry is often more interested in the exceptions than the rules. Physicists like rules. So, physicists have a tendency to iron out the exceptions in favor of their laws. This is a little like the difference between constituted law, and common law. The exceptions, the anomalies, can be informative. They provide precedents which can shape future judgments. We should be prepared to consider what anomalies teach us. We should be prepared to see new patterns emerge from the anomalies if they start to get out of hand. In short, we should celebrate the anomalies!

So, to return to the question posed by my inquisitor: I am a researcher who loves anomalies. I can appreciate why a reductionist approach to knowledge works most of the time, but not always. In fact, I love the way the Universe seems to defy our desire to reduce it, to fit it into a box. I am open to the possibility that something really weird is going on out there, and that weirdness is part of the Universe's inherent charm. That is not to say there are not important physical laws. But in the same way that biology builds upon chemistry, the Universe likes to make use of its fundamental laws in increasingly complex ways.

To do justice to the complexity of this subject, the material I present in this book necessarily needs to be robust. I have tried my best to reflect the sense of the scientific work I have drawn upon, not just the words used. All academic disciplines encode their knowledge within increasingly complex forms of language. Jargon is often unnecessary, e.g. that used within many business environments. In academia, however, new knowledge necessarily calls for new classifications and a new vocabulary. That can facilitate technical clarity for those who have learned the language of the discipline, but it can also create an almighty barrier for anyone who has not.

Scientists often 'translate' their work into plain English when putting together a press release, or giving a media interview. Through this they demonstrate the ability to provide a more accessible synopsis of their work. In this book, I sometimes cite press releases for this very reason, alongside the source material in the scientific literature. But that desire to be understood by the public need not extend to academics' scientific writing. I have tried to make that technical material more understandable, pitching my writing at a level which hopefully strikes an appropriate balance. If I have failed to capture a nuanced aspect of someone's academic writing, I apologize. It is my earnest desire to demystify scientific research, and I do so in good faith.

It is fortunate that the diversity and complexity of this investigation lends itself very well to a book format. I have tried to structure the book into a sensible arrangement of chapters. This has not been easy, however, because many of these diverse strands are interconnected. They defy a straightforward linear ordering. I think of the contents of this book as a gestalt: There is an underlying pattern of thought that hopefully emerges from the whole.

The first chapter of this book sets the scene and introduces the various categories of objects in the solar system. It also essentially provides a glossary of terms which will prove helpful later in the book. The second chapter discusses the history of Planet X. From this foundation, the complexity of the science about Planet X grows steadily. The first half of the book deals with the latest evidence about Planet X, for and against. It also sets out my own ideas about what the underlying problems are, and offers some novel solutions in an attempt to address those seemingly intractable problems.

Some of the chapters about Planet Nine may prove a challenging read in places. I considered shifting some of the more scientific material into an appendix. However, this would run the risk of losing important threads which later weave into arguments further into the book. Please bear with me if you find some material heavy-going. It does get better!

The second half of the book broadens out to look at how the existence of a rogue planet affects different parts of the solar system. I have devoted each of these chapters to readily compartmentalized topics: The Moon, Mars, Pluto, and so on. Some of the areas covered in **Dark Star**, which remain more or less current, have been afforded less space in this new title. In these cases, I have referred back to it within the text. As noted, this book is not an update, or even a re-write, but instead a completely new offering which complements, rather than replaces, the former book.

I hope you will find this book stimulating. I truly believe that our outer solar system holds within it a great mystery. It is my great pleasure to share my investigation into this mystery with you.

The Incomplete Solar System

When it comes to the solar system, the conceptual framework that most people rely upon can be summarized by a single diagram from a high school textbook. A slither of the Sun's surface adorns the left flank of the diagram. The eight planets are dutifully arranged in order left to right: Mercury, Venus, Earth, Mars, Jupiter, Saturn, Uranus, and Neptune. They are presented according to size. First in line are the four small terrestrial planets of the inner solar system. Then we meet the four much larger gas or ice giants of the outer solar system. Between these two sets of four is a sketchy band known as the asteroid belt.

Once upon a time, little Pluto made an appearance at the end of the group of eight. But in response to the emergence of a whole new class of Pluto-like worlds out there, Pluto has been re-classified as a dwarf planet. Its place at the back of the planetary queue has been replaced by a second asteroid belt, known as the Kuiper belt (or sometimes, the Edgeworth-Kuiper belt). Pluto is now the most famous representative of this belt, and is simply one of many Kuiper Belt Objects (or KBOs) which expand out from Neptune in an extensive disk.

For the most part, these objects (planets, dwarf planets, asteroids) revolve around the Sun on the same level. This 'floor' of the solar system is known as the plane of the ecliptic. This orderly series of circuits reflect the early beginnings of the solar system, when the Sun emerged from a condensing disk of rotating material, known as the pre-solar nebula. So, in that sense, the linearity of the planets from our high school textbook is correct.

What the textbook fails to describe, however, is the sheer scale of the space occupied by these planets. How could it? If the page were drawn to scale, we would likely not be able to even see the dots representing the planets upon it. The Sun would be a small solitary disk printed onto the left-hand side of the page, with the microscopic dots orbiting it largely left to our imagination. Furthermore, the steady, linear ordering of these familiar worlds on the page belies the exponentially expanding space between them.

So, for clarity, it is necessary to misrepresent the scene.

Fig. 1-1 The Solar System, Dwarf Planets, Asteroids, The Sun

Then, there's the size of the numbers involved. Objects in the solar system are many millions of miles apart from one another. For simplicity's sake, astronomers helpfully provide us with a unit of measurement which I shall adopt throughout this book. The distance between the Earth and the Sun is about 93 million miles (~ 150 million km), and is presented by one unit; one "Astronomical Unit" (AU). An AU is not to be confused with Australia! The Astronomical Unit is a very useful unit of measurement indeed, especially as we start to explore the outer solar system and beyond.

The Structure of the Solar System

The four terrestrial planets of the inner solar system are ordered relatively neatly: Mercury whizzes around the Sun at a distance of just two fifths of an Astronomical Unit. Venus, in many ways our sister world, is located three quarters of the distance of that of the Earth. This relative proximity when we are on the same side of the Sun is the reason why it Venus is often seen as such a bright star, either in the evening or in the morning. Then, beyond the Earth, Mars is half as far again from the Sun, at ~1.5 AU.

The asteroid belt covers a broad swath of space between roughly two and three AU. Then things start to stretch out. Jupiter is 5AU away from the Sun. Despite this distance, it often appears almost as bright as Venus. This is because of Jupiter's immense size, which helps to compensate for the much greater distance by reflecting back much more of the Sun's light towards us.

Glorious Saturn, with its beautiful rings, lies 9.5AU away, about double that of Jupiter. Then we double the distance again to get to Uranus, at about 19AU. You have to cover roughly the same distance again to get to distant Neptune's orbit, at about 30AU.

When grappling with the outer solar system, I find it useful to center the numbers on Saturn. It is roughly ten times the distance of the Earth from the Sun. This is one order of magnitude; 10AU to 1AU. From here, it is simple. Jupiter's orbit is about half that of Saturn's. The orbit of Uranus is roughly double that of Saturn. Neptune's is roughly triple that of Saturn.

Distributed amongst these four outer planets are found additional minor bodies known as "Centaurs". Their orbits cross those of the major outer planets, and so are dynamically unstable. They are generally asteroid-like, but sometimes behave like comets if they get close enough to the Sun. Similar cometary bodies, which generally have orbits within 10AU and lie on the ecliptic plane, are known as short-period comets.

Another category of minor bodies in this zone are those that accompany a shepherding planet along its orbit. They are known as "Trojans". They either precede the planet by 60°, or follow the planet by 60°. Trojans sit in stable positions in the orbital path of the planet known as LaGrangian points. Jupiter has literally thousands of Trojans (with the group preceding Jupiter being sub-classified as "Greeks"). Distant Neptune has far fewer, but makes up for it with its influence over the Trans-Neptunian Objects in the Kuiper belt beyond.

The Kuiper Belt

This is where the conventional view of the solar system classically ends. Neptune rubs up against the right-hand flank of the page, with perhaps a quick mention of diminutive Pluto, whose irregular orbit usually extends beyond that of Neptune's. In this conventional view, the solar system consists of these eight planets, neatly assembled, and neatly arranged like ducks in a row. But, in reality, this is just skimming the surface of the Sun's true domain. It describes a tiny core of real estate beyond which lies an almost completely unexplored realm of immense proportions.

First out is the afore-mentioned Kuiper belt, theorized separately by two astronomers in the middle of the Twentieth Century (Edgeworth 1943; Kuiper 1951). The objects in this belt generally extend out from Neptune's territory at 30AU, to about 50AU. You will have noted, then, that the Kuiper belt's depth is about twenty times that of its inner cousin, the asteroid belt. Given how much further it is away, then it is clear that the Kuiper belt covers a huge disk-shaped area across the plane of the ecliptic.

The Kuiper Belt Objects are thinly spread, and incredibly difficult to detect. They are often referred to as Trans-Neptunian Objects (TNOs) by astronomers; a term which encapsulates not just their position in the solar system, but also the relationship they often share with the outer ice giants. I will be referring to these objects a lot during the course of this book and will be using both terms (KBOs and TNOs) interchangeably. In many ways, the planet Neptune is their Shepherd of the Deep. It orders them into stable, "resonant" orbits. This means that their orbital periods are related to Neptune by a ratio of small integers. For example, Pluto is in a 2:3 resonance with Neptune, meaning that for every two orbits Pluto makes, Neptune makes three.

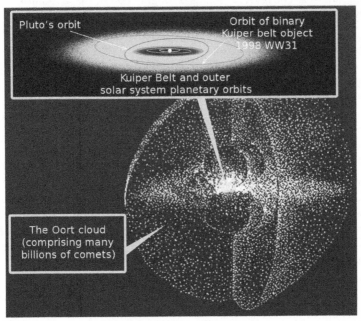

Fig. 1-2 The Kuiper Belt Objects

Bodies in the solar system quite often establish these settled, stable relationships between themselves over the course of time. This is because as these objects dance around the Sun, they exert regular patterns of gravitational influence over one other. Resonance patterns provide stability by preventing the bodies from getting too close to one another. So, when Pluto crosses Neptune's path during the perihelion passage of its elliptical orbit (the closest position to the Sun), this "mean-motion orbital resonance" with the ice giant ensures that it will never get too close to Neptune. If it did, the larger planet would "perturb" Pluto's orbit, nudging it somewhere else.

Although many TNOs have these kinds of resonance relationships with Neptune, some do not. Their orbits remain stable because their perihelion distance is sufficiently far away from Neptune's orbit to avoid the gravitational influence of the ice giant, even if they happen to line up. Some of these are scattered out beyond the main Kuiper belt. These "Scattered Disk Objects" (SDOs) hold great significance for the discussions which will follow later on. They are often referred to as "Extreme Trans-Neptunian Objects" (ETNOs).

If we tried to incorporate the Kuiper belt into our school textbook, we would have to create a pull-out page to fit it in.

The Heliosphere

The next major feature of the solar system is the "heliopause", which is the outer edge of the Sun's "heliosphere". If you imagine the planets fitting within a page of a school textbook, then this boundary represents the edge of a large table that the book is sat upon. Up until now, most of the objects we have discussed move around the Sun in the same plane: The "ecliptic". The "heliopause" is a boundary condition which resembles the surface of an outer bubble. The zone within that bubble is known as the "heliosphere". In fact, it is not actually a sphere. Instead, it is shaped like a teardrop - or possibly even croissant-shaped, according to astrophysicists from Boston University (A & G 2015; Opher et al. 2015).

The heliosphere incorporates the domain of the solar wind. It is intrinsically linked to the power of the Sun, and its relationship to the surrounding galactic environment. It marks the point where the pressure created by plasma driven out by the solar wind meets an equal and opposite pressure exerted by the interstellar medium beyond.

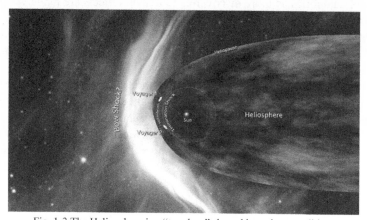

Fig. 1-3 The Heliosphere is a "teardrop" shaped boundary condition.

Because these internal and external pressures can vary, the "heliosphere's" position is not fixed. Just to complicate matters further, there are a couple of additional zones within the "heliopause" where conditions vary slightly. These are known as the "termination shock", where particles emitted from the Sun decelerate below the speed of sound, and the "heliosheath" beyond.

Because of the irregular shape of the "heliopause", its distance from the Sun is not uniform, but an accepted figure of 122AU marks the point where the Voyager 1 spacecraft passed through this boundary in 2012, and went on to enter interstellar space.

If we tried to fit the heliosphere into our school textbook, the pull-out page would now have to be several pages in width. The solar system is already starting to feel like a hugely extended environment within which the planetary zone resembles just the core. This distance is now two orders of magnitude greater than our starting position on Earth. Yet, even though we have officially entered interstellar space beyond the "heliopause", we are far from finished. This is because the Sun's gravitational influence extends well beyond the heliosphere.

The Oort Cloud

There appears to be something of a gap beyond the heliosphere. We have to travel some distance to reach the next feature: The "inner Oort cloud" (sometimes known as the "Hills cloud"). This truly colossal disk of objects begins about 2000AU away. It contains a reservoir of icy objects. These sometimes become dislodged and fall down towards the Sun, then becoming long-period comets.

The torus-shaped repository of comets that makes up the inner Oort cloud may have a relationship with the "scattered disk" of Extreme Trans-Neptunian Objects located beyond the Kuiper belt (Fernández 2004). So, there may turn out to be some muddying of these categories as more discoveries of distant objects are made. As you can see, we are now very much beyond the comfort zone of our school textbook, both in terms of distance and terminology. We are now over three orders of magnitude away from our starting position, and just beginning to enter the realm of the comet clouds.

Or, at least, that is what is generally thought. It should be noted that the Oort cloud is first and foremost a theoretical construct. It is named after its proponent, the Dutch astronomer Jan Oort. We know that long-period comets emerge from this immense area because their trajectories can be tracked backwards to it. But what is less clear is what this area represents or, perhaps more importantly, where the objects within it originally came from.

The main bulk of the Oort cloud - should it exist - is the mind-bogglingly vast outer Oort cloud. It is an incomprehensibly immense spherical shell of distant comets extending from 20,000AU to 50,000AU - and then possibly upwards to anywhere between 100,000AU and 200,000AU. We are now four or five orders of magnitude out there, and our poor old high school has had to pay for a thousand or more text books to help us simulate the limits of the 'real' solar system!

At these distances, the numbers are increasingly burdensome, and it becomes necessary to shift to a different scale of measurement: The light year.

Estimates of the extent of the Sun's gravitational influence vary, but for the sake of convenience some like to alight upon the simple figure of one light year as the outer gravitational boundary of the solar system. This marks a generally recognized point where the Sun's influence over the comets in the outer Oort cloud is balanced out by outside influences, like the galactic tide, and that of passing stars.

One light year is equivalent to 63,241AU, and many astronomers consider 50,000AU as a realistic estimate for outer boundary for the Oort cloud. Others argue for 200,000AU, which is over 3 light years distance. This is well on the way towards the nearest star: Our nearest stellar neighbor, the red dwarf Proxima Centauri, is 4.25 light years away.

The difference in opinion about these distances tells you a lot about how little is understood about this region, and the trillions of objects which are thought to occupy it. They simultaneously occupy interstellar space, whilst feeling the shepherding tug of the Sun. As stars pass each other, like ships in the galactic night, these shoals of comets get pulled around. In this way, many of the Sun's outer comets could have originated from clouds of comets around other passing stars. Alternatively, many more may have been captured from interstellar space as free-floating, independent minor bodies. The sizes of these captured objects may vary significantly. We just don't know.

It should be pretty clear by now that our knowledge of the solar system is not really as well tuned as we might think. There is a vast realm of space out there in the Sun's cosmic backyard that we know next to nothing about. Yet, you could be forgiven for thinking that astronomers have got it licked, well and truly. At least, that is the impression they give when discussing possible planets beyond the confines of the generally recognized outer solar system.

If we tried to provide the proper sense of the limits of the comet clouds in the outer solar system, then our school textbook pull-out would now extend out by about 1600 page widths. When we start to examine this 30m long pull-out to see what we actually know about the contents of the outer solar system, we find that the vast majority of it is simply blank. **Here be Dragons!**

The Problem with Reflected Light

There is another problem with trying to find distant objects in the outer solar system. Cosmic objects may be self-luminous, like stars. The light we see from stars is emitted by them directly. However, the light from other objects, like planets, is the reflected light from their sun. In the case of the objects in the solar system, the reflected light we are talking about is, of course, the Sun's. So, before we can see the light from any given object in the solar system, be it planet, moon, asteroid or comet, the light must first have traveled from the Sun to the object, and then been reflected back towards us.

The brightness of the reflected light we observe is a function of the distance from the source of light. This follows an inverse square relationship. The greater the distance, the more rapidly the luminosity of the object fades. The brightness of an object falls away to the second power as you steadily move away from the light source. So, light intensity changes by a factor of one divided by the square of the distance between the light source and the observer. The intensity of the light we see diminishes very quickly with distance.

A consequence of this is that when you double the distance, light intensity decreases to a quarter of the original value. Let us imagine that an astronaut who is standing on one of Jupiter's moons pointed an incredibly strong flashlight towards the Earth. An astronomer on Earth might note the brightness of that light source, or its light flux, in her notebook as a certain value. Then, let us say our intrepid astronaut heads off to one of Saturn's moons and does the same again. Saturn's moon is nearly twice the distance away, but the astronomer on Earth will register the light flux she observes in her telescope as only a quarter of the earlier value.

This relationship actually provides astronomers with a very good method of working out the distance that a luminous object is from us. If she knows how powerful the light a certain class of star emits is, relatively speaking, then the intensity of the light an astronomer observes will allow her to calculate the star's distance from us, using this inverse square law.

That is the mathematics regarding the light coming to us directly from the Sun, or from other self-luminous objects, stars. However, viewing conditions are significantly worse when you consider the light being reflected back at us from all the other 'dark' objects in the solar system. Unlike our astronaut's flashlight, solar system objects are not self-luminous. Why is this worse in the case of reflected light? Because the relationship of rapidly diminishing brightness has to be applied twice: First, when the light from the Sun travels to the object and, second, as the light travels back towards us. The culmination of essentially doubling the overall distance the light is traveling is that reflected light diminishes by the power of four with distance from object to observer.

So, if our astronaut were to do the same experiment again, but this time with a mirror rather than a flashlight, the light reflected back from the mirror held up on one of Saturn's moons is now one-sixteenth the value reflected back from the same mirror held up on one of Jupiter's moons. As a general principle, if you move any planet out to a position twice as far from the Sun, then it would be one-sixteenth as bright because the sunlight not only has to get all the way out there – but also has to come back. The inverse square rule is applied twice, factoring the diminishing of reflected light with distance to the power of four.

This has a fundamental effect upon the detectability, or otherwise, of objects in the outer solar system. Actually, things are made worse still by reflectivity. In many cases, very little of the light landing upon solar system objects is reflected, or radiated, back towards us. Solar system bodies are not like mirrors, reflecting back perfect parcels of light. Instead, they absorb much of the light as heat. This adds color to the fraction of light reflected back at us. Furthermore, being largely spherical, solar system objects do not present us with a uniform surface, like a mirror does. Reflected light and radiated heat spill out into space, meaning that only a fraction of the total detectable light and heat actually head in our direction.

Some objects absorb more of the sunlight than others. This reflective quality is known as the object's "albedo". If a solar system object is mostly covered with ice, then its albedo is going to be high, and the fraction of the Sun's light reflected back is high. But if the object is covered in dark, rocky surfaces, organic materials or, in the case of the Earth, oceans of water, then its albedo will be low, and it will absorb most of the sunlight, becoming warmer as it does.

The object will then radiate some heat back out into space. As a result, astronomers often have more chance of discovering these distant, dark objects in infrared light than they do in visible light.

Finding Planets

These issues create tremendous problems for astronomers seeking out new bodies in the outer solar system. These mathematical relationships also explain why it took so long so discover Uranus (1781) and Neptune (1846), despite the fact that these planets are both massive ice giants. (As an aside, Neptune was actually spotted before 1846, by none other than Galileo Galilei, in 1613 (Browne 1993). Having seen Neptune through his small telescope, Galileo dutifully recorded its position. Unfortunately, he did not realize that it was a planet).

Uranus and Neptune are both contained within the main page of our school textbook. As you move further out, scanning the extensive pull-out, the viewing situation deteriorates rapidly. The light we hope to observe, which has been reflected back towards us from outer solar system objects, becomes negligible. We require larger and larger telescopes, and more and more luck.

This is why we can readily see the light from distant stars - and even extremely distant galaxies - when we look up at the night sky, but conversely why astronomers need huge observatories full of computerized kit to even hope to observe incredibly faint outer solar system objects located at a tiny fraction of the distance. This creates a false conception of how well we know the universe. As far as distant, self-luminous objects go, we are doing great. However, that confidence about our knowledge of stars and galaxies can distort the very different approach we need when appraising our knowledge of dark objects.

Pluto was discovered in 1930 by American astronomer Clyde Tombaugh, who was actually hunting for Planet X. The name was suggested by a British school girl, and was given the nod by Tombaugh because the first two letters were the initials of Percival Lowell, the founder of the Lowell Observatory, where the decades-long Planet X hunt had been conducted. Although Pluto was clearly not the substantial Planet X body the late Percival Lowell had initially hoped to find, it was granted planet status nonetheless. (I have previously discussed this quirky aspect of Planet X history in my first book "Dark Star: The Planet X Evidence" (Lloyd 2005)).

The 'first' Kuiper Belt Object, 1992 QB1, was discovered much, much later – as recently as 1992 (Jewitt & Luu 1993). It is a testament to the sheer magnitude of the task that over sixty years had passed by between the discovery of the dwarf planet, Pluto, and 1992 QB1. Since then, many more KBOs have been discovered, gradually filling out what had been up until then only a theoretical second disk of asteroids.

Our Cosmic Backyard

If you are hoping to find a new planet out there in the outer solar system, then its distance from the Sun - as well as its albedo - will have a tremendous bearing upon how difficult your task will be. The fact is, Planet X has still not been discovered, despite well over a century of searching (although, admittedly, the appetite among astronomers for seeking out such an object has ebbed and followed considerably during that time). But, that does not mean that it is not there. There are significant barriers to discovery involved.

Let us compare the solar system to a farmhouse set in rolling countryside, on a warm autumnal night. It is pitch black outside. The lights are on in the house, with the curtains open, and some of the light spills out into the garden. You and I are sat on the patio, enjoying a few beers with our friends.

Fig. 1-4 The Solar System with Planet X lurking somewhere in the darkness.

We can see each other clearly enough in the light from the windows. Just for fun, we pretend we are the planets in the solar system (you're right, it hasn't been the liveliest party ever, I'll admit). You are the Earth, and I am Mars; so I move my chair slightly further away from the house. You can see my face clearly reflected in the light from the farmhouse window.

Your partner pretends to be Venus, and a friend of theirs wants to be Mercury. They shift their chairs towards the patio door. We now see them half-silhouetted by the light from the window. At the edge of the patio, another friend is tending to what is left of the barbecue. He is Jupiter – which is just right for the alpha-male taking charge of the burgers. Let us say that the waft of smoke drifting across the patio is the asteroid belt. The patio area, then, is the inner solar system, illuminated reasonably well by the Sun.

Now, a few of our friends have decided to go and have a "vape" on the lawn beyond the patio. As they progress away from the farmhouse, they start to disappear into the gloom of the garden. Our friend pretending to be Saturn is twice as far into the garden as the barbecue king, and is still reasonably visible. The unfortunate bloke who has drawn the short straw with the next planet, is twice the distance again. Then, in the darkness at the end of the garden, the friend who is pretending to be Neptune accidentally steps into the fishpond. That's the god of the sea and freshwater for you!

Anyway, the last two of our friends are pretty much invisible in the darkness - although the swearing is apparent enough. When they "vape", though, the tiny light emerging from the e-cigarette is more clearly visible than they are. The e-cigarette is self-luminous, albeit faint. Most of the light from the house which has reached our friends is absorbed as heat by them and their clothes, depending upon their overall reflectivity, or albedo. So, only a small amount of light bounces back towards us.

Beyond the garden are the farm's fields. They extend out in all directions, further into the darkness of the night. If one of our friends were to climb over the hedge (the Kuiper belt) at the bottom of the garden and wander off, we would quickly have no idea at all where they were. Yet, they are still in the 'solar system'. It might take them all night to walk to the light of the nearest city (representing Proxima Centauri), especially if they were drunk.

This book is about the hunt for Planet X, and possible associated anomalies within the solar system. This 'missing friend' is very likely to be out in the cosmic fields somewhere. Why do I think that? Although we have not yet seen the culprit, we have heard something or other moving around out there. Unless it was our imagination (and this is a moot point), we have growing evidence that there is a stranger in the cosmic fields…worrying the celestial sheep.

He may be a friend; he may be a foe. Really, none of us can be sure. What we all agree on is that trying to see him moving around out there in the darkness is proving very tricky indeed!

    Timeless Voyager Press

The History of Planet X

There is one object missing from our solar system inventory that is centrally important to this book. That is "Planet X". It sounds like a title of a 1950s 'B'-movie (which is actually true). But what is it really? For the purposes of this book, I will define the term to mean an undiscovered planet orbiting the Sun which is at least as big as Mars. Often, I will imply that such a body is considerably larger than this. There may be many planet-sized terrestrial bodies out there, awaiting discovery. The real prize is an object of similar proportions to the outer known planets.

The astronomical community has had a rather on and off relationship with the concept of Planet X. For the most part, it's definitely "off". The subject has become toxic for a number of reasons, many of which will be fully explored in the pages of this book. But it is not just the speculative fringe associated with the name Planet X that puts professional astronomers on the defensive (and sometimes on the offensive).

Peer-pressure plays a part in Science. Some of the professional astronomical community's biggest hitters are (or were) vocal Planet X skeptics, like E. Myles Standish Jr. of the Jet Propulsion Laboratory (JPL) in Pasadena (Standish 1993), or the late Brian G. Marsden of the Smithsonian Astrophysical Observatory (Browne 1993). Skepticism is an important part of the scientific method, but as with any academic discipline, knowledge can become embedded and immutable. There needs to be a healthy balance between the thorough picking apart of new ideas and data, and openness to them.

The idea of discovering a new planet can seem fanciful to some, and become an obsession for others. For many years, Planet X was deemed to be a less than serious subject for proper scientific research. Young astronomers seeking this astronomical Holy Grail in the outer solar system were quietly discouraged from such a foolish venture. I am in contact with interested astrophysicists who are still very, very careful who they share their ideas on this subject

     Timeless Voyager Press

with, despite the fact that Planet X is currently enjoying a more positive reception in the astronomical community overall.

Planet Formation

Then, there is the matter of the theoretical underpinning of a size-able planet beyond Neptune. Theoretical considerations played a part in the declining interest in Planet X in the second half of the twentieth century. Theories about how planets form during the early stages of a star system have always been a problem for advocates of Planet X.

Stars form from condensing clouds of molecular gas, collapsing inwards due to their own gravity. The cloud the Sun formed from is known as the "pre-solar nebula". Planets form from a disk of pre-solar nebula materials left over from the star's formation. This "protoplanetary disk" rotates around the young star. Its composition may vary with distance from the Sun, but the density of the protoplanetary disk beyond the point where Neptune is thought to have formed was threadbare.

If the accretion of dense material to form proto-planets provides the building blocks for planet building, as is believed, then this stuff has to come from somewhere. The Sun's early protoplanetary disk is currently considered insufficient to drive the process of planet formation beyond Neptune. This is not as clear-cut as one might think, however. Protoplanetary disks observed around young stars vary considerably in size. Their radius can extend out to 1000AU. As such, the formation of wide-orbit planets beyond Neptune seems a realistic possibility – as long as the Sun's own protoplanetary disk was larger than is generally thought.

There are other ways planets can find themselves out beyond Neptune. As we have seen with comets located in the Oort cloud, there are good reasons to suppose that at least some of the population of outer solar system objects has been derived from elsewhere. One might then argue that the Sun could have captured a planet from a passing star (or even a free-floating planet moving through interstellar space). This would remove the need to explain how a planet might have formed so far out. However, astrophysicists have often long argued that this is a highly unlikely scenario. The odds against such a planet being drawn into a sufficiently stable orbit around the Sun are vanishingly small, they argue. This has been the mantra of Planet X skeptics for decades.

One of the purposes of this book is to explore the possible origin of Planet X. I will present some new ideas about how conditions beyond the Heliosphere might be driving planet-forming processes. Before that, let us take a look at the rather erratic history of Planet X. Although we are currently in a period of relative optimism about the imminent discovery of a new planet, this has not always been the case.

The Haphazard History of Planet X

It is perhaps surprising that astronomers have been interested in finding Planet X for well over a century. The prospect of finding this body was first advanced in the nineteenth century – before Neptune itself was discovered! A German astronomer, Peter Andreas Hansen, who was grappling with fluctuations in the orbit of the planet Uranus, considered it likely that not one, but two unknown planets were tugging at it (Grosser 1964). Two years after Neptune had been discovered, French astronomer Jacques Babinet claimed that the new planet was also showing signs of disturbance, indicating the presence of another planet lurking beyond it.

This concept enjoyed intermittent popularity in the nineteenth century among astronomers, but failed to yield a successful discovery. For instance, theoretical work on the possibility of two trans-Neptunian planets was published by the learned Glaswegian electrical engineer and astronomer, Professor George Forbes (1880). He had been making calculations based upon the assumption that the paths of comets are affected by planets in the outer solar system. His calculations led him to confidently predict the existence of two extra worlds beyond Neptune. The first missing planet was located about 100AU away, with an orbital period of about 1000 years. He argued that the second was further out, at 300AU, giving it an orbit of about 5000 years.

The term Planet X was first coined by Frenchman Gabriel Dallet, around the end of the nineteenth century (Orrman-Rossiter & Gorman 2015). Rather than being the Roman numeral standing for planet number ten (after all, at that point there were definitely only eight planets), the term instead encapsulated its mystery. This is the same sense of mystery that makes the phrase "Planet X" so perfect for the title of a Fifties 'B'-movie.

There was a lot of interest in the subject in the years leading up to the Great War, with several astronomers offering various orbital arrangements for Planet X (Schlyter 1997). It wasn't just Neptune and Uranus which offered tantalizing anomalies to study, but also the ranges of various comets in the outer solar system which seemed to hint at the presence of another planet (Crockett 2014). This continued effort to locate Planet X drew in the polymath businessman Percival Lowell. His interest in the subject led to a substantial financial investment culminating in the construction of a whole new observatory in Flagstaff, Arizona. Some years later, the Lowell Observatory became the place where Clyde Tombaugh discovered Pluto.

It is interesting to note that Lowell's own earlier search for Planet X – conducted between 1913 and 1915 – had actually turned up Pluto. Twice! Like Galileo before him, Lowell and his team had not realized that they had found something significant. This was despite the candidate object being located within the area of the sky that Lowell had confidently predicted would contain Planet X. Perhaps he was just expecting something rather grander than poor little Pluto?

(As an aside, I would argue that this same principle is probably going to prove true for the search for Planet X in modern times, too. Once it has been observed and its location officially verified, I suspect there will be many regretful realizations of prior observations that had been shrugged off).

Despite the success with finding Pluto in 1930, not one extra trans-Neptunian object was discovered over the next sixty years. Theories predicting planets beyond Neptune came and went during that time. Many of the calculations pointed to an object located between 60AU and 80AU (Schlyter 1997). Each sought to address orbital anomalies observed in the movement of outer solar system objects, but none was capable of scoring a direct strike. The elusive Planet X continued to weave its dark magic across the skies.

Nibiru/Marduk

In 1976, a journalist and independent scholar named Zecharia Sitchin argued that ancient Mesopotamian mythology could readily be decoded as a cosmogonic allegory. In particular, he argued, the ancient Babylonian Enuma Elish creation myth could be interpreted as describing the formation of the solar system, including planets which the ancient Babylonians had absolutely no business knowing about, like Neptune and Pluto (Sitchin 1976).

Sitchin's controversial interpretation of Enuma Elish, and various older Sumerian texts, led him to propose the existence of an additional planet in the solar system's family, which he described as 'Nibiru/Marduk'. The name Nibiru stemmed from early Sumerian cosmological descriptions, and Marduk was the name of a Babylonian deity. Sitchin argued that their identity was one and the same, wrapped up in the existence of a missing planet.

Fig. 2-1 Brown dwarf flare image.

His description of a fiery red interloper, making a disorderly entrance into the early solar system, has evoked decades of conjecture about the potentially catastrophic "return of Nibiru". Sitchin's description of Nibiru/Marduk's highly eccentric, elliptical orbit is consistent with a Planet X object following a comet-like path. He argued that this Planet X/Nibiru body's orbital length was 3,600 years. 3,600, or 1 "Sar", was an important number in the ancient Mesopotamian sexagesimal numbering system that Sitchin decided held cosmological significance. We shall return to aspects of Sitchin's writings in due course.

Around about the same time, astronomers Tom van Flandern and Robert Harrington of the U.S. Naval Observatory became interested in the subject of Planet X. They hoped to explain stubborn anomalies in the orbits of Uranus and Neptune. Projected parameters for their Planet X (Harrington & Van Flandern 1979) seemed to sit well with those suggested by Zecharia Sitchin (Sitchin 1990a).

However, van Flandern was skeptical about Sitchin's proposal that Nibiru/Marduk had caused a planetary catastrophe following its entry into the solar system. Sitchin had claimed that the "Celestial Battle" between the god Marduk and the Babylonian watery monster goddess "Tiamat" described a cosmic clash between planets. In this interpretation, Tiamat was a large watery planet that went on to become the Earth.

Van Flandern did not consider it possible to reconcile a catastrophic collision between Nibiru/Marduk and another planet with the current layout of the solar system. According to van Flandern, the near circular orbit of the Earth was inconsistent with such an event ever having happened (Alford 1998).

Improved Calculations of the Orbits of the Outer Planets

Harrington, on the other hand, seemed more open to Sitchin's thesis, and participated in a two hour video interview with the controversial author. Harrington spent years trying to find Planet X, but to no avail. Six months before his untimely death, new calculations were published about the masses of the outer planets of the solar system. These stemmed from improved ranging data supplied by the two Voyager probes as they flew past the outer planets. The new data shifted the numbers Harrington had worked with in the wrong direction. As a result, there seemed less need to invoke the presence of a Planet X body (Browne 1993).

Following these recalculations, it was claimed that the movements of the main planets could more or less be accounted for naturally. It is perhaps more accurate to say that there remain anomalous discrepancies in the orbits of Uranus and Neptune. The detailing of these astrometric 'errors' continues to be explored to this day, particularly as new ranging data is returned by spacecraft like Cassini and New Horizons (Iorio 2017). However, at the time, these minor anomalies were dismissed by powerful voices in the astronomical community. They were considered to lie within the scope of experimental error (Standish 1993). The more accurate measurements of the positions of the outer planets were gleefully touted as ringing the final death knell for Planet X in the latter part of the 20th century.

A few years prior to the premature announcement of 'the death of Planet X', John Anderson, a research scientist working at the Jet Propulsion Laboratory in Pasadena, had also suggested the existence of a Planet X object. He argued that it must have an eccentric, elliptical orbit which was inclined to the ecliptic plane. Anderson's interest lay in the various probes sent out into the outer solar system, and he subsequently became a central player in the scientific effort to explain the so-called "Pioneer anomaly" (Anderson et al. 2002). This anomaly involved the unexpected slowing of the Pioneer spacecraft in the outer solar system.

Anderson argued that the nature of Planet X's inclined elliptical orbit meant that its effect upon the known bodies of the outer solar system was fleeting (Wilford 1987). In other words, only when Planet X comes close to the planetary zone would it begin to exert any real gravitational influence. During the current observational era, he argued, the planet is far too distant. As a result, the gravitational influence it has upon other bodies in the outer solar system is less easy to detect.

The intermittent influence exerted upon the outer solar system by Planet X seems to be an important factor to consider. Anderson argued that the unseen planet must have crossed the plane of the ecliptic during the nineteenth century, at which point it created the discrepancies noted at that time in the orbits of Uranus and Neptune. Since then, Planet X shifted ground away from the ecliptic plane sufficiently for these fleeting, minor perturbations to fade away.

Nemesis

The media's reporting of the Planet X hypothesis around this time was complicated by another theory that had emerged just a few years before. Two separate groups of scientists had proposed the existence of a sub-brown dwarf "failed star" in the outer solar system (Whitmire & Jackson 1984; Davis, Hut & Muller 1984). This massive object was given the rather catchy name "Nemesis". With so many competing theories, it was starting to get tricky to discern just what was meant by Planet X (Wilford 1987). But Nemesis was a different beast altogether. Instead of a terrestrial planet lying beyond the outer fringe of the Kuiper belt (the classic Planet X concept), the failed dwarf star, Nemesis roamed moodily through the outer reaches of the Oort cloud.

Rather than having an orbit of several hundred, or perhaps several thousand years, Nemesis circumnavigated the Sun over about 27 million years. Its distance from the Sun would be measured in tens of thousands of astronomical units. If we return briefly to our cosmic landscape analogy in Chapter 1, instead of being somewhere out in an adjoining field, like Planet X, Nemesis is miles and miles away, across hills and dales.

The other difference was its size. Nemesis was proposed during a time when brown dwarfs were mere conjecture, and more commonly known then as "black dwarfs". But, just to complicate matters, Nemesis – if it exits at all – is likely to be smaller and lighter than a brown dwarf 'star'. Hence, it falls into the category of a "sub-brown dwarf". Weighing in at several Jupiter masses or more, you would think such an object which should command some respect. Instead, Nemesis is more often than not the subject of intense ridicule.

Extinction Cycles

There was a good reason why two teams of serious astronomers had independently proposed the existence of Nemesis. It was in response to another theory published in the scientific press that year. This particular theory owed more to palaeontology than astronomy. It involved the scarily regular occurrence of mass extinctions that had taken place during Earth's geophysical history. Palaeontological studies provided evidence for a non-random, cyclical pattern of mass extinction events (Raup & Sepkoski 1984).

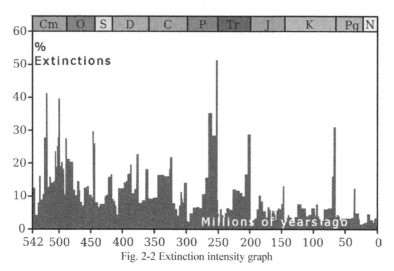

Fig. 2-2 Extinction intensity graph

If you look at a graph showing the numbers of extinctions of species over geological time periods, there are a number of sharp spikes in the fossil record. Moreover, there appears to be a sequential regularity to these events. A mass extinction event on the Earth appears to take place in tune to a beat running every 26-27 million years, or so. It should be noted that the case for a periodic sequence of extinction events is hotly contested. However, a more recent study seems to reinforce Raup and Sepkoski's original data, and even extends the extinction sequence back further in time from 250 million to 500 million years ago (Melott & Bambach 2010).

The reason why astrophysicists got involved in a debate about periodic extinctions in the Earth's fossil record is that the time scales are so vast that the seemingly orderly nature of the events calls for nothing less than a cosmic cause. The catastrophic impacts of comets are high on the list of potential mega-destructive events befalling the Earth. If such events have a periodicity to them, then it is natural to consider what might be inducing regular swarms of comets. This is where Nemesis comes in.

If a massive planet were to sweep through a particularly dense cloud of comets during its slow orbital path around the Sun, then the disruption it causes might explain the sudden flurry of destructive activity. Nemesis might follow a somewhat eccentric orbit itself, for instance, although the stability of any such orbit was questioned quite early on (Hills 1984). As a result, its "perturbing" effect upon distant comets becomes a periodic function of its ungainly motion around the Sun. It is easy to see how the periodic extinction cycle led to the conceptualization of Nemesis.

However, there are alternative explanations for the still-contested extinction event cycle. Like all the stars in the Milky Way, the Sun moves around the center of the galaxy in a similar manner to that of a carousel horse on a fairground ride. As it slowly rotates around the center of the galaxy, it bobs up and down through the galactic plane. This occurs every 35 million years or so. This may have repercussions for global climate conditions on Earth as it does so (Russell 1979). This is because there are many, immense molecular clouds concentrated in the galaxy's spiral arms. As the solar system encounters these clouds, the Earth's atmosphere may be inundated with dust blocking light from the Sun, causing dramatic global cooling (Pavlov et al. 2005). There is also the potential for periodic bombardment by cosmic rays (Medvedev & Melott 2006).

The problem with any hypotheses involving the Sun bobbing up and down through the Galactic plane is this – the cycle is not exactly the same as the reported extinction cycle seen in the fossil record. Either party may be wrong in their calculations, of course, in which case there may well be a determining relationship between the Sun's carousel-like motion and extinction events on Earth. But 27 into 30 (even 35) does not fit as things currently stand.

This is why Nemesis maintains its attraction as the astrophysical force majeur to explain the extinction cycle. The planet's theoretical orbital period can be altered to fit the evidence available in the fossil record, unlike the Sun's calculated motion through the galaxy.

Tyche

A watered-down version of the Nemesis theory emerged in 1999. Professor John Matese, working with one of the original advocates of Nemesis, Daniel Whitmire, published findings about long-period comet activity. Because of the spherical nature of the Oort cloud, dislodged comets should emerge from this massive reservoir more or less randomly. They should appear from all points in the sky. Studying the data, Matese and Whitmire, and their colleague Patrick Whitman, instead discerned a non-random patterning. This was in terms of both the direction and frequency of in-coming long period comets. This led them to argue that the long-period comets, which visit the planetary zone of the solar system from the outer Oort cloud, were being perturbed by a substantial unseen body (Matese et al. 1999).

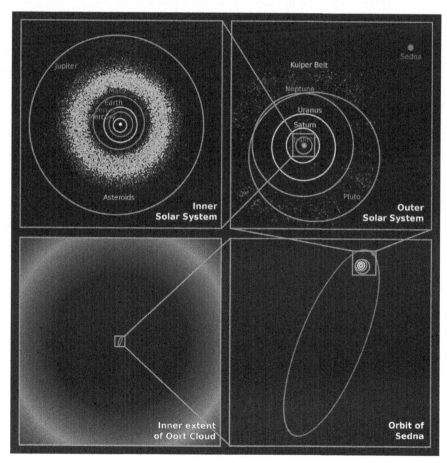

Fig. 2-3 Oort Cloud Sedna orbit image

Later, after the 2003 discovery of the rather curious scattered disk object named '90377 Sedna', Matese and his colleagues linked Sedna's anomalously eccentric orbit to their proposed giant planet, which by this time they had nicknamed "Tyche" (Matese et al. 2006). They argued that Tyche is a gas giant planet of between one and four Jupiter masses. Its projected atmospheric temperature (-70°C) is considerably warmer than that of Jupiter (-140°C). This is despite its much greater distance from the Sun: The 'warmer' temperature of Tyche relative to Jupiter is due to the former's greater mass and density.

Tyche is a sub-brown dwarf, lying somewhere along the spectrum of planetary mass objects between Jupiter and the brown dwarfs. It should be noted that astronomers use several different names for this type of object. The terms in use include sub-brown dwarfs, "planetary mass Y-dwarfs", and "planemos" (unsurprisingly, this latter term never really caught on).

You may be wondering: What is the difference between Tyche and Nemesis? Tyche was derived from the original Nemesis concept, but has been brought closer in. Its distance may be about 15,000AU, giving it an orbital period of a couple of million years. Therefore, it would not be responsible for the proposed cycle of extinction events on Earth. It may have been captured into a wide binary orbit by the Sun, being derived from a sister star within their mutual birth cluster (Matese & Whitmire 2011). The inner limit for an object this size has potentially shifted into the inner Oort cloud, which would necessitate something of a rethink about this torus-shaped grouping of comets (Matese et al. 2006; Fernandez 2010).

A very similar scenario had been proposed around the same time as Tyche, this time by the English environmental and planetary scientist John Murray. He also advocated a body of several Jupiter masses to explain the same dataset of long-period comet trajectories (Murray 1999). I've discussed both sets of work quite extensively in my previous book (Lloyd 2005). They were catalysts for my own research into this subject (Lloyd 1999).

Suffice to say that the merits of the case for a sub-brown dwarf lurking among the comets had been adequately covered in peer-reviewed academic journals. These were hypotheses, argued from empirical data. Counter-arguments were made that the evidence provided was insufficient to draw a strong conclusion about the existence of such a massive planet out there. After all, the object had not actually been observed by astronomers, and a great deal of skepticism about the existence of Planet X remained. Some media reports about these theories led the reader to think that NASA had located a new, distant planet. This was not actually the case (Plait 2011).

However, it is not entirely true to say that the discovery of a Tyche-sized object had never been discussed by NASA. In fact, like a shadowy figure of a suspect caught on an old CCTV image, the sighting of a cold, Jupiter-sized planet lodged in the outer solar system had been claimed by NASA, in 1983 (O'Toole 1983). Its position had twice been pinpointed by scientists analyzing data from the infrared space telescope IRAS. At the time, it was thought to lurk some 500AU away, to the west of the constellation of Orion. The IRAS scientists seemed skeptical, even uncomfortable, about their own announcement, and the story seemed to quickly disappear without trace. This fleeting announcement has become the subject of many a conspiracy theory since.

The size and distance of this hotly debated object, announced briefly by scientists in 1983, fits the scenario I originally depicted in my own Dark Star Theory: A Jupiter-sized world caught somewhere in the no-man's land between the Kuiper belt and the inner Oort cloud. This is closer than the distances claimed for Tyche, and certainly Nemesis, whose distances are marked out in the tens of thousands of astronomical units. Any massive object located within the inner extent of the inner Oort cloud would represent a particular challenge: How could it have evaded detection for so long? In the next chapter, we will explore the detectability of brown dwarfs and sub-brown dwarfs in more depth.

Sub-Brown Dwarfs in the Infrared

As many reasoned skeptics are apt to point out, it's certainly not impossible that another major planet might be located in the outer solar system – even if it might prove difficult to work out quite how it could have got there. The problem is that it perennially evades detection. Objects of this size may be cold by Earth's standards, but they are still relatively warm compared to the cold of space. Their heat signature, however faint, should be detectable during infrared sky searches.

The infrared sky survey by IRAS was pretty much unique for its time. Infrared telescopes necessarily need a cold environment in which to operate; ideally, out in space. This is because the Earth's atmospheric water vapor absorbs infrared radiation. This effectively creates a heat blanket over any ground-based observatory, causing a great deal of interference. Even in space, the infrared telescopic equipment relies heavily upon coolants, and these eventually dissipate.

Such was the case with NASA's Wide-field Infrared Survey Explorer (WISE), which enjoyed a relatively short working life. Launched in December 2009, NASA's WISE mission conducted intermittent sky searches (2010-2013) which, when taken together, lasted little more than one year. Nonetheless, it managed to scan much of the sky in that time, in various wavelengths, and at a resolution which was far more detailed than its predecessor IRAS had achieved.

WISE was the first infrared search of the sky since IRAS, and exceeded its scope by several orders of magnitude. Theoretically, WISE was capable of detecting a second Jupiter as far as one light year away. Surely, then, this latest attempt to map the sky in infrared should have easily found Planet X, especially if such an object was a gas giant or sub-brown dwarf?

Some of the scientists involved in the WISE program were actively on the lookout for brown dwarfs, and cooler sub-brown dwarfs, in our stellar neighborhood. The team included Davy Kirkpatrick of Caltech, who is an acknowledged expert on brown dwarfs. Although they found a limited number of so-called "Y-dwarfs", none was found closer than nine light years away (NASA/JPL 2011; Kirkpatrick et al. 2011). At least, so it had appeared.

Fig. 3-1 WISE Artist's conception of Brown dwarf

Work on the WISE data by astronomer Kevin Luhman of Pennsylvania State University, who might be considered the WISE guru of astrophysics, unearthed subsequent brown dwarf discoveries from the mass of data the sky search had provided. Prominent among them was the discovery of a binary pair of brown dwarfs located just 6.5 light years away. Known as Luhman 16, this pairing became the third closest set of stellar objects to the Sun (Luhman 2013). These brown dwarfs are similar in size, and orbit around each other every 27 years, or so.

Looking back through other earlier sky searches revealed previous sightings of Luhman 16 AB, including by IRAS back in 1983! There were more of these "pre-coveries" in other infrared surveys, too, but despite their relative proximity, the binary pair of brown dwarfs had gone unnoticed at the time. It goes to show that such objects are not always identified correctly (or at all), even when imaged frequently over many years.

Then, in 2014, the coldest sub-brown dwarf yet was also dug up out of the WISE data. Known somewhat awkwardly as WISE J085510.83–071442.5, it lies just over seven light-years away and weighs in somewhere between three and ten Jupiter masses (Luhman 2014). It is the fourth nearest extra-solar system, although it would appear to be a free-floating planetary mass object, rather than a dwarf star.

Although these discoveries are impressive in their own right, this harvest of nearby ultra-cool sub-brown dwarfs and brown dwarfs actually remains rather disappointing. The perceived lack of such objects in the Sun's neighborhood undermines previous projections that there should be at least as many brown dwarfs as stars. Instead, the results that came out of WISE suggest that there are about five stars for every brown dwarf (Clavin 2012). Perhaps that merely indicates that finding these objects on our doorstep is not as straightforward as assumed. We will return to this point later.

For many astrophysicists, and skeptics from outside of the profession, the failure of WISE to detect a backyard sub-brown dwarf once again doomed Nemesis, Tyche and Planet X to oblivion. There was little recognition at the time that WISE had its own limitations, mostly borne out of its short shelf-life. Instead, NASA proclaimed the question of a Nemesis or Tyche-like Planet X body in the solar system to have been decisively settled (Clavin & Harrington 2014). WISE had not found it; therefore, it was not there.

Questioning the Wisdom

To spot faint, relatively nearby objects it is necessary to track their motion across the sky over several observations. There is the need to conduct what are called "parallax measurements". These are observations of an object in space from different positions along Earth's orbit. As the Earth changes position, a relatively nearby object will be seen to change its position against the background field of stars. This effect is known as *parallax*.

The nearer the object is to the Earth, the greater this visual shift in the object's position in the sky. Optimally, astronomers try to observe an object at six month intervals, when the Earth is located on either side of the Sun from the Sun/object axis. This enables astronomers to triangulate the object's position, and in doing so accurately work out its distance from Earth. Parallax measurements can be crucial for finding minor bodies in the solar system. Photographic plates taken at different time intervals will expose how unidentified points of light change position through the sequence. This is like a "spot-the-difference" puzzle. Nowadays, computer programs can sieve through huge numbers of images to pick up the changing positions of these tiny specks of light.

The same principle applies in infrared as it does in visible light. The problem for sky searches in infrared is that they are relatively short-lived, and so their capacity to compare images over time is curtailed. Spotting a tell-tale heat signature of a distant object is one thing; spotting a moving warm object quite another. Given the short shelf-life of the WISE telescope, making those parallax measurements consistently for all data points is bound to be a less than perfect procedure.

Professor Matese alluded to this when considering why there seemed to be a short-fall in local brown dwarf discoveries by WISE. This "blind-spot" for infrared sky searches raises the possibility of a far-too-hasty declaration for the non-existence of a massive Planet X body (Matese & Whitmire 2011a). After all, larger brown dwarfs emerged from the data later on, indicating they had previously been missed.

There are other reasons why objects might evade detection. Although the results from WISE seemed to rule out a Jupiter-sized gas giant companion object out to about 26,000AU (Luhman 2013a), it is quite possible that one might also have es-caped detection by WISE if it was located close to a bright star. The light from such a star would have overwhelmed the candidate object (Redd 2014).

Skeptics seemed overly hasty to dismiss the case for Planet X, despite indirect evi-dence continuing to accumulate that something was still out there. Although ostensi-bly the hunt for Planet X seemed to be over, as a result of the negative data coming back from WISE, people in the know were keeping their options open.

More Stars than Dark Stars

For some time, astrophysicists have argued amongst themselves over how many sub-brown dwarfs there might be in the galaxy. Opinions vary. We saw earlier how the WISE survey, conducted between 2011 and 2013, seemed to demonstrate that the number of these objects might be much lower than initially thought. However, other published evidence, this time concerning the stellar populations of open star clusters, points towards there being more brown dwarfs and sub-brown dwarfs than stars in our galaxy. This is at odds with the WISE findings and, again, highlights po-tential failings in the infrared sky search methodology to pick out all of these objects successfully.

When I use the term sub-brown dwarf, I'm generally referring to gas giant planets/ ultra-cool dwarf stars which are several times more massive than Jupiter. This may be up to ~13 times as massive as Jupiter. At this mass and above, objects swiftly burn their limited stocks of deuterium in nuclear reactions, emitting light for a short while (Spiegel et al. 2011). Objects above this mass are classified as brown dwarfs, the lowest 'stars' on the stellar spectrum, even though this early period of active light emission may be very short-lived. We will return to this later on.

Denser than Jupiter

Most examples of ultra-cool brown dwarf objects (those more than a few million years old) are essentially dark. By contrast, very young sub-brown dwarfs may give out a little light, because they still retain enough heat from their formation. There remain significant gray areas in brown dwarf classification.

It is a curious quirk of nature that sub-brown dwarfs are actually the same size, or smaller than Jupiter, despite being heavier. This is because of a degenerate state of matter which is created at very high pressures within massive planets. This quantum mechanical effect kicks in around the mass of Jupiter. The result is that multiple-Jupiter mass objects become denser rather than larger, as the accreting young planet piles on the weight. The planet's density builds until it reaches the temperature and pressure where nuclear fusion is triggered, becoming a brown dwarf 'star'.

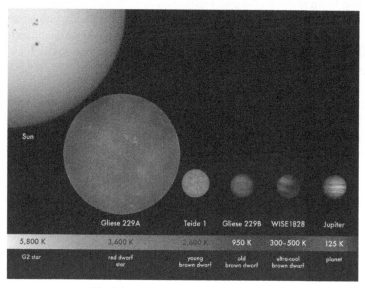

Fig. 3-2 Brown dwarfs comparison image

Because these objects are so small, and so dim, they are extraordinarily difficult to observe. Some have been found, but they are usually either extremely young (and therefore still burning brightly), or are "exoplanets" discovered orbiting parent stars (and so detectable through gravitational "wobble" effects, or other clever means of discovery).

Their mass, lying between that of Jupiter and the deuterium-burning limit at about 13 Jupiter masses, seems to single sub-brown dwarfs out as rather special objects. It looks likely that they form, like planets, within proto-planetary disks. As a result of the chaotic forces at work in these environments, they may migrate outwards, or become scattered into distant orbits, like primordial comets. Some may even break free of their parent star to escape into the interstellar space between stars (Sumi et al. 2011), becoming "free-floating" sub-brown dwarfs.

Counting Dark Stars

It has been my contention for some time that the population of these objects is significantly underestimated. WISE is partly to blame for this. Many ultra-cool dwarf stars may be free-floating objects moving through interstellar space. These independent objects may have been ejected from young star systems as the fledgling planets in those systems jostled for position. Alternatively, larger brown dwarfs may have formed independently, condensing out of relatively small molecular clouds in the same way as stars do.

Opinions about their numbers vary greatly among astrophysicists. Some evidence suggests that there may be twice as many of these objects as stars, including studies involving gravitational micro-lensing surveys of the galactic bulge (Sumi et al. 2011). However, results from other studies contradict this conclusion. Instead, they predict that there may be as few as one sub-brown dwarf per 20-50 stars within a given star cluster (Scholz et al. 2012).

This enormous discrepancy is important. The difference between these two estimates is statistically very significant - perhaps as high as two orders of magnitude. This ultimately affects our understanding of how many free-floating sub-brown dwarfs we can expect to find out there.

If the number of free-floating sub-brown dwarfs is on the high end of expectation, then it means that there are also likely to be far more of these objects in wide, distant orbits around their parent stars. This, in turn, increases the likelihood of there being a similar Dark Star object (or perhaps more) in our own immediate solar neighborhood.

Recent observations of binary star systems indicate that wide binary objects tend towards very eccentric orbits over the course of hundreds of millions of years (Kaib et al. 2013). So, if a companion object were found to exist at the kinds of distances discussed above, then one would anticipate that its orbit should be eccentric. This would fit with ideas about a large Planet X object maneuvering along an elliptical orbit around the Sun.

It has been shown that sub-brown dwarfs can orbit around their parent stars at these kinds of extreme distances. The infrared camera aboard the Spitzer Space Telescope spotted a sub-brown dwarf, of seven Jupiter masses, orbiting its parent white dwarf star WD 0806-661. The distance between the sub-brown dwarf and the white dwarf star it is orbiting around is a staggering 2,500AU (Luhman et al. 2011).

Fig. 3-3 Brown dwarf image

A similar object found moving around our own Sun would find itself embedded in the inner Oort cloud. This provides us with a remarkable precedent for the kind of wide-orbital binary that we might imagine for our own solar system.

Despite this successful discovery, free-floating sub-brown dwarfs are extremely difficult to locate, particularly in visible light. Within our own galactic neighborhood, astronomers stand a better chance of imaging them using near infrared sky searches, like WISE and 2MASS.This is because the sub-brown dwarfs' size and density should produce sufficient internal heat to allow them to stand out from the frigid background of space. Even so, their age (1-10 billion years old) makes then exceedingly faint. Taken together, it is perhaps no surprise, then, that the random trawl across the sky by WISE for these local minnows was not particularly successful (Clavin 2012).

Guessing the Number of Sweets in a Jar

How can we know how many of these objects should be out there? It turns out that sub-brown dwarfs may not be distributed completely randomly across interstellar space. Studies of open clusters and mass groupings of young stars have received much attention in recent years. These clusters may well host the brightest examples of low-mass sub-stellar objects in the solar neighborhood.

The youngest of these 'local' associations, within several hundred light years, is TW Hya (which is usually known simply as "TWA"). This star cluster, located about 100 light years away, contains a few dozen 10-million-year-old stars, all moving through space together, like a shoal of fish (Carnegie Institution for Science 2017). It is known that many of TWA's low-mass objects have yet to be observed, despite careful study by astronomers for over a decade. These objects were too faint to have been detected by ESA's Hipparcos mission, which accurately determined the astrometry (accurate positioning and distance) of a huge number of stars during the lifetime of its mission.

A census of the TWA star cluster has made use of computer simulations to try to fill in some of the gaps in the sub-stellar populations of this open star cluster. This is a bit like working out how many minnows there might be lurking within a large school of fish. The scientific team was led by Carnegie's Jonathan Gagné and researchers from the Institute for Research on Exoplanets (iREx) at Université de Montréal.

They seemed particularly interested in working out whether the open cluster might contain additional hidden objects in the ~5–7 Jupiter masses range. In other words, sub-brown dwarfs. Their interest had been prompted by previous discoveries using another infrared survey. Known as 2MASS (Two-Micron All Sky Survey). This one looked for objects in the near-infrared. Appropriately enough, 2MASS found two objects in this mass range within TWA. The implication was that there could be more. In order to determine whether there might be more free-floating sub-brown dwarfs in TWA, the team performed an astronomical calculation known as an "initial mass function". This function can be used to determine the overall distribution of mass in the cluster of moving stars. Once it has been determined, scientists can infer the number of undiscovered objects that might lie hidden within it helping to hold the cluster together (Sci News 2017).

The result of this calculation was that there are likely to be at least an additional 10 objects, in the ~5–7 Jupiter masses range, lurking unseen within the TWA cluster. It is possible that there are more than 20. This large number of potential objects surprised the research team because it surpassed the numbers of low-mass objects expected, given the populations of stars in the census of TWA (Gagné et al. 2017).

The new estimate is in keeping with other studies of galactic interstellar populations of sub-stellar objects of this size. It was consistent with recent estimates for the space density of objects at the deuterium-burning limit (Gagné et al. 2015), the afore-mentioned discovery of a cold sub-brown dwarf seven light years away (Luhman 2014), as well as the results from micro-lensing surveys of the galactic bulge (Sumi et al. 2011).

Hidden Populations

It looks like the number of free-floating sub-brown dwarfs may well be more than the number of actual stars in our galaxy after all! This inevitably leads us back to our question: If sub-brown dwarfs are so numerous, and our near-infrared sky surveys have become ever more powerful, then why have we not been detecting more of them?

It looks like WISE was not as good at detecting these objects as many hoped. It is not that they are not out there. Their multiple dark presences have been discerned through micro-lensing techniques, and through astronomical and statistical analysis of star groupings.

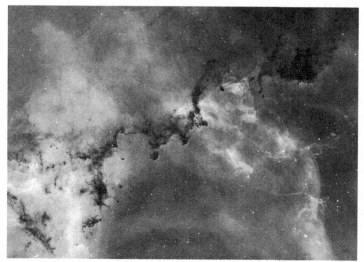

Fig. 3-4 Star cluster image

So, why have these objects not been observed en masse, as theoretically predicted? Is there an intrinsic property of these free-floating interstellar objects which is blinding them to us in infrared and visible light? These are questions I will attempt to address in detail in forthcoming chapters, as I propose a possible scenario to explain this paradox. Suffice it to say for now that infrared searches for sub-brown dwarfs are only revealing part of the picture. That means that negative, knee-jerk reactions about the WISE data were misplaced.

I would argue that there remains a good possibility that at least one massive object remains undiscovered in the outer solar system. Perhaps that object is indeed Nemesis, or Tyche, or something similar. Perhaps it is the sub-brown dwarf, or Dark Star, I described in my previous book (Lloyd 2005). But it might be a lesser gas giant, or an ice giant of a similar mass to the planet Neptune. It may be an ejected super-Earth, taking the form of either a massive terrestrial world, or a mini-Neptune. Or, it may 'just' be a regular planet, of a similar size to Earth, or Mars. There are many possibilities vying for the title of Planet X.

A Doomed Prediction

In **Dark Star** (2005) I confidently predicted the presence of a sub-brown dwarf in the outer solar system. I had seen a correlation between the questions raised about extinction events by astrophysicists, and the mythical representation of a "doomsday planet" (Lloyd 2001). I considered it possible that this planet existed, and perhaps had even moved through the planetary zone of the solar system during prehistorical times. Although I was confident that this planet poses us absolutely no concern at the present moment, I did think it possible that the mythology associated with Planet X might have some basis in truth. There may have been a celestial phenomenon that had once been seen from Earth in prehistoric times.

The mythical descriptions of the doomsday planet seemed to tie in well with a new stellar entity known as a brown dwarf (Lloyd 1999). Brown dwarfs span the uncomfortable gap between stars and planets. Although these smoldering, red objects represent a category of the smallest possible 'stars', and actually stop emitting bright visible light quite early on in their lifetimes, they are still ominously magnificent as planets. I referred to these objects as *Dark Stars*.

As more factual information emerged about the brown dwarfs, a new category of objects was created by astronomers to describe the ultra-cool objects at the lower end of the mass spectrum. These were the sub-brown dwarfs, and they were more like planets than failed stars. It became clear to me that the mythical doomsday planet that I had been trying to authenticate fell more into this category. Although more massive than Jupiter, sub-brown dwarfs are actually the same size, or even smaller, than the familiar gas giant. That made them denser and warmer, and so more likely to be picked up in an infrared search of the sky.

As we have seen, several infrared searches have been made of the sky in the last few decades. Although they have not turned up as many sub-brown dwarfs and brown dwarfs in the Sun's neighborhood as predicted, they have at least found some (Luhman 2013; Luhman 2014). But they have not found any whose distance lies between here and the nearest star. As a result of the search conducted using WISE, it has been generally concluded that there are no objects of this sort in the Sun's own backyard (Luhman 2013a).

The non-discovery of Planet X by WISE seemed to doom this fabled world. But, around the same time that NASA was noisily ringing the death knell for Planet X, evidence emerged from a different quarter that would breathe sensational new life into the subject.

The Extended Scattered Disk

I'll be honest with you: When WISE didn't turn up the *Dark Star* as I had predicted it would, I became quite despondent about the whole thing. Despite the caveats expressed in Chapter 3, it really did seem to me that I had been wrong all along. If a sub-brown dwarf was out there, then the range that WISE was said to cover should have pushed it out to the very fringes of the Sun's gravitational field of influence.

Nemesis (or Tyche, for that matter) could not have been observed during the time of modern humans. At tens of thousands of astronomical units distant, these hypothetical bodies would lie too far out to have made an incursion into the planetary zone in the last 100,000 years or so. I was hoping that such an object would be closer to home. Perhaps there might still be a Nemesis object, I thought, but my own proposal that a massive planet might lurk closer in, perhaps within the inner extent of the inner Oort cloud, seemed increasingly wide of the mark.

The Habitable Solar System

However, I still believed that in order to generate sustainable conditions for life in the outer solar system, a substantial body like a sub-brown dwarf was needed (Lloyd 1999). Its relative warmth and, in particular, gravitational influence would be sufficient to warm a moon, or orbiting terrestrial planet. This is in the same way that Jupiter's presence allows liquid water oceans to exist beneath the pristine, billiard-ball surfaces of some of Jupiter's Galilean moons. Since I wrote **Dark Star** (2005), it has become apparent that Saturn's moon, Enceladus must also have a global sub-surface ocean below a cracked layering of ice. This tiny moon is losing water and other volatiles into space via immense geysers, graphically displaying the incredible forces at work beneath its surface.

Not wishing to miss the party, Saturn's largest moon Titan not only has oceans and lakes of liquid hydrocarbon below its thick nitrogen and methane atmosphere, but also the potential for hidden subsurface oceans of water. These subsurface oceans seem to be becoming the new norm among the moons of the gas giants. Distant Neptune's moon, Triton, has a surface which is geologically active, with cryovolcanoes and geysers spurting out nitrogen gas into space. This provides the moon with a very thin layer of nitrogen across its surface. Essentially, Triton maintains a tiny atmosphere, albeit a very cold one. Remarkably, the same is true for Pluto.

If these conditions are possible upon the moons of gas giants and ice giants, and even tiny binary dwarf planets, then what kind of conditions might prevail on a moon orbiting a sub-brown dwarf? Perhaps such a moon might be warmed sufficiently to maintain a viable atmosphere capable of supporting life? This seems a very reasonable supposition, although the orbiting lunar planet would clearly need to be much, much closer to the parent sub-brown dwarf than the Earth is to the Sun. It would need to look more like Jupiter's system of moons.

The projected habitable zone of a moon around a sub-brown dwarf would make the smoldering parent 'star' appear very large in the moon's sky. The moon would very likely also be tidally locked, meaning that, like our own Moon, it would spin at the same rate as it rotates around the parent body. This creates the effect that the same side of the moon always faces the parent planet (Lloyd 2010).

Even more speculatively, this scenario might offer an explanation for ancient myths about human-like "gods" visiting Earth from the heavens (Lloyd 2009). We might then begin to incorporate Zecharia Sitchin's theory about the mythical Nibiru into the mix, accounting for how a race of beings might conceivably live on a planet in the outer solar system (Sitchin 1976). However, even if there might be some truth to this, Sitchin's Nibiru could not be a single terrestrial planet. I remain convinced that there has to be a combination of planets within a system for his theory to work. This is needed to facilitate the internal heating required to make any such world habitable in the first place.

But all of this conjecture rested upon the need to find such a system in the Sun's backyard. No such planet had been forthcoming, and skeptics were gleeful in their condemnation of the whole subject. This was not helped by the continual claims in the tabloid media about an in-coming Planet X object which would strike the Earth. This is such an unlikely scenario in so many ways, however I believe it preys upon something deep within the human psyche.

One might argue that a prior appearance of this planet in prehistorical times had such an almighty effect upon the humans living at that time that it became ingrained as an archetype within our collective unconscious. Or, perhaps it just makes a great sensationalist story that sells tabloid newspapers. Either way, this *fake news*, as one might describe it these days, has made Planet X ever more toxic. It is little wonder, then, that it invites such vehement skepticism from so many quarters.

Despite all of this, the potential for the existence of Planet X continues to actually grow. Quite incredibly, strong evidence suggestive of the presence of such an object emerged after the dark night of the soul that was WISE. This evidence rests upon the anomalous orbits of scattered objects in the outer solar system, and the seemingly compelling correlations between them.

The Problem with Sedna and Friends

As has been briefly noted before, not all of the objects discovered in the Kuiper belt behave as expected. Actually, that's something of an understatement. The minor planet Sedna (originally 2003 VB12) quite literally threw astrophysicists a curve ball. It was way out in left field, completely ignoring the rules of the game.

Sedna is a curious red-colored dwarf planet. It is a rare and rather strange occupant of the so-called scattered disk beyond the Kuiper belt. Its discovery, by astronomer Dr. Michael Brown in late 2003, sparked great controversy, due in part to the object's highly unusual orbit. Its closest approach to the Sun is 76AU, well beyond the outer edge of the Kuiper belt.

Sedna became the latest member of the so-called extended scattered disk located beyond the Kuiper belt (Gladman 2002). However, Sedna's orbit could not be easily explained. It was not possible to explain its very long perihelion distance by the usual mechanism of chaotic diffusion of outer solar system bodies. Sedna was a mystery.

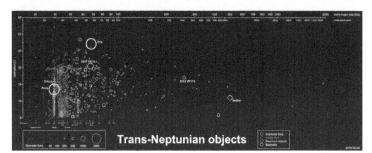
Fig. 4-1 KBOs and resonances diagram

Competing theories have been put forward to explain the elongated orbits of Sedna and other extended Scattered Disk Objects with very distant perihelia (>75AU). This sub-grouping of objects is referred to as the "Sednoids". Many of these theories about the Sednoids involve the influence of an unseen planetary object beyond the Kuiper belt (Gomes et al. 2006; Lykawka & Mukai 2008). Something is thought to be pulling these objects away from the main Kuiper belt. The further away the influencing planet is, the more massive it has to be to create the same observed effect (Matese et al. 2006).

But the existence of Planet X is not the only possibility. Other theories invoked such mechanisms as the prior disruptive influence of a passing star (Kenyon & Bromley 2004), or the capture by the Sun of extra-solar planetessimals from passing brown or red dwarfs (Morbidelli & Levison 2004). Academic papers arguing the case for Sedna's capture from a passing star were later successfully tested by computer simulations, helping to strengthen the case for this kind of mechanism (Jilkova et al. 2015).

Migrating Neptune

Another possibility is that Sedna was pushed outwards as a result of a highly eccentric migration of the planet Neptune. It might come as a surprise that the farthest ice giant might be capable of moving around to this extent, given its present stable orbit around the Sun.

For a while, various anomalies in the structure of the solar system have led astrophysicists to contemplate migration of outer solar system planets to explain the present patterning of minor objects. The internal structure of the Kuiper belt, for instance, contains a "kernel" of clustered objects whose patterning is difficult to explain within the context of the present observable structure of the solar system. This collection of tightly-bound objects moves within the plane of the ecliptic, like the planets. Yet this clustering is clearly distinct from the more unpredictable spread of the other Kuiper Belt Objects.

Astronomers wonder whether this distinctive pattern of behavior points to a particular incident in the early system which shepherded them, en masse, to their current location. Attempts to model the orbits of the kernel objects using collisions between larger KBOs proved fruitless. Instead, is has been argued the structural nature of the kernel objects' orbits implies some kind of primitive resonance with the planet Neptune, which early on had migrated across the solar system (Nesvorný 2015). This may have resulted from an interaction with a missing fifth massive planet (Nesvorný 2011).

A Hidden Planet X

The problem created by the existence of Sedna was recognized before I wrote my book *Dark Star*, and I was able to outline the headache Sedna was causing at that time (Lloyd 2005). Astrophysicists specializing in the composition of the outer solar system were already contemplating the presence of a perturbing influence to account for this object's bizarre orbital properties. It seemed to me that if enough of these anomalous Sednoid objects could be pinned down, then it might be possible to triangulate back to the Planet X body causing all the trouble.

Even I could see that this was easier said than done. The mathematics involved would require super-computer simulations involving the kind of randomized "Monte

Carlo" calculations often used by astrophysicists to model complex systems. It was certainly beyond my fairly basic level college math, although several readers of my book asked how I was getting on with my calculations! One or two even tried themselves, using computer programs like Sandbox.

The number of variables in such a calculation is huge. Nonetheless, from the position of my cozy armchair, I advocated this mathematical approach. I predicted that a massive planet would be found to be the influencing factor at play. I suggested that it was located somewhere beyond the heliosphere, but within the inner Oort cloud (i.e. between about 200AU and 2000AU). The orbit of this massive Planet X object would be inclined to the ecliptic, highly elliptical, and have an orbital period measured in the tens of thousands of years (Lloyd 2005). Although it would be at its most influential when it was closest to the planetary zone of the solar system (during its perihelion passage), it would very likely be currently located further out towards the comet clouds (nearer aphelion).

This predicted scenario has continued to build in strength. As we shall see, the indirect evidence for such a planet has continued to grow. Arguably, it has reached the point where the existence of such an object is highly probable, statistically. The enigma is that it remains hidden from view, despite all of the hope and expectations vested in WISE. Against the odds, no one has managed to find the elusive Planet X.

Weighing up the Options

By the end of the nineties, a number of different scenarios had been put forward about the existence of Planet X. As mentioned previously, the larger the perturbing body, the further away it would have to lie to create the same effect. So, a Planet X body smaller than the Earth, and located just beyond the Kuiper belt might do the trick (Lykawka & Mukai 2008). Alternatively, a larger, Neptune-sized world within 2000AU, or even a Jupiter-sized planet within 5000AU, might also account for the anomaly (Gomes et al. 2006).

You will recall that the inner extent of the inner Oort cloud is located around 2000AU. That meant that a perturbing body located within this boundary – in the no man's land now incorporating the anomalous extended scattered disk – could be any sized world up to that of Neptune, which weighs in at about seventeen Earth masses.

If you want to propose a larger body to explain the anomaly, like a top end gas giant, then it would have to sit out in the inner Oort cloud. Clearly, such a presence would create a potential effect upon the comets located there. The inner Oort cloud is torus-shaped, contrasting with the spherical immensity of the outer Oort cloud beyond. If a massive Planet X was inclined to the plane of the ecliptic, then its effect upon the torus-shaped inner Oort cloud might only be significant when its orbital path crossed the plane.

Fig. 4-2 Disk image

This, then, might be the kind of driver for non-random comet swarm activity initially advocated by Matese et al. (1999). The hypothesis about the proposed sub-brown dwarf known as Tyche shifted ground because of Sedna. Taking into account the Sedna anomaly, Professor Matese's new preferred distance became a planet of five Jupiter masses located around 7850AU from the Sun (Matese et al. 2006).

This proposed object is in keeping with other calculations regarding realistic parameters for a range of possible planets located at various distances, and their various effects upon comet clouds (Fernandez 2010). These calculations set an upper limit to the mass of any massive solar companion to several Jupiter masses, no matter how far away it might be from the Sun.

Science vs. Sensationalism

If my *Dark Star* proposal were to work, then this body would need to be located in the inner Oort cloud, somewhere between 2,000 and 8,000AU. However, this was further out than I had proposed in 2005, and in keeping with this, the *Dark Star's* orbital period would have to be significantly larger, too.

Scientific evidence which might support the *Dark Star* hypothesis was moving steadily away from Sitchin's theory about Nibiru. It was becoming clear to me that either they were not related at all, or there were two bodies involved. After all, Nibiru was known to the Sumerians as the 'Planet of the Crossing' (Sitchin 1976). It was also sometimes translated as 'the ferry'. So, it seemed possible that Nibiru was a transiting object between the Sun and the unseen companion. We will return to this concept later in the book.

This idea made some sense to me, but by this point my original brown dwarf idea had mutated across the Internet into a colossal Doomsday Planet. A myriad of books were published about an in-coming brown dwarf set to destroy the Earth within months. YouTube videos spread the word, fueling countless articles on websites and in tabloid newspapers. The application of reasoned scrutiny showed that these claims lacked factual foundation.

Fig. 4-3 Incoming comets

I found myself becoming a skeptic of the increasingly angry doomsday prophets, while simultaneously advocating a Planet X solution to the growing anomalies emerging in the outer solar system. However, any nuanced approach on my part was largely lost in the fray of counter-cultural fervor. The Four Horses of the Virtual Apocalypse had already bolted while I was left behind to bang the stable door shut.

For the astrophysicists, however, such popular media considerations were of no relevance whatsoever. For them, any link between their proposed outer worlds and this "Nibiru" nonsense was to be discredited immediately. My interest in potential habitability might have provoked the odd raised eyebrow among them, but even generic discussions about alien life at the microbial level can cause stomach ulcers among astronomers.

Astrophysicists interested in the outer solar system continued to concentrate simply upon the evidence emerging from the growing number of extended Trans-Neptunian Objects (ETNOs) now being almost routinely picked out of the darkness of the night sky. Some of these objects, like the Sednoids, defied any explanation based upon the standard model of solar system formation. In response to this conundrum, academic papers arguing for a Planet X solution to explain the ETNOs, and other anomalies, grew in number (e.g. Gomes & Soares 2012; Iorio 2012).

A Pattern Emerges Among the Sednoids

An important case in point was the discovery of a Sedna-like object, named 2012VP113. Its perihelion is located well beyond the Kuiper belt, at 80AU, and it was grouped in amongst the Sednoids. 2012VP113 is a member of the extended scattered disk which appears to have somehow become detached from the main Kuiper belt. Because of this disconnect, the Sednoids are not influenced by Neptune in the same way regular TNOs in the Kuiper belt are.

The discoverers of 2012VP113, astronomers, Chad Trujillo and Scott Sheppard, wondered whether this steadily growing scattered disk could be a missing link in the outer solar system. Perhaps the Sednoids were dynamically linked to the more distant inner Oort cloud (Trujillo & Sheppard 2014)? In other words, rather than them being TNOs which had been scattered outwards, these extended scattered disk objects might have instead been scattered inwards from the inner Oort cloud beyond.

As intriguing as this notion was, another even more remarkable observation appeared in their landmark paper. Trujillo and Sheppard had discerned a distinctive pattern emerging among the orbits of the various members of the extended scattered disk. This included Sedna, 2012 VP113 and ten others, all of which generally resided beyond 150AU.

Common Arguments of Perihelion

All twelve of these objects move through their perihelion passages around the same time as they cross the plane of the planets (Crockett 2014). In astrophysical jargon, there was a noticeable alignment of their "arguments of perihelion" (sometimes also referred to as the "argument of periapsis" (ω)).

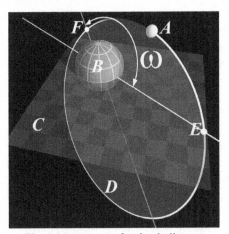

Fig. 4-4 Arguments of periapsis diagram

The accompanying diagram shows the elliptical orbit described by object A (in this case, a Sednoid). The Sun (denoted by B in the diagram) resides in one of the two focal points of object A's elliptical orbital path. The orbits of the planets sit upon a common plane known as the ecliptic (C).

If you draw a line through the longest points of an elliptical orbit (i.e. through the perihelion (F) and aphelion points), then this line is the main axis of the ellipse (the red line in the diagram). If an object's orbit (D) is not aligned with ecliptic, and is tilted away from the ecliptic in 3-dimensional space, then you could also draw a second line through the points (or "nodes") where its orbit intersects the ecliptic (the green line in the diagram). The argument of perihelion (ω) is the rotation angle of the object's main axis with this second line drawn towards the "ascending" node (E).

This is not an easy concept, so don't worry if you can't picture this. What is important is that the calculated value of ω for all twelve of these Sednoids was approximately zero.

This alignment seems way too much of a coincidence. Although things could have started out this way, over the course of the lifetime of the solar system this neat arrangement should have naturally scattered out, and become more random in nature. This is what is observed, for instance, among the objects in the Kuiper belt. The range of arguments of perihelia among the regular KBOs is broad; but this is not the case for the Sednoids beyond.

Why would the unleashed ETNOs, free of Neptune's influence, demonstrate a stronger alignment of properties compared to their shepherded cousins in the scattered disk? This did not seem to make a lot of sense.

The Sednoids and Planet X

Contemplating this mystery, Trujillo and Sheppard concluded that something was holding the perihelia of these twelve extended scattered disk objects in place (Trujillo & Sheppard, 2014). They proposed that the orbital patterning of the Sednoids likely pointed to the presence of a substantial Planet X object. Their version of this planet was between two and fifteen times the mass of the Earth. Its average distance from the Sun (or "semi-major axis") is projected to lie somewhere between 200AU and 300AU (Jenner 2014).

However, this range was considered too short by the Italian astrophysicist Lorenzo Iorio. He argued that a Planet X body causing this effect among the Sednoids should have an average distance of between about 500AU for the smallest size, to perhaps more than a thousand astronomical units for the largest (Iorio 2014). His calculations took into account more up-to-date information about the exact positions of various outer solar system planets, and the gradual changes observed in their orbits over time.

By this point, WISE seemed to have ruled out the presence of additional gas giants in the outer solar system (Clavin & Harrington 2014). However, a terrestrial-sized object seemed to satisfy most commentators: It might just be small enough to have evaded detection by infrared sky surveys, particularly WISE (Hand 2016). WISE had conducted a short survey, set at lower wavelengths, which might just have been sensitive enough to detect a distant planet this small. However, this particular scan had covered only 20% of the sky.

Arguments against taking this patterning too seriously were made by planetary scientist Dave Jewitt of UCLA. He pointed out that most of the twelve sednoids entered the Kuiper belt at their perihelion points. So, they may have been close enough to Neptune for it to be the not-particularly-mysterious organizing influence.

There was also the potential for this small sample being prejudiced by the way astronomers seek these objects out – a case of "observational bias" (Crockett 2014). However, this last criticism was discounted by Spanish astronomers Carlos de la Fuente Marcos & Raúl de la Fuente Marcos, of the Complutense University of Madrid (2014). Instead, they proposed a two-body solution to explain the clustering effect. In other words, there might be two Planet X objects involved, pulling in different directions.

Having opened up a whole new vista of Planet X astrophysical research, Trujillo and Sheppard seemed content to let things stand there. Scott Sheppard remarked in a later interview that studying the dynamics of what was going on was not really their forte (Hand 2016). Nonetheless, their pioneering work nudged others to explore the matter further.

Caltech's "Planet Nine"

The pattern of arguments of perihelion among the Sednoids piqued the interest of charismatic planet-hunter Mike Brown of Caltech, who, you will recall, found Sedna back in 2003. After his discovery of yet another anomalous outer solar system object, now known as 90482 Orcus, Brown discussed publicly his belief that it was simply a matter of time before a major discovery was made in the outer solar system (C.I.T. 2004).

Intrigued by Trujillo and Sheppard's findings, Brown approached his departmental colleague Konstantin Batygin, a dynamicist (one who researches and investigates dynamics) and computer modeling expert, to help crunch the numbers. Together they investigated Trujillo and Sheppard's hypothesis by plotting out the positions of the twelve extended SDOs, and created a theoretical model of the outer solar system. They then used computer simulations to test whether the perceived patterning of these Sednoids might provide clues to finding new objects in the outer solar system, like undetected planets.

As they plotted out the objects, they realized that there was an additional dimension to this clustering that Trujillo and Sheppard had not spotted (Hand 2016). A subset of six of the Sednoids made closest approaches to the Sun in a manner that was physically clustered in space. Combined with the alignment of their arguments of perihelion, the data strongly implied that they were being affected by another, as yet unseen planet.

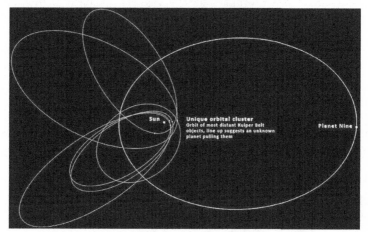

Fig. 4-5 Planet Nine orbit diagram

The Caltech team crunched some numbers to work out the probability that such an alignment could simply be coincidental. At the time of writing their first "Planet Nine" paper in 2016, that chance stood at 0.007%. The only way this set of circumstances could have arisen, they argued, was that the objects have a common exterior force which was demonstrably influencing their orbits. The more similarly-aligned objects are added to a cluster of this sort, the more the probability of this being a coincidence disappears towards zero. Six objects seemed to create a sufficiently strong pattern to make the heady claim that Planet X had to exist (Batygin and Brown 2016).

Others were not so sure. After taking into account the Caltech argument about the clustering in physical space, Sheppard estimated the probability that the mystery planet exists at about 80% (Kaufman 2016). This cautious appraisal was still strongly suggestive of the existence of Planet X.

However, Jewitt argued that Caltech's halving of Trujillo and Sheppard's original cluster of twelve down to six weakened their overall argument (Hand 2016). He wanted to see more Sednoids align with Brown and Batygin's six before he would accept that something suggestive of the existence of Planet X was really going on out there.

Nonetheless, Dr. Brown remained bullish. This clustering of six Sednoids occupied one hemisphere of the outer solar system for the majority of each of their orbits. The computations Batygin and Brown had made indicated to them that the perturbing body must lie in the opposite hemisphere. Planet X was in an "anti-aligned" position,

Timeless Voyager Press

they argued. Furthermore, its orbit was highly inclined to the ecliptic, or plane of the planets.

According to the Caltech model, Planet X's perihelion lies in the same direction as the six objects' aphelion. The orbits of the six cross through the orbit of Planet X, but the timing of their circuits mean that they do not actually meet.

Caltech created a rather elegant image to accompany the press-release about this 'new' hypothesized planet. Combined with Brown's considerable reputation as a planet-hunter, this announcement in January 2016 became big news among the world's media. This may have been helped considerably by the Caltech team's decision to re-brand Planet X. Their proposed new body would henceforth be called 'Planet Nine'. This may have reflected Brown's self-confessed part in the demise of Pluto as a recognized planet (Brown 2010).

As noted in Chapter 2, the original "X" in Planet X had not stood for "ten". The concept of Planet X had been around long before Pluto's discovery. So, this re-branding was not strictly necessary. Either way, Batygin and Brown felt that a change in nomenclature was required, and Planet Nine was born.

The Properties of Planet Nine

Unfortunately, the orbital properties of Planet Nine were vague at best. It was not possible to pinpoint the object's exact position in space from the available data. It certainly had not been actually observed. Instead, the team had determined that the effect that was observed in the clustering could be attributed to an object moving along a set of orbital paths. Their dynamical computations provided a number of possible solutions to the problem, rather like probability waves in quantum mechanics. The clustering provided a series of constraints for the possible position, mass and orbital trajectory of the planet. For the time being at least, the "discovery" of Planet Nine has remained a theoretical construct encased within a set of parameters.

If this sounds woolly to you, then you're right. This was less a discovery than a complex probability solution. Still, the Caltech duo seemed to have gone further than any previous research to confirm the existence of Planet X... even if only theoretically.

Brown and Batygin proposed that their Planet X object orbits, on average, twenty times farther from the Sun than Neptune, at an average distance (its semi-major axis) of about 600AU. The shape of the planet's orbit would be highly elliptical. The new planet's orbital period would lie somewhere between 10,000 and 20,000 years. This shifted the average distance for Planet X more than twice as far away as Trujillo and Sheppard's original work (2014), having taken into account Iorio's corrections (2014).

This new Planet X body had an orbit several times greater than that specified by Zecharia Sitchin for his 'Nibiru' (Sitchin 1976). Planet Nine did not appear to be Nibiru.

However, somewhat incredibly, the orbital description for Planet Nine fits very well with my back-of-an-envelope proposals over ten years before (Lloyd 2005), which I later incorporated into my novels *Ezekiel One* (Lloyd 2009) and *The Followers of Horus* (Lloyd 2010). As a result, I was very excited about the announcement of Planet Nine, although I recognized that my sub-brown dwarf had been scaled downwards considerably to make it fit.

As we have seen, the parameters for the mass of Planet X can vary considerably. In order to be causing a gravitational effect upon objects relatively close to the Kuiper belt, the range of masses expands upwards with distance. The closer Planet X is, the less massive it should be to account for the observed effect. At an average distance of 600AU, Brown and Batygin provided a range of masses lying somewhere between five and fifteen times that of the Earth.

If the orbit's average distance (its semi-major axis) shifted outwards beyond 700AU, taking the planet's aphelion position closer towards the inner Oort cloud, then that range of mass values could climb considerably higher. Brown and Batygin settled upon ten Earth masses as a reasonable median within their range of values. This proved to be a straightforward, media-friendly solution.

At the upper end of their scale, the new planet had a mass edging towards that of the so-called ice giants Uranus and Neptune (14 and 17 Earth masses respectively). Uranus is a small, very cold gaseous planet with what is thought to be an icy inner core. Neptune's thick, gaseous exterior also gradually transitions down towards an icy, watery mix within. Additionally, Neptune is thought to have a rocky core at its center. Was Planet Nine an ice giant? Perhaps.

At the lower end of the proposed mass range, Planet Nine might instead be classified as a "super-Earth" object. "Super-Earth" is the name given to the most common known form of exoplanet. They are several times more massive than the Earth, but not quite as massive as the ice giants. It is assumed that these worlds are similar to Earth, only bigger and denser. Even though no one really knows for sure, it remains unclear what a hybrid terrestrial/ice giant planet of, say, ten Earth masses, might look like (Fesenmaier 2014).

To complicate matters further, the object in question would be located in the frigid backwaters of our own star system. This habitat is quite unlike the super-Earth exoplanets that have, to date, mostly been discovered because of their close proximity to their parent star. So, even if we discover a great deal about the super-Earth exoplanets in the current exoplanetary catalogue, this might not give us a clear picture of what our Planet Nine would look like, or what it might be composed of.

Nonetheless, whether it was a super-Earth or a small ice giant, this proposed new object was definitely a substantial new planet. Just the thought of its discovery caused quite a stir.

Re-Branding Planets

The press release about Planet Nine came out on 20th January 2016, quickly making global headlines. At the same time, Batygin and Brown's academic paper (2016) was made available Online. I had been keeping a watchful eye on the extended scattered disk anomalies, and had previously reported on Trujillo and Shepherd's paper (2014) in my *Dark Star* blog (Lloyd 2014). Furthermore, as we shall see, another serious Planet X related 'discovery' had been publicized in late 2015. These recent developments placed the Caltech announcement into an emerging narrative (Lloyd 2016).

The ALMA 'Sightings'

In the months leading up to the announcement of Planet Nine, a great deal of excitement had been generated about Planet X within astronomical circles.

Fig. 5-1 Ancient Constellations over ALMA image

Two announcements had been made by teams of scientists working at the Atacama large millimeter/sub-millimeter array (ALMA) in Chile. In December 2015, they had jointly, but independently, announced the serendipitous discoveries of two quite separate candidate objects in the outer solar system (Liseau 2015; Vlemmings 2015).

Mike Brown, and other leading outer solar system experts, quickly questioned the scientific strength of this unpublished work, which had been announced quite noisily in the media at the time (Billings 2015). Brown's concerns were well founded. It turned out that the two papers had not yet undergone peer review for journal publication. This proved to be a serious error of judgment. The two ALMA scientific teams were criticized for having released their data prematurely into the public domain.

However ridiculed the subject of Planet X might be, one thing should never be doubted: The scientific stakes involved are considerable. One might speculate that the ALMA scientists were anxious to get the story out before someone else beat them to it. If they had gone through the due peer-review process first, concerns probably would have been raised prior to publication, saving embarrassment. As it turned out, both teams had made critical errors in their arguments.

Both of the ALMA teams hastily withdrew their papers in light of some public criticism by noted scientists. Because skeptical scientists have a general tendency to offer rather more guarded responses about the work of other scientific teams in public, the tone of Brown's critique in particular (Kaplan 2015) indicated to me that something was up. I sensed movement in the Planet X waters.

Symbiosis and Science

Planet X research is often a fringe topic, laced with romanticism and not a little obsession. There are many unqualified researchers actively engaged in the subject, more often than not predicting the end of the world. They can quickly gain a wide audience. Then there are serious academic scientists who become quite convinced that another planet exists, and get the Planet X bug. In contrast, most of their colleagues consider Planet X to be nothing more than a myth.

As a researcher, I don't fit into either of these camps. I have been privileged to receive a scientific education, and am familiar with scientific jargon. I can sometimes recognize important details that professional scientists quietly bury in their work while at the same time analyze trends in the progress of the scientific work. Often I can make loosely-based predictions that sometimes turn out to be correct years later. Among many other things, I had predicted a clustering relationship over ten years before: I had argued that it should be possible to extrapolate the position of Planet X from the emerging relationship between scattered disk objects directly affected by it (Lloyd 2005).

But, I'm not an academic astrophysicist, and so cannot describe myself as an *expert*. It is difficult to know whether serious scientists pay the slightest bit of attention to the writings of non-experts in their field. On the most part, I suspect they don't.

During the days following the announcement of the likely existence of Planet Nine, an odd thing happened. I remembered that there had been a B-movie called **Plan 9 From Outer Space**, starring the ill-fated Bela Lugosi (Wood 1959).

Fig. 5-2 "Plan 9 from Outer Space poster image

My chronically warped sense of humor got the better of me, and I decorated my blog with an image of the original theatrical release poster of the movie. Underneath the image, I wrote the tongue-in-cheek question "Planet 9 from Outer Space?"

Within a couple of days, the same movie poster image (but this time cleverly altered to read "Planet 9 From Outer Space") appeared on Mike Brown's blog. Later, Batygin entitled his 2016 Watson Lecture at Caltech with the same phrase. We seemed to share a collective sense of humor.

The Internet has opened up academic writing to a wider audience, and therefore public scrutiny. Many scientists enjoy sharing their work with non-experts. I think that is a good thing. It makes science more accessible, and invites young people to get involved. This connectivity can work in both directions. Potentially, it can provide an interface that enables people to question scientific arguments, suggest new ideas, and even to get involved in research.

That argument can only extend so far, however. The problem is that you cannot really democratize science. Science relies heavily on expert knowledge and is necessarily didactic in its approach to the dissemination of knowledge. Nonetheless, there is a place for creative minds to share and explore ideas. As a result of the Internet, those ideas can be disseminated widely, and then go on to influence others in curious ways. I suspect that my popular blog piece about the announcement of Planet Nine had been read by someone in the astrophysics camp. This humor cluster seemed to transcend coincidence. Without doing the probability calculation, I would say that this indicated healthy connectivity.

Planet X and Pluto

Mike Brown is an important and celebrated astronomer who claims to have been instrumental in the reduction of Pluto's longstanding status as planet to that of a minor planet (Brown 2010). The decision to demote Pluto was made at the International Astronomical Union's General Assembly held in Prague in 2006 (IAU 2006). This was a highly contentious decision which still stirs up strong feelings among astronomers and members of the public alike. Perhaps this is because Pluto was the only planet of the original line-up which had been discovered by an American, lifting its political status among many opposed to it demotion.

That notwithstanding, Pluto's planetary status was becoming untenable on a practical level. Its demotion to dwarf planet in 2006 was based upon the increasing number of other Pluto-like objects found within the Kuiper belt. If you count Pluto as a planet, then why not include other, similarly sized bodies, like Eris? A number of solar system objects are larger than little Pluto, among them a widening selection of KBOs. Forget Planet X; how about Planet Y, Planet Z, etc.? We could quickly run out of letters.

It seems likely to me that this rather dry academic argument about nomenclature was never meant to attract the amount of public irritation that it did. I suspect that the public ire generated by their collective decision caught everyone at the IAU by surprise. Perhaps for Mike Brown, being the person who discovered the 'real' ninth planet would restore a certain cosmic balance.

There is a more serious issue here, as well. As we have seen, this re-branding of Planet X flies in the face of well over a century of tradition. The original hypotheses put forward for another massive planet beyond Neptune were proposed well before Pluto had been discovered. The proposed object was referred to as "Planet X" even then.

Fig. 5-3 Comparison of Kuiper Belt objects image

The original name was not an indicator of its place in the existing order, but rather an indication of it being an unknown quantity. The name may now be toxic. Beyond its quirky association with old B-movies, like **The Man from Planet X** (Ulmer 1951) and **The Strange World of Planet X** (Gunn 1958), it is now strongly associated with the idea of a doomsday planet, partly due to the writings of Zecharia Sitchin (1976).

For serious astronomers looking for additional major planets in the solar system, these associations can make it difficult to look their more skeptical colleagues in the eye. There is a strong ridicule factor around the subject. I am aware of strong peer pressure exerted upon young scientists to avoid getting wrapped up in "that Planet X nonsense".

Brown and Batygin have managed to break through a barrier, asserting some authority upon a subject which is fraught with ambiguities. In that sense, their re-branding of Planet Nine has had some very helpful consequences. But the new evidence for Planet X's existence really represents an important step along the long winding path towards its eventual discovery. Planet Nine had been cleverly presented in a very media-savvy way, and had been backed by an important, ground-breaking paper.

The Caltech press release was supported by some beautiful imagery, including an elegant, multi-colored clustering of sednoids, and a very distinctive, artistic impression of a deep blue planet floating in the void. The public imagination had been successfully captured.

The essential groundwork for the Planet Nine story had been laid by Trujillo and Sheppard, who had originally spotted the anomalous clustering of the ETNOs (2014). They determined that the arguments of perihelia of the most distant of the scattered disk objects were commonly around zero degrees. This strongly implied the existence of a super-Earth object located around 250AU from the Sun.

Batygin and Brown's additional realization about the clustering in physical space extended the argument further, and made the existence of an additional planet more likely still (2016). The Caltech scientists had provided the world with a robust case for a substantial Planet X object. Their case was built upon statistical probability. Having assumed that the objects in question could be stable over the long-term, Batygin and Brown showed that "with some fine tuning, this scenario could be feasible" (de le Fuente Marcos 2017).

Although many astronomers had inferred Planet X from their datasets of outer solar system objects, the Caltech team was able to argue that the odds of it being out there had improved dramatically. With the announcement of Planet Nine, scientific reputations had been put on the line for an object that had spent many years consigned to the wilderness. Suddenly, it was okay for scientists to talk about Planet X again.

The planet-hunters from Pasadena had boldly laid claim to Planet X, hoisting their deep blue colors over the Planet X vista, and re-naming the territory in honor of the fallen planet Pluto. The hunt for Planet Nine was on.

The Hunt Begins

As we have seen, Brown and Batygin's theoretical work served two purposes. Firstly, it made the case for the existence of a substantial Planet X body much stronger. It had achieved this by examining the close relationship between a cluster of objects lying beyond the Kuiper belt, and showing that the probability of this relationship being attributable to mere chance was vanishingly small.

Secondly, their work helped to narrow down the range of orbital values for their Planet Nine. It still wasn't possible to triangulate its exact position, though, as I had naively hoped for a decade before. Instead, computer simulations had come up with bands across the sky where this Planet X object might be, and helped to eliminate other huge swathes of the sky from any future search. With these sets of parameters, Brown professed confidence that the discovery of Planet Nine was at hand. It should pop up within a couple of years if a concerted effort was made to find it.

Fig. 5-4 The Maunakea Observatory image

Brown hoped to conduct at least part of that search himself, using the powerful 8.2-meter optical-infrared Subaru telescope at the summit of Maunakea, Hawaii. The telescope is operated by the National Astronomical Observatory of Japan. In the autumn of 2016, following the release of their paper about the likely existence of Planet Nine, the Caltech team secured a sizable chunk of valuable time on the Subaru telescope.

The astronomers favored an area close to Orion as their prime hunting ground. They had less than a week to cover the roughly 2,000 square degrees of sky they wished to search. Really speaking, this area required twenty nights of uninterrupted observation to properly do it justice. But, if luck was on their side, and they happened to search in the right chunk of sky during that time, then maybe they'd turn up the find of the century. If they were not so lucky, then they could at least eliminate parts of the sky from their ongoing inquiries. Unfortunately, their search turned up nothing.

You might wonder why the Caltech team went public with their theoretical findings at such an early stage. After all, they could have sat on their computerized correlations, adding data from new extended scattered disk objects as they emerged from the darkness of the outer solar system. The more data gained, the better the likelihood of pinpointing the position of Planet Nine.

But that would have taken time, and risked their unpublished theoretical work being surpassed by a chance discovery of Planet X by others. Even though the claims of discoveries of massive Planet X objects made by the ALMA teams in December 2015 had proven erroneous, the very fact that these claims could emerge so suddenly from large teams of international researchers showed how high the stakes were.

As the old English proverb goes, a bird in the hand is worth two in the bush. Better to get what you've got out there and published, than sit on it, risking obsolescence. Besides, invoking another idiom, many hands make light work. If the theoretical prediction for the presence of Planet Nine gained traction within the astronomical community, then more teams might engage in the hunt. More importantly, perhaps, a greater array of instruments would be used by those various teams, and more precious telescope time would be turned over to the hunt for Planet X.

I think that the Caltech duo took what was a calculated risk to release the results of their computer modeling in such a public way, relying upon the strength of the theoretical data. Would their claim for the re-branded Planet Nine become a laughing stock among their skeptical peers? Or would it galvanize other astronomers and astrophysicists into a concerted effort to locate this object once and for all?

In the months following the release of the original Planet Nine paper, there was a flurry of Planet X-related research activity. An unusually large number of papers were published in the scientific press on the subject, mostly by scientific teams with a longstanding interest in the subject. Astronomer, Scott Sheppard has likened the current search for Planet Nine to the run-up to the discoveries of other outer solar system planets – Uranus and Neptune – and the dwarf planet Pluto (Kaufman 2016). There was a palpable expectation that something big was about to pop out of the woodwork.

Constraining Data

Without being able to observe the planet directly, astronomers were reliant upon its perturbing effects upon other bodies in the solar system as a guide to its possible position in the heavens. Because of the very great distances involved in the case of Planet Nine, compared to the relatively neighborly locations of Uranus and Neptune, the gravitational effects it engenders upon other solar system bodies is tiny. As a result, the measurements of the positions of other bodies in the solar system have to be very precise if there is any hope of detecting the kinds of tiny changes which might be attributable to a distant missing planet. Otherwise, any discernible effect would be lost within experimental, or observational, error.

Some of the papers which were published after the Planet Nine announcement considered how such an object would impact upon solar system "ephemerides". There are the positions of natural or artificial objects (like satellites and space probes) in space at any given point in time. A particularly strong candidate for this inquiry is the planet Saturn, whose position and movement is now known as accurately as the inner planets, thanks to the space probe Cassini (Iorio 2017).

Analysis of the Cassini ranging data provided valuable information about Saturn's ephemeris, and this was used to further constrain the parameters of Planet Nine's proposed orbital path (Fienga et al. 2016; Holman & Payne 2016a). This work was also taken into account in subsequent computer simulations trying to pinpoint the current location of Planet Nine. These simulations produced another series of zones in the sky where the putative planet was most likely to reside (de la Fuente Marcos & de la Fuente Marcos 2016).

If Cassini had been allowed to continue its mission for a few more years, then the future data that would have emerged about Saturn's ephemeris would have been even more helpful in this regard. Unfortunately, the decision was made to finish Cassini off, culminating in a controlled fall into Saturn's atmosphere in September 2017.

But Saturn is not the only planet in town. Careful study of the position of the dwarf planet Pluto, and possibly fellow TNOs, could also help to reduce the parameters for the proposed Planet Nine's location in space. These objects are located much further away from us than Saturn, and their loosely bound orbits provide fertile hunting ground for gravitational perturbation effects caused by an unknown body beyond. After the long-period comets, TNOs are among the most loosely bound bodies we know of in the solar system. They are more readily susceptible to Planet Nine's charms than the other planets bound more closely to the Sun, and the more we learn about them, the closer we get to nailing Planet X.

Because of its tiny size, and immense distance, Pluto's position was not as well documented as Saturn's. Fortunately, new data emerged from a fresh analysis of historic photographic plates of Pluto, interposed upon modern stellar catalogs (Buie & Folkner 2015). This re-examination had been undertaken to provide more accurate information to better support the New Horizons probe's flyby of Pluto.

Fig. 5-5 Kuiper Belt Object artist concept image

Fittingly, it was conducted by astronomers at the Lowell Observatory, the site of Pluto's original discovery. Interestingly, one of the authors, Marc Buie, spoke of a possible massive Planet X back in 2002, in response to findings about another anomaly. This anomaly involves the truncation, or cutting short, of the Kuiper belt, which occurs around 50AU. This border zone is known as the "Kuiper Cliff". Buie considered it likely that an unknown planetary body had been sweeping out the zone beyond the Kuiper belt (Couper & Henbest 2002). It seems that most people who study the outer solar system end up considering Planet X at some point or another!

Anyway, astrophysicists from the Harvard-Smithsonian Center for Astrophysics took a look at this new data about Pluto's position. Based upon their analysis, they concluded that Planet Nine is likely to be either larger than the proposed super-Earth object of around ten Earth-masses or failing that, the object would have to be much closer than had been previously suggested (Holman & Payne 2016).

Far from ruling out a Planet X object, these new papers demonstrated that other anomalies are present in the outer solar system that might be explained by the presence of an unknown planet. It could even turn out to be larger than expected. This begs the question of how Planet X has managed to remain hidden for so long.

Two Planets for the Price of One?

The domain of Planet Nine seems to lie in the substantial gap between the outer edge of the Kuiper belt (the Kuiper cliff, at about 50AU), and the inner edge of the inner Oort cloud, which starts from about 2,000AU. The predicted approximate values for Caltech's Planet Nine place it from about a third to half the distance through this zone. Batygin and Brown predicted a semi-major axis (its average distance) of 700 AU. It would be inclined to the ecliptic by about 30°, and orbit the Sun in a highly elliptical orbit (value = 0.6), but it would not come near to the rest of the planets (2016).

As we have seen, the original work by Trujillo and Sheppard, which preempted the Planet Nine paper, placed Planet X beyond 200AU (Trujillo & Sheppard 2014). The Caltech work set the planet much farther back, perhaps three times as far. The variability of Planet Nine's orbital elements creates some confusion, and different research groups propose different sets of values on the unknown planet's orbital properties.

A Planet X body lying close to the Kuiper belt might be Mars-sized. Push it a bit farther out, and you would need an Earth-sized world. Another 500AU or so, and Planet X becomes a 'super-Earth' of about 10 Earth masses. Once Planet X is among the comets, it ranges from a Neptune-sized ice giant, up towards a gas giant, and eventually a sub-brown dwarf lying at the edges of the solar system. This planetary spectrum creates a wide variation in the possible identity of Planet X.

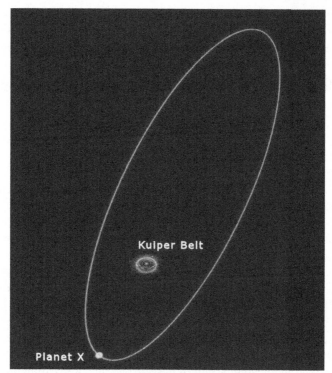

Fig. 5-6 Orbit of Planet X

Other research groups have estimated the distance of an undiscovered Planet X body, based upon other perceived correlations. A Spanish team of astrophysicists have conducted a great deal of work on the subject of Planet X over recent years. After the Planet Nine paper was published, the team conducted numerical N-body simulations using powerful computer programs, to test the dynamic stability of the Planet Nine cluster over time (de la Fuente Marcos, de la Fuente Marcos & Aarseth 2016).

Their calculations provided additional constraints upon Planet Nine, and led them to propose that there must be more than one Planet X body out there. One Planet X body alone seems insufficient to maintain the clustering of ETNOs (de la Fuente Marcos & de la Fuente Marcos 2016a). They argued that, instead, a multiple planet scenario was needed to properly explain the data from the extended scattered disk cluster (Spanish Foundation for Science and Technology 2016).

The Spanish team had proposed this before (2014), prior to the Caltech announcement about Planet Nine. Their work on ETNOs followed Trujillo and Sheppard's paper on the extended scattered disk object 2012 VP113 (2014). It included an analysis of how the orbits of these objects may have been shepherded by Planet X. The team considered the existence of two Planet X bodies, located at 200AU and 250AU respectively. These additional planets would clear the area of objects through the action of their combined resonances. They argued that, between them, these distant, undiscovered planets explain the permanent clustering of ETNOs, and their anomalous orbital properties, better than, say, early interactions with the Sun's primordial siblings in its original birth cluster.

Planet Nine and the Riddle of the Sun's Obliquity

At about 30°, the large tilt of the orbit of Planet Nine could explain other unexplained features of the solar system, as well as the observed clustering of extended scattered disk object beyond the Kuiper belt.

It is an odd, undisputed fact that the Sun does not spin on its axis exactly perpendicular to the plane of the planets – as you might expect it to – instead it is tilted by six degrees. Therefore, there is a precession effect involving the 6° tilt between the invariant plane of the giant planets, known as the ecliptic, and the solar equator.

The Sun's axial tilt is an anomaly which Brown and co-workers attribute to Planet Nine (Yuhas 2016). According to their model, the putative super-Earth is massive enough, and has an orbit which is so significantly inclined from the plane of the other planets, that the solar system has slowly twisted out of alignment with the Sun (Bailey, Batygin & Brown 2016). They think that Planet Nine offers an elegant explanation for the spin-orbit misalignment of the solar system. Its high inclination off-sets the overall angular momentum vector of the Solar System away from the ecliptic.

Fig. 5-7 Large image of Planet Nine

The team used an analytic model to investigate interactions between the highly inclined Planet Nine body they have proposed, and the remaining giant planets. Starting from a flat initial state for the solar system (as you would expect from a conventional protoplanetary disk), and adding in an additional Planet X body with the kinds of parameters currently considered for Planet Nine, the model shows that the distant planet was capable of causing the observed obliquity and pole position of the Sun's spin axis.

The key to this quite dramatic effect is the high inclination of Planet Nine's projected orbital plane with respect to the rest of the solar system. This seems to be a more important factor than the object's mass. The high tilt to the plane of the ecliptic is key. This point was also made by Alessandro Morbidelli, of the Côte d'Azur Observatory in Nice, France (Boyle 2016).

Over very long time periods, computer simulations independently deployed by both teams show that Planet Nine might generate the observed solar system anomalies naturally. Significantly, this would imply that Planet Nine would have taken its place within the ranks of the planets quite early on in the lifetime of the solar system. This, in turn, implies that its orbit is a stable feature of the solar system, rather than the result of a relatively recent capture of an exoplanet.

It should be pointed out that there are other possible explanations for the tilt of the rotational axis of the Sun, and the relative obliquity of the Sun's equatorial plane compared to the invariant plane of the ecliptic. But the existence of a Planet X object at a high inclination may well provide the most straightforward explanation for these effects, as well as the anomalies of the extended scattered disk beyond the Kuiper belt.

However, Planet X might need to be considerably larger than a super-Earth to succeed in this endeavor. Notably, earlier work by Batygin considered whether the obliquity effect might have been caused by the early presence of a binary companion to the Sun (Batygin 2012). Such a massive body, if suitably inclined, would have had the gravitational capacity to warp the solar system into its present configuration. This concept is more in keeping with my own view about a massive Planet X, and suggests that there is some leeway to shift Planet Nine's mass upwards to properly capture all of these effects.

The Hunt for Planet Nine

As we have seen, it is not possible to pinpoint the exact location of the missing planet through triangulation. The physical clustering in space of a set of extended Trans-Neptunian Objects (ETNOs), otherwise known as the Sednoids, insinuates the presence of a shepherding body way beyond the Kuiper belt. However, this is insufficient to prove its existence, much less to nail down where it is. Instead, it creates a broad set of bands across the sky where the object might be found. Hopefully, it eliminates the rest.

Despite this ballpark approach, different research groups favor different locations. The Caltech team's favored location for Planet Nine is in the vicinity of the constellation of Orion, edging towards the zodiacal constellation of Taurus (Batygin & Brown 2016). They judge this to be the zone where Planet Nine reaches its farthest distance from the Sun (aphelion). It also happens to be a rather lovely part of the sky, which is readily identifiable by most people with just a notional interest in the sky.

It seems likely that Planet Nine should be towards the back end of its eccentric orbit. If it was at its closest approach to the Sun, there is more chance that it would have been located by now. Brown and Batygin consider the closest approach (perihelion position) of Planet Nine to take place somewhere within an equally broad expanse of sky in the vicinity of the zodiacal constellations Scorpius/Sagittarius (Batygin & Brown 2016). This is in the diametrically opposite part of the sky. It could be there, too, but they think this unlikely. If it was, this part of the sky would present its own set of problems, as we shall see.

The Music of the Spheres

The Caltech duo's conclusions are built upon the assumption that the cluster of six ETNOs is anti-aligned to the orbit of the perturbing planet. They argue that the bulk of Planet Nine's orbit lies in the opposing portion of the sky to the bulk of the orbits of the Sednoids. Although it would be wonderful if we could now simply say "X marks the spot", it is not possible to simply triangulate the position of Planet Nine from the position of the objects it is affecting. This is because all of these objects are constantly moving, along different orbital paths, at different speeds.

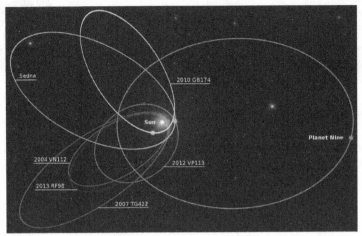

Fig. 6-1 Planet Nine orbit diagram

The relationship between Planet Nine and the ETNOs amounts to a stable pact between them, built up over many millions of years of interaction. Kathryn Volk and Renu Malhotra of the University of Arizona have suggested that there may be some resonance effects at play within these orbital pacts (2016). These long-term relationships serve to harmonize orbits. You could almost argue that these integer relationships create a "music of the spheres", which is a rather poetic concept from older times. If such a set of relationships did exist between Planet X and the Sednoids, it would help to narrow the planet's location down. Bear in mind, though, that it takes a super-computer to untangle these kinds of puzzles: The number of variables here are vast. At the moment, the available data can only provide probabilities as to the whereabouts of Planet Nine.

The search for Planet Nine is like a massive missing person's inquiry – except that no one has ever met the missing person. Instead, lots of people have vague fragments of memories of this person existing and interacting with them. By gathering all those vague memories together, a stronger picture begins to emerge: Patterns of behavior; favorite locations to visit; most recent sightings. From these fragments it may prove possible to piece together a routine, and then use that to infer the location of the missing person at any given time. But in the case of this particular planet-hunt, not enough is known about Planet Nine. As a result, the search area remains uncomfortably broad.

As Viewed from Earth

There have been short-lived announcements in the past about the discovery of Planet X. The most famous examples resulted from a previous infrared sky search. In 1983, IRAS Scientists separately highlighted both Orion (O'Toole 1983) and Sagittarius (Redfern & Henbest 1983) as locations containing massive planets deep in the solar system. I have discussed these short-lived announcements at length in my previous book (Lloyd 2005).

Most mainstream astronomers show little or no interest in these historical claims. They have been dismissed by skeptics, arguing that the 'sightings' were simply misidentified dense molecular clouds or distant galaxies (Plait 2008). We will return to the misidentification of "fuzzy objects" later in this book. Nonetheless, it is interesting to note that both of these two constellations synchronize well with the projected long-axis endpoints along Planet Nine's eccentric orbit. This seems a rather odd coincidence.

Fig. 6-2 Sun Halo image

Astronomers show zero interest in the many popular claims of a planet or brown dwarf star hidden behind the Sun. As far as Planet X is concerned, there continue to be many 'recent sightings' posted by members of the public on the Internet. These are not taken seriously by the astronomy community. YouTube videos often depict a double Sun in the sky. These images and videos are readily attributable to sun-dogs, lens flares, and various other lighting artifacts which manifest when you point a camera directly towards the Sun. Atmospheric conditions, like water crystallization in cold air, can generate similar effects.

Reports of this ilk have been around for decades. If you were to take even a selection of them seriously over that time, you could infer that the mystery planet has somehow remained hidden behind the Sun for many years. The only way that could happen is if the hidden world is in an orbit which is diametrically opposite that of Earth, with an orbital period of exactly one year.

Let us say the Earth really did have a twin 'star' located in the (not terribly stable) L3 Lagrangian point, on the opposite side of the Sun from the Earth. We could never see it directly from Earth, but we would have seen it from probes sent out into the solar system. It is conceivable that some minor rubble sits undetected in Earth's L3 Lagrangian point: But not a planet.

A Planet X body approaching the Sun presents even greater problems for those who believe that we are all doomed, and that their videos can prove it. Planet X has a vast orbit, and its perihelion passage is a relatively sedate affair compared to Earth's comparative speedy jig around the Sun. As the Earth changes position around the Sun during the year, its perspective of the broader solar system shifts significantly. An in-coming planet cannot hide behind the glare of the Sun for long – a month or two at best. But Planet X's perihelion passage would take longer than this. Like a comet, it would be visible to the naked eye. It would be seen across a range of sky as it progressed through the heavens, not just behind the Sun.

Then there are the telescopes. There are many amateur astronomers on the look-out for new comets. An in-coming Planet X would stand out a mile for them, as it would for powerful sky surveys conducted by professional astronomers. The kind of grand conspiracy required to keep a lid on all of this boggles the mind. On this basis of these reasoned arguments, I can very confidently assert that Planet X is not currently located in the planetary zone of the solar system.

A Cluttered Sky

Pinpointing the location of Planet X still involves a lot of guess work, given the paucity of firm data available. The anomalous clustering of the ETNOs create strong evidence for the existence of another planet out there, but provides little concrete data on its actual whereabouts amid the broad swathe of possible orbital paths. The suspected parameters of Planet Nine's orbit are dependent upon many variables, and these create vast sections of sky in which to search for this mysterious object. There is an approximately twenty degree width for the long band of possible orbits across the sky. This is because of uncertainties in its projected inclination, and other orbital values.

Astrophysicists have been trying to make these search areas smaller. They have been working on novel ways to "constrain" the properties of Planet Nine's orbit. This is rather like ruling out suspects from a line of inquiry. If Planet Nine is located at a particular distance and location, and has a certain mass, it may necessarily affect the exact positions of other outer solar system bodies. This can help astrophysicists to rule out further areas of the sky. They can constrain Planet Nine's properties, and cut back its possible locations. The bands across the sky then become segmented, as various areas are ruled out of the search.

Some search areas are deemed more probable than others in different ways. If a given search area has been thoroughly investigated by many prior sky searches then it may be considered less likely to harbor an undiscovered object than areas which have been relatively neglected. This may be because some constellations are darker than others, unaffected by, say, the presence of the Milky Way galaxy in the background. Scientists hunting for solar system objects have tended to hunt for minor bodies in the solar system in darker areas, to cut down on the background noise which would hamper their search. This has resulted in an historical observational bias towards darker regions in the sky.

Conversely, constellations which are literally overrun with stars and other celestial features have generally been avoided by sensible astronomers hunting minor bodies. Their field of view is simply too cluttered. The constellation of Sagittarius is a prime example. To my mind, then, Sagittarius, and its environs, remains the most fruitful area to search for Planet X, as I believe this area of the sky contains the planet's aphelion position. There are other reasons for this, too, as discussed in my previous book (Lloyd 2005). However, although the Caltech team also highlight a zone close to this constellation, they consider this to be the 'wrong' side of the sky to search for Planet Nine.

If Planet Nine is located in this zone, immersed in the star field near to the center of the galaxy, then it is either at perihelion (which means it should be much easier to find), or it is aligned with the ETNO cluster.

This represents a worst-case scenario for the Planet Nine duo. The object's location in the galactic plane would explain why it is proving difficult to spot (Billings 2018). However, it would also mean that the anti-aligned pattern they initially advocated was wide of the mark (by exactly 180°). However, if Planet Nine turned out to be in the wrong place, that is still a significantly better outcome to not turning up at all…

Planet Nine and the Kuiper Belt

In 2018, the Caltech team, working in conjunction with a scientist from the University of Michigan, published a new paper which appeared to hint at a change in direction (Khain, Batygin & Brown 2018). As we shall consider in more depth in Chapter 10, the formation of Planet Nine is problematic, based upon standard models of planetary and solar system formation. Whatever processes placed it in its proposed current position would have significantly affected the layout of the Kuiper belt within its overarching orbit.

Computer simulations of the early Kuiper belt reveal the complex interplay between it, Planet Nine, and the objects in the extended scattered disk. The calculations showed that the broad early Kuiper belt generated a bimodal structure of scattered objects associated with Planet Nine – some aligned, some anti-aligned with its position in space (Khain, Batygin & Brown 2018). The averaged distances for the anti-aligned objects is similar irrespective of the shape of the early Kuiper belt. However, the simulations show a difference in the aligned objects, whose distributions vary greatly with initial Kuiper belt conditions.

The stability of the aligned objects and the anti-aligned objects seems more complex than previously thought. Different configurations of the early Kuiper belt appear to shape the ETNO cluster in different ways, allowing astronomers to potentially work backwards from future discoveries to determine how Planet Nine came to be.

There appears to be a subtle shift in emphasis about whether the objects which are aligned or anti-aligned to Planet Nine are stable. I detect movement on this issue, which may have a significant impact upon the projected location of the planet itself. I continue to suspect that the aligned/anti-aligned bodies in the initial Planet Nine cluster are the wrong way around, and that this will become increasingly apparent as more of these scattered objects are discovered.

Narrowing Down the Location of Planet Nine

Fig. 6-3 ALMA Observatory antenna image

Various research groups have added constraints to the possible orbital tracks, and available locations of Planet Nine. We have already considered how a Planet X body gravitationally affects other bodies in the solar system, creating tiny unexpected perturbations in their orbital paths. An observed absence of these effects can constrain the possible size, distance and location of an unseen planet. These constraints serve to create a patchwork of high probability areas, low probability areas, and no chance areas across the sky. Armed with this knowledge, astronomers can concentrate their precious telescope time on these areas.

Other anomalous effects can constrain the position of Planet X, as well as providing evidence for its presence. For example, one of the anomalies in the solar system that Planet Nine might provide an answer for is the "obliquity" of the Sun. We are used to the idea of the Earth being tilted with respect to its orbit around the Sun, currently 23.27°. This tilt, or obliquity, creates the seasons. As described in Chapter 5, the Sun's own rotation is tilted with respect to the plane of the planets which rotate around it (the ecliptic). The reason for Sun's obliquity, of about six degrees, is a mystery. Its rotational axis should have been perfectly perpendicular to that of its protoplanetary disk, from which the planets emerged. But that is not the case. Instead its axial tilt is 6°. This implies that the entire solar system has been warped out of shape over the last 4.5 billion years. No one knows why.

Planet Nine may offer a solution to this problem, if its own orbit is sufficiently inclined to the ecliptic. It could create the long-term gravitational tug which caused the warp in the plane of the planets. This possible relationship adds further constraints to the likely parameters of Planet Nine (Gomes, Deienno & Morbidelli 2016). Some of these new constraints are compatible with the orbital configurations for Planet Nine that allow it to affect the cluster of ETNOs. If the obliquity argument has any merit, the search area for Planet Nine can be narrowed down further.

Historical sky searches can also be of help. In an article he wrote shortly after the publication of the Caltech paper, Brown indicated the areas which these had effectively already covered (Brown 2016). These sky searches were each capable of finding objects down to a particular magnitude. The observed luminosity of the stars in our sky is measured by their magnitude. Planets like Jupiter and Venus are the brightest objects in the sky, after the Sun and Moon, and have high relative magnitudes. Bright stars like Sirius have magnitudes of a similar order to that of the visible planets. Many stars (and outer planets) are much less bright, requiring telescopes to see them. Planet Nine would have a very low magnitude, due to its great distance, making it difficult to observe among so many other similarly faint objects in the night sky.

Its great distance may cause another problem for automated sky searches. Because Planet Nine is so distant, its lateral movement across our skies is so small as to be almost negligible. As a result, many of the sky searches would automatically negate this object, assuming it to be a background star. Brown cites the more helpful searches for faint solar system objects conducted by the Catalina survey, the PanSTARRS survey, and moving object surveys – listed in decreasing orders of magnitude. In conjunction with the powerful WISE infrared search, which covered busy portions of the skies these other surveys sensible tended to avoid, Brown mapped out the "best bet" zones for Planet Nine (Brown 2016).

This early sky mapping for Planet Nine was adjusted by other researchers analyzing astrometry data of various planets. This further constrained these regions and other parameters for the planet's properties, like its mass and distance (Fienga et al. 2016; Holman & Payne 2016 & 2016a; Iorio 2017). However, despite these modifications, there remains a significant patchwork of sky left for astronomers to examine.

A Combobulation of Locations

It is not clear to me whether there is any official cross-over between all these studies in terms of a common focus for this search, although Caltech has hosted informal conferences for interested parties to share their latest Planet Nine research.

Attempts have been made to bring some of these different threads together. For instance, there are four areas of the sky favored by the de la Fuente Marcos brothers, who carried out computer simulations in an effort to constrain the values for Planet Nine, and work out its most probable location. They included comparisons with previous work by Batygin and Brown (2016), and Agnès Fienga, of the Observatoire de la Côte d'Azur, et al. (2016). The zones of interest were:

1. Most probably Hydra, away from the main bulk of the Milky Way.

2. Equally probably between Scorpius and Lupus, towards the Galactic bulge, but relatively far from the galactic center. These first two are the most probable, say the authors, but have values inconsistent with the "pseudo resonant scenario" described by Brown and Batygin.

3. Less probable, they say, is Taurus, a calculation which uses values favored by Brown and Batygin, and also Fienga et al., and is also consistent with anti-alignment (2016). A similar randomized 'Monte Carlo' simulation approach to Fienga's values additionally provides the constellation of Cetus.

4. Least likely, is a zone between Microscopium and Sagittarius. (de la Fuente Marcos & de la Fuente Marcos 2016)

Again, these become arguments about probabilities. It is interesting to note, though, that based upon this analysis, the endpoints of Planet Nine's projected long-axis are the least probable locations for its current location. Astrophysicists have widely differing opinions on where Planet Nine probably lurks.

So, what of the 'new' favored locations? Matthew Holman and Payne, astrophysicists from the Harvard-Smithsonian Center for Astrophysics, explored possible correlations between the Cassini data, and a broader range of values for Planet Nine's possible orbit (Holman & Payne 2016a). As a result of a theoretical model they developed to analyze this spread of possible values, they argued that they have been able to significantly constrain the parameters for Planet Nine's orbit, and indicate a preferred area of the sky within which Planet Nine might reside.

It turns out to be a broad sweep of sky between the constellations of Eridanus and Cetus. This sky zone correlates well with recommendations arising from Fienga's et al.'s randomized computer simulations (2016). However, this search area of the sky still remains thousands of times larger than the full Moon.

Ranging Data

I asked Lorenzo Iorio whether a consensus has emerged about the probable location of Planet Nine, based upon all of this work. His reply sums the present situation up neatly:

"No, to my knowledge. Its parameter space is still too large, even including the possible mass. The only thing it could be stated with a certain degree of security is that it should not be close to the perihelion. It will be important to have all the Cassini ranging data available after its end [in September 2017]. And, who knows? Perhaps, New Horizons' telemetry could give us some more info." (Iorio 2017a)

The Cassini mission came to end with a controlled fall into Saturn's atmosphere on 15th September 2017. The spacecraft's closer proximity to the gas giant during its final hours may have provided a finer level of detail as to its astrometry, allowing astrophysicists to fine-tune their calculations of Planet Nine's potential perturbing effect.

However, it should be noted that Cassini mission managers at JPL do not think that Cassini has been affected by any unforeseen tugs, however small. From the measurements taken between 2004 and 2016, no untoward data emerged, according to William Folkner, a planetary scientist at JPL (Dyches 2016).

The spacecraft would not itself be unduly affected by Planet Nine anyway, but if Saturn's own position and trajectory varied because of Planet Nine's distant presence, then this should have been picked up in the Cassini ranging data while the probe was orbiting the ringed planet. Arguments made to keep Cassini going, in the hope of further constraining the position of Planet Nine (Fienga et al. 2016), were not, in the end, heeded.

Fig. 6-4 Radar Dish image

The New Horizons spacecraft flew past Pluto and its moons in the summer of 2015, and continues to travel deeper into the solar system. The communications that take place between the probe and Earth offer the opportunity to measure its distance from us accurately, and provide evidence for any anomalous shifts in its trajectory over time which may be attributable to an unseen massive planet beyond.

Other sky searches, like the Dark Energy Survey, have been getting involved in the hunt, too. Scientists working on this project were already obtaining and crunching data for areas of the sky highlighted as possible locations for Planet Nine (Hall 2016).

As far as I can tell, Brown continues to favor the Orion area. I wrote to him to ask whether he was considering other zones in the sky, given these various constraints, but I received no reply.

Planet Nine Hunters Turn Up Brown Dwarf

In 2017, efforts to discover Caltech's proposed Planet Nine object brought together tens of thousands of enthusiastic, eagle-eyed volunteers. They diligently sifted through page upon page of infrared images of the sky (Zooniverse 2017) which had been taken by NASA's Wide-Field Infrared Survey Explorer. The citizen volunteers used Internet-based platforms to work through "flip-books" of infrared sky-images amalgamated from the WISE data.

Old school human processing has its uses, even today. Computer programs designed to spot astronomical needles in haystacks can get bogged down in details that humans readily dismiss as irrelevant. However, the sheer amount of data provided by WISE was way too much for even a sizable group of professional researchers to sift through. The substantial number of volunteers coming forward to search for Planet Nine allowed a highly efficient analysis of the available data. Any anomalies that were spotted were sent to them through professional astronomers for further analysis, and possible investigation using powerful telescopes. The uptake of this voluntary work was extremely impressive, allowing professional astronomy teams to delegate out a great deal of work.

The driving force was to find a Planet X object in the outer solar system, but there was also the potential for the discovery of other objects, too, like new Kuiper belt objects, comets, asteroids, and even distant brown dwarfs. The project didn't reveal the location of Planet Nine, but it did result in the discovery of a new brown dwarf located some 100 light years away, named WISEA J110125.95+540052.8. This provided something of a consolation prize (Kuchner et al. 2017).

Temperatures of brown dwarfs vary depending on their age and mass, ranging from a few thousand degrees down to room temperature. The warmest brown dwarfs are known as L Dwarfs, the next warmest group as T Dwarfs, and the coolest (many of which are categorized as sub-brown dwarfs) as Y Dwarfs. The known Y dwarfs range from 3 to 20 Jupiter-masses (Leggett et al. 2017), so this category rather confusingly includes both sub-brown dwarfs planets and ultra-cool brown dwarf stars.

You might wonder what these letters stand for. There is no particular sense to this set at all. They are simply some of the letters left over from the more mainstream categorization of star types.

The first Y-dwarf was observed as recently as 2011, and to date only a couple of dozen of these cool objects have been observed and studied. However, this small number represent the very tiniest tip of the Y Dwarf iceberg, with estimates based upon statistical analyses putting the galactic population of these objects at about a billion (Yates et al. 2017). However, as we have already noted, estimates still vary considerably. The small number so far discovered reflects their intrinsic dimness.

Not So WISE

Arguably, the potential for discovering more brown dwarfs may have been the true driving force behind the creation of the Planet Nine citizen search. Beating the bushes to find the legendary Planet X is a more glamorous prospect than seeking out brown dwarfs, after all. The motivating factor for the professional astronomers at WISE may have been an Easter egg-hunt for small brown dwarfs, especially the highly elusive Y dwarfs.

There had been disappointment among several astronomers hoping to find a massive Planet X body using WISE. They did not consider the job to have completed with any great certainty. It has been noted that two WISE searches were carried out while the space telescope was operating, and each search had spotted objects missed by the other. The implication was that there are likely to be other objects lurking in the solar neighborhood that had been missed by both (Prigg 2008). However, bids to extend scientific work on NEOWISE data, to locate more local Y dwarfs, were turned down by NASA on cost grounds (Howells 2014a).

Reading the abstract to the science paper announcing the brown dwarf's discovery (Kuchner et al., 2017), one might be forgiven for thinking that the WISE astrophysics team is taking advantage of the popularity of the Planet Nine subject to enhance their prospect of discovering sub-brown dwarfs. Given the context, who can blame them? There also remains the possibility that Planet X is an old, low mass Y Dwarf.
So, connecting the two searches in this way seems to me to be a sensible proposition.

The human eye appears to be superior to the computational methods of spotting moving objects within the WISE images, providing the opportunity to find brown dwarf objects of about a single point of magnitude fainter than those perceived by WISE's automatic processes. This indicates not only the potential benefits of bringing in outside help to comb the celestial beach for brown dwarf driftwood, but also the recognized limitations of WISE itself. These recognized limitations cast doubt on claims that WISE was able to entirely eliminate Planet X from its line of inquiry several years ago (Clavin & Harrington 2014).

In turn, the same may be argued for more up-to-date efforts to find Planet Nine within the WISE and NEOWISE databases (e.g. Meisner et al. 2017).

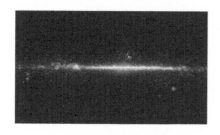

Arguments Against Planet Nine

The crux of the argument for Planet Nine hinges upon the validity of a clustering effect of extreme Trans-Neptunian Objects (ETNOs) in the extended scattered disk (Batygin and Brown 2016). As samples go, the six objects that Brown and Batygin have selected is a relatively low number. We have already seen how this whittling down of selected objects, from Trujillo and Sheppard's original population of a dozen ETNOs (2014), has encountered criticism.

Based upon statistical analysis, Brown and Batygin have argued the following: The probability that each of these six objects randomly exhibits the same common property is just 0.007%. The chance that this clustering could be attributed to mere coincidence is of the order of 1 in 10,000.

In mathematical terms, this provides a value for statistical significance of 3.8-sigma. This value relates to standard deviation of the data points from the mean. In this case, 3.8-sigma is comfortably above the 3-sigma rule of thumb required in scientific work, which is equivalent to a 99.7% certainty. However, the 'proof' for the existence of Planet Nine did not reach the 5-sigma level that would have clinched it, according to astrophysicist Dave Jewitt of UCLA, an expert on comets and other primitive bodies in the solar system (Hand 2016).

Successful inclusion of a seventh body into this cluster would elevate the statistical significance to the 5-sigma level. However, other ETNOs selected by Trujillo and Sheppard were not included in this particular set. That is not to say that every object out there should behave in the same way, but it does raise questions about which objects to include or exclude, and this can skew the statistical analysis.

Advocates of the existence of Planet Nine expect these correlations to strengthen with the discovery of new ETNOs. Further objects have indeed been discovered since the announcement of Planet Nine's existence. It turns out that some fit, and some do not. These new objects appear to have muddied the celestial waters considerably. Skeptics have criticized the whole concept of

Planet Nine on the basis of the data points which do not fit. On the other hand, the planet's advocates have highlighted new relationships between these growing numbers of objects which, they claim, strengthen their case still further.

The Bigger Picture

The scientific battle-lines have been drawn for a while, with reputations at stake. In this chapter, we will look at some of the skeptical arguments laid at Planet Nine's door, and how the proponents of this object have fought back.

First, it's important to recognize that the Planet X field of study has never been founded on any single piece of evidence. Instead, it tends towards a patterning of anomalies, which collectively cause those studying the subject (scientists and laypersons alike) to conclude that there is an important missing piece in the cosmic jigsaw puzzle. The difference, perhaps, with the case for Planet Nine is how robust the supporting mathematical evidence is.

Second, even if these patterns continue to emerge from the data, that does not necessarily mean they might be attributable to a Planet X body. It is true to say that the existence of a substantial Planet X body would explain a great many anomalies within the solar system, as described in my previous book (Lloyd 2005), but some or all of those anomalies might also be explained by other phenomena, too. So, the fact that there are anomalies does not prove the existence of Planet X. However, it does give us good reason to consider the perturbing planet's presence as a common factor.

The OSSOS Objects

Astronomers from Outer Solar System Origins Survey (known as OSSOS) have been hunting for new Kuiper Belt Objects. In 2017, they identified eight new ETNOs, four of which are of the type used to make the initial case for Planet Nine (Sokol 2017). If the case for Planet Nine's existence is to be believed, these new objects should align with the extended ETNO cluster (Shankman et al. 2017).

Although there is potential for a fit with some of these new objects, one of them (2015 GT50) is clearly out of sync with the rest of the greater cluster. Additionally, the orbits of a few of the objects lie on the opposite side of the sky from the original clustering.

These eight objects represent an independent dataset of ETNOs of a similar size to the original Planet Nine cluster. The authors conclude that the underlying angular distribution of this OSSOS sample is boringly randomized. In other words, there is no external influence reconfiguring the orbits of these scattered objects into a discernible pattern. If you were independently presented with the eight new ETNOs discovered by OSSOS, you would not have noticed any unusual patterns emerge in

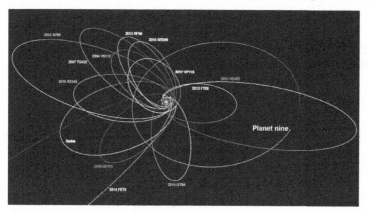

Fig. 7-1 Planet Nine ETNO diagram

the first place, and would not have concluded that a mysterious planet was involved (Shankman et al. 2017). The OSSOS team argued that this casts the existence of Planet Nine into doubt: The observed clustering effect among the six original Sednoids is likely to be due to observational bias.

Observational Bias

This alleged bias relates to practical considerations in astronomy: Which parts of the sky receive the most attention from ground-based telescopes; the cumulative effect of atmospherics affecting locally derived datasets compared to those obtained elsewhere in the world, and so on. These seemingly subtle differences can lead to observational bias, which results in the emergence of non-random patterns in the data, which are then falsely attributed to an invisible external influence.

If data is obtained from just one telescope, for instance, then there is a strong likelihood that these localized factors will have an impact upon the spread of results. If the data is derived from multiple sky surveys, conducted using different telescopes located in different parts of the world, the potential for observational bias lessens. So, perhaps the observed clustering effect is simply a product of observational bias to do with the positioning of, and atmospheric conditions around, key observatories?

However, the problem for this skeptical argument is the fact that the ETNOs were discovered by different surveys and observatories. This should preclude such bias. While the OSSOS team recognizes this, they argue that observational bias still plays a part, despite the varied provenance of the original cluster's discoveries. Additionally, they argue that one might reasonably expect statistical variations to arise from a small set of detected objects like the Planet Nine cluster, irrespective of whether the data has been derived from a variety of sources or not.

The key object for their argument is the off-kilter object known as 2015 GT50. For skeptics, 2015 GT50 represents the fly in the Planet Nine ointment. Yet, it should be noted that despite their skepticism, the OSSOS team could not definitively rule out the presence of Planet Nine (Brennan 2017).

Planet Nine Advocates Return Fire

Shortly after OSSOS published their data about the new ETNOs, astronomer, Scott Sheppard pointed out that three of the four new objects do actually have clustered orbits consistent with a distant Planet X object. The problematic object 2015 GT50 does not necessarily represent a knockout blow for Planet Nine, he argued. Sheppard had always expected that there would be some new objects which would not fit (Sokol 2017).

The Caltech team was also skeptical that the new OSSOS sample tolled the death-knell for Planet Nine. After examining the new data provided by OSSOS, Batygin published an article analyzing the impact of the discovery of these new ETNOs upon the case for Planet Nine. The short version of his rather technical analysis is that although the objects are, on the face of it, randomly distributed, a more detailed analysis shows that they are largely consistent with Caltech's original thesis. They are either anti-aligned to the purported Planet Nine body (as per the original cluster), or aligned with it in what Batygin describes as a "meta-stable" array (Batygin 2017).

Batygin calculated that the fly-in-the-ointment, 2015 GT50, turns out to have a resonance relationship with a set of values for the unseen Planet Nine object, as described in the original Caltech paper (Batygin & Brown 2016). 2015 GT50 sits smack bang on one of these resonant orbit paths.

Others may sense a circular argument here: A cluster of objects points to a set of parameters for the mystery perturber whose theoretical orbital characteristics can then be used to explain a data point that fits neither the anti-aligned nor the aligned cluster. Perhaps Scott Sheppard is right to merely shrug 2015 GT50 off instead?

Brown then presented calculations showing the probability that observational bias wrecked the cluster's anomalous properties is marginal, at best. He looked at the ten ETNOs with the largest semi-major axes. The chances that these exist within a broader population exhibiting an overall uniform longitude of perihelion, is about 1%. When combined with the additional common trait to do with their orbital poles, the chances drop still further, to 0.025% (Brown 2017). Statistically, argues Brown, the existence of Planet Nine remains highly probable.

TNOs with Retrograde Orbits

Batygin notes that the cluster effect is just one of several lines of inquiry pointing towards the existence of a super-Earth Planet X body. Even if the cluster argument is weakened by the new OSSOS dataset of objects (he does not agree that it is), he argues that there is still compelling evidence for Caltech's Planet Nine object. This body of evidence includes the mystery of the retrograde objects known as "Niku" and "Drac" (Batygin & Brown 2016a).

The discovery of the unusual, highly inclined TNO "Niku" was announced in August 2016 (Ying-Tung Chen et al. 2016). Niku has a retrograde orbital path: Essentially, it is moving the wrong way around the Sun, in comparison to nearly every other object in the solar system. In this, and other respects, Niku is similar to another retrograde TNO, 2008 KV42. This object had been nick-named "Drac", for its ability to climb walls.

Brett Gladman, of the University of British Columbia and one of the discoverers of Drac, thinks that this object began life in the inner Oort cloud, and migrated inwards after receiving a gravitational nudge some time ago. He considers this to be an object in transition, and that its destiny is to become a Halley-type comet (Hecht 2008).

Is it possible that a distant Planet X object might have provided the nudge which first brought Drac inwards from the comet clouds? I put this thought to astrophysicist Matthew Holman, of the Harvard-Smithsonian Center for Astrophysics, suggesting that this might imply that, to achieve this perturbing effect, a Planet Nine body would need to have a much wider orbit than suggested by Batygin and Brown. Here was his response:

> "I really have not figured out the explanation for the common plane, pro-grade and retrograde, that these objects seem to share, but it's not clear to me that if something like a Planet Nine is perturbing them that it would need to have such a large aphelion. The gravitational influence of the disk and bulge of the galaxy are strong enough [to] reduce the pericenters of inner Oort cloud objects into the region of the planets."
> (Holman 2016)

So, if Niku and Drac did originate from the inner Oort cloud, then it is not necessary to infer the presence of Planet X. But, clearly, the external presence of such an object lodged in the vicinity of the inner Oort cloud would help...

There is a third retrograde object in this category, 2013 SY99, which is sometimes referred to as L91. This extended scattered disk object, which has a 20,000 year orbit, was first discovered by OSSOS in 2013, but details of its orbital path were published in 2016 (Witze 2016). L91 comes in as close as 50AU at perihelion - around the Kuiper Cliff zone - but manages to sweep out as far as 1,400AU at aphelion. This is farther away than both Sedna and 2012 VP113.

Explaining this bizarre orbit is problematic. It may have been banished by Neptune towards the inner Oort cloud to start with, before gradually moving back towards the Sun after further gravitational nudges by passing stars, or the influence of the galaxy's tidal forces. This is the preferred solution of astrophysicist Michele Bannister of Queen's University, Belfast (Mann 2016). It is somewhat in keeping with the idea that these retrograde objects begin life in the inner Oort cloud, but the difference in this case is that 2013 SY99 sits within the plane of the solar system, rather than being tilted at high angles.

This difference also casts into doubt Planet Nine's influence, according to Dr Bannister (Witze 2016). But the complexity of the mechanism proposed by OSSOS to explain the changing orbit of 2013 SY99 has also been challenged – this time by the Caltech team. They think the intervention of Planet Nine provides a far simpler explanation for the orbits of these objects. These retrograde bodies may have all originally been extended scattered disk objects, under the influence of Planet Nine. Rather than being objects drawn down from the inner Oort cloud, they consider these objects to have originally belonged to the Kuiper belt, or other areas of the solar system closer to home. The Caltech astronomers consider these three anomalous objects to provide yet more evidence for the existence of Planet Nine (Batygin & Brown 2016a).

Patterns in the Nodal Points

Jupiter has a powerful effect upon the orbits of many comets, including the position of their nodal points along their orbital paths. As we have seen, "nodes" are points along an inclined orbital path crossing the ecliptic, or plane of the planets. For a standard object moving around the solar system in the usual direction (prograde), the ascending node is where it crosses the ecliptic moving from the southern celestial hemisphere into the north.

Among other effects, Jupiter affects the nodal positions of comets, creating a bimodal distribution (Rickman, Valsecchi & Froeschlé 2001). Because of the presence of Jupiter, the points where comets cross the ecliptic tend to congregate around two maximum values, equivalent to two humps in a distribution graph. Centaurs also show this distribution pattern, again tied in with Jupiter's influence. This is despite the neighboring presence of Saturn.

Could similar effects be evident among other, more distant outer solar system bodies? There does indeed seem to be a correlation between nodal distance and orbital inclination (i) for objects in the extended scattered disk beyond the Kuiper belt.

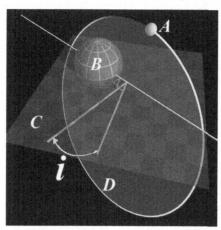

Fig. 7-2 Inclination in elliptical orbit diagram

Spanish astrophysicists argue that it may be possible that the 22 ETNOs they looked at are being similarly affected by the perturbing power of a Planet X object (de la Fuente Marcos & de la Fuente Marcos 2017a). Their calculations show that the correlation produces a best fit with a Planet X object located several hundred astronomical units away. This object would take the form of a super-Earth planet weighing in somewhere between ten and twenty Earth masses. Significantly, as a result of this observed correlation, the de la Fuente Marcos brothers argue that observational discovery bias can be ruled out for these objects.

As far as this nodal distribution effect is concerned, the brothers conclude that there is one unknown planet causing this effect, and not a combination of two, as they have previously suggested. But in the same way that Jupiter is the dominant driving force for this effect amongst comets approaching the inner solar system, and amongst Centaurs (despite the presence of neighboring Saturn), it seems reasonable to assume that another lesser Planet X object could still be in the frame without contributing significantly to this nodal effect. All the same, one large Planet X object is all that is required in this instance.

Collective Gravity, and Caju

You might also wonder whether other factors might have caused these effects in the outer solar system – far removed from the influence of powerful players like Jupiter. Objects located beyond Neptune are more loosely bound to the Sun than Centaurs and short period comets, and are therefore more susceptible to outside influences.

Computer simulations, conducted by scientists working at Boulder, Colorado, suggest that the orbital paths of Trans-Neptunian Objects may become more eccentric and extended simply as a result of gravitational interactions from within the Kuiper belt itself (Strain 2018). If true, then there is no need to seek out Planet Nine. Al-

though a series of small-scale perturbations may provide a model to explain anomalies, this does not seem to offer an explanation for the definitive clustering of the
ETNOs, however. One would expect such a series of small-scale events to randomize any effects, not modulate them.

It may also not explain the highly inclined nature of some of these objects. A new
object which has been added to the Planet Nine cluster of ETNOs lies perpendicular
to the plane of the planets. 2015 BP519 (named "Caju") was detected by the Dark
Energy Survey. It is currently located some 55AU away, at the periphery of the Kuiper belt. But Caju's semi-major axis is some 450AU, and its inclination from the
ecliptic is a staggering 54 degrees.

Like the rest of the objects in the extended scattered disk, this is no ordinary KBO.
Its properties not only suggest that it sits well with the Planet Nine cluster, but it also
literally elevates that cluster to a whole new level. Objects with this kind of orbital
property were predicted to exist if Planet Nine lurks out there somewhere, influencing this distant minor bodies (Batygin & Brown 2016). Without taking into account
the presence of an additional planet beyond, it has so far proved impossible to simulate Caju's spectacularly complex orbit (Becker et al. 2018).

Some Objectivity

There is a healthy scientific debate going on about the possible perturbing influences
that are shepherding these distant scattered disk objects. The various possibilities
include galactic tidal forces, passing stars and undiscovered planets of various sizes
lurking unseen in the outer solar system. Objectively, it does not seem possible to yet
say which of these theories is the most valid.

I asked Dave Jewitt, an astronomer at University of California at Los Angeles who
specializes in primordial solar system bodies, his opinion about whether the new
ETNOs discovered by the OSSOS team had made the existence of Planet Nine more
or less likely. Here was his response:

> "The significant new evidence I've seen is (1) detection of new objects
> which are probably in the purported cluster and others which are definite-
> ly not. The claim has been made that even the latter can be explained by
> a Planet 9, but to me it looks like the clustering evidence weakened. (2)
> Shankman et al. [2017] published the first survey in which bias is accu-
> rately treated, and they conclude that bias could easily produce a grouping
> like the one others believe is caused by a planet.

> "So, the case didn't get any stronger, in my eyes. The trouble is that, like
> cold fusion and life after death, Planet 9 is something a lot of people want

to believe is true. Well, to be fair, Planet 9 has a better chance than life after death!" (Jewitt 2017)

It seems that the Caltech team has quite a ways to go before they convince many of their colleagues in the astrophysics community that Planet Nine is for real. One of the points they make is that Planet Nine's effects upon the solar system are many and varied. The frequent need to bring in a large undiscovered planet to solve various anomalies and solar system riddles is an argument I have often used myself down the years. Individually, each problem which can be solved by the distant presence of Planet X does not, in itself, provide proof. But when you start to mesh together multiple anomalies, then a more comprehensive picture emerges calling for another planet.

However, the temptation is to then try to explain just about everything in terms of this object. This is where Sheppard's more nuanced approach may be wise – an occasional shrug when there is not a completely perfect fit. Anomalies can work in both directions.

Are They Digging in the Wrong Place?

There's another important possibility that I would like to offer here, which involves the fit between the objects discovered by OSSOS, and the original Planet Nine cluster. The orbits of two of the OSSOS objects were located in the opposite hemisphere to the original ETNO cluster.

The original Caltech computer simulations predicted that there should be a long-term, stable cluster of bodies with orbits that are anti-aligned with respect to the major axis of the undiscovered planet. In other words, the orbits of the cluster sit on the opposite side of the sky from that of Planet Nine. This stable cluster is what had been originally observed. But the model also predicts a weaker, metastable cluster of objects that are aligned with the orbit of Planet Nine (Batygin & Brown 2016).

Objects in the stable cluster are those that were scattered into distant orbits billions of years ago, and were locked into a permanent, resonant pattern. Conversely, objects in the opposing, meta-stable zone were jilted out of the Kuiper belt by Neptune more recently (perhaps hundreds of millions of years ago), and are now interacting with Planet Nine. It will one day eject them from the solar system (Batygin 2017).

Let's say the extended cluster does indeed point towards a Planet X body. The original cluster was very one-sided, and the Caltech team interpreted that as being the dynamically stable "anti-aligned" orientation to Planet Nine.

In other words, the bulk of Planet Nine's eccentric orbit (arrow #1– [1 elliptical orbit] in the diagram) lies in the opposite hemisphere of the solar system than the bulk of the ETNO orbits within the cluster (arrows #2– [4 elliptical orbits] in the diagram).

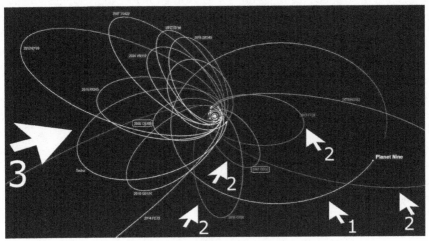

Fig. 7-3 Anti-aligned ETNOs Planet Nine diagram

It is natural from their point of view to assume that the observed objects are more likely to populate the stable zone, as opposed to the meta-stable zone with respect to the position of Planet Nine. But this may have been attributable to observational bias.

Now that we have data points which fall into opposing zone (arrow #3 – [10 elliptical orbits] in the diagram), it seems perfectly possible to me that as more discoveries are made, the population on this side of the fence will grow (and with it the position of the #1 arrow [elliptical orbit] denoting Planet X). Who is to say that the dominant hemisphere will not shift from one side to the other? In which case, it may be that the Caltech team have misinterpreted the initial data set as anti-aligned, when they were actually looking at the meta-stable aligned grouping.

I argued for this configuration after the announcement about Planet Nine in 2016. I believe that Planet X is located in the exact opposite part of the sky to Caltech's pre-ferred location in Orion (Lloyd 2016a). This claim harked back to my previous work, which had successfully predicted much of the Planet Nine scenario. This included the main axis of the planet's orbit, its approximate inclination, and its approximate distance. I had placed its aphelion location at the opposite end of the long-axis argu-ing that Orion was the perihelion location, not the aphelion location (Lloyd 2005). As a result of this, I believed that Planet Nine should lie in the same hemisphere as Caltech's original ETNO cluster.

I now think it is possible that observational bias provided astronomers with the rarer members of the meta-stable, aligned cluster. It is too early to say whether opposing objects start to shift the balance towards my preferred location. But if more opposing objects are found, then, in the immortal words of Indiana Jones, we can conclude that "they're digging in the wrong place!"

How Did Planet Nine Get There?

One of the earliest objections to Planet Nine is that a planet could not form in the sparse regions of space beyond Neptune. There was never the necessary amount of material out there to construct such a world, astrophysicists argue. One might counter that a planet could have started life elsewhere, within the bosom of its star, before being heaved out of the cosmic nest to fend for itself. Like offspring who never really leave home entirely, Planet Nine might be hanging around, having not been completely ejected during the early solar system's chaotic adjustments. Alternatively, it may have been captured from another star in the birth cluster, establishing its new home on the outskirts of the Sun's planetary family.

In the light of the 2016 Planet Nine paper, these scenarios, and others, have been put to the test through computer simulations to determine the probabilities of them occurring. It turns out that the odds are pretty long for many of these formation scenarios. Worse still, the odds of Planet Nine surviving in its proposed orbit are not that great either – certainly during the early period of the solar system when stellar flybys were more common (Li & Adams 2016).

That is not to say that Planet Nine could not have formed; just that its formation via the most likely routes is improbable, as is its subsequent survival as an eccentric outlier in the outer solar system. Taken together, it probably should not be there. Given how tough it seems to be to find, it is little wonder that skeptics favor the "No Planet Nine" option. Still, the anomalies that suggest its presence persist. As with the evolution of humans, many argue that the odds of that happening seemed to have been pretty long. Yet, we're still here.

So Where Is It?

One of the strongest arguments against the existence of Planet X is its continued absence in the eyepieces of astronomers' telescopes. Conspiracy theorists put this down to a worldwide campaign of silence among astronomers on the subject. However, as we have seen, there are a number of academic astrophysicists and astronomers who have staked their reputations on this topic, and would very much like to see it found. Over the last couple of years, Brown and Batygin have spent many an autumn and spring night on the summit of the Hawaiian dormant volcano Mauna Kea, searching for Planet Nine. They make use of the 8m Subaru telescope, booking precious slots to seek the object out. These times of the year are ideal for hunting in the constellation of Orion, which remains high on their list of cosmic haystacks.

Sometimes, their efforts are thwarted by the elements (Müller 2018). But Brown remains ever optimistic that Planet Nine will turn up. Planet X has proven an elusive object for nigh on 150 years of searching, and it seems that it is not about to give its location away just yet. That is assuming, of course, that it does actually exist in the first place.

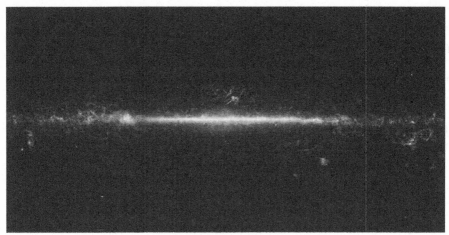

Fig. 7-4 WISE sky image

Although much of the sky has been effectively ruled out as a potential location for Planet Nine, the areas where it may yet lurk remain vast. There is no guarantee that Brown's favored spot – which doesn't seem to have changed despite various published papers further constraining Planet Nine's position – is correct. Looking for a needle in a haystack involves first locating the right haystack. There are hints that the favored search location may be shifting, which I think would be the right move. Once you know which haystack to search, then your best bet is to invest in a decent metal detector!

Without a candidate object, it is possible that Planet Nine is set to become a cautionary tale about how not to announce a new scientific finding. An article published in Scientific American in March 2018 emphasized the current lack of progress on finding the new planet – particularly in light of the claims which had been attached to it just two years before. The article noted that Planet Nine was supposed to be in the bag by this point. In his defense, Brown argued that the theoretical foundation provided by himself and Konstantin Batygin remains sound. The chances of Planet Nine being out there are 99.99% positive, he argued (Billings 2018).

This continues to beg the question: With so much academic and media interest in this search, how has it managed to evade detection? Unless this point can be adequately addressed, Planet Nine will return to the periphery of astronomy, no matter how well supported it is by theory.

Planet Nine and the Nice Model

In the previous chapter, we considered some of the challenges Planet Nine faces. It is hardly surprising that a re-branded Planet X should evoke substantial criticism from peers and skeptics alike. We should expect nothing less. "Extraordinary claims require extraordinary evidence", the late Carl Sagan famously once argued. Nuclear physicist Stanton Friedman countered dryly that there is no such thing as 'extraordinary evidence', just evidence. But extraordinary claims surely do require a higher standard of proof.

The seemingly watertight case for Planet Nine, set out by an eminent astronomer with a track record of remarkable discoveries in the outer solar system, potentially drives the case for Planet X towards that higher standard. It is no wonder, then, that many scientific teams took up the gauntlet to test this exciting new hypothesis. Some looked back through archived surveys to try to find the object. Some analyzed fine-tuned astrometry data to constrain the planet's orbital parameters. Others, with a more theoretical interest, set about modeling Planet Nine, its effect upon bodies in the solar system over time, and how it might fit into the bigger picture.

Fig. 8-1 Blue planet image

Following publication of the Planet Nine hypothesis (Batygin & Brown 2016), a team of scientists tested the idea by simulating the solar system over a 4 billion year period; first, as it appears to us now, and, second, with the addition of a super-Earth located well beyond the Kuiper belt (Lawler et al. 2016). They found very little difference in the outcome with respect to the resultant architecture of the outer solar system, except that the extended scattered disk beyond the Kuiper belt would be significantly denser in the presence of a Planet Nine body.

The hidden planet was presumably either drawing some of these objects out of the Kuiper belt, or pulling comets into the scattered disk from the inner Oort cloud. This was what one might expect. However, notably, the computer simulated populations of ETNOs were not shown to exhibit similar properties to one another. They did not create a clustering of objects with anomalously similar properties, as has been observed with the aligned arguments of perihelion.

A lack of synchronicity is to be expected where there is no Planet X object present. The arguments of perihelion for the ETNOs should randomize out over hundreds of millions of years due to the perturbing action of the known planets. But as there is a synchronicity across a cluster of objects, then this implies that something else is going on.

As we know, various teams of researchers have independently concluded that a substantial Planet X is probably necessary to maintain the observed distributions of objects in and beyond the Kuiper belt (Gomes et al. 2006; Lykawka & Mukai 2008; Trujillo & Sheppard 2014; de la Fuente Marcos & de la Fuente Marcos 2014; Iorio 2014; Batygin & Brown 2016). However, because the inclusion of a Planet X body within a set of simulations has not thrown up this observed effect, these computer simulations may cast some doubt over whether Planet X is really capable of achieving this effect.

The Monte Carlo Method

This assumes, naturally enough, that the way complex solar system dynamics are modeled is always accurate. One could argue that you get out of these calculations what you put into them. Working backwards can also create circular arguments. In order to carry out what are known as N-body "Monte Carlo" simulations with super-computers, a number of underlying assumptions are made about the configuration of the early solar system.

The Monte Carlo method involves computational algorithms making use of a huge number of randomized elements. It was given the name because one of the inventor's relatives had a gambling problem and frequented the Principality's famous casino. Many random variables are set in motion through these astrophysical simulations – generally the orbits and masses of minor bodies and planets. Multiple simulations are carried out with altered variables each time.

Fig. 8-2 Monte Carlo Casino image

This is a trial and error method, and many of the results of the simulations prove wide of the mark. In other words, over a series of simulations, many will fail to achieve what is observed. So, how do you know when "many" becomes "all"? At what point do you conclude that you have completely ruled out a scenario because every time you try to emulate the observed data through computer modeling, you fail? How many calculations do you have to perform to satisfy this negation?

As you can no doubt see, there is no answer to this. It is difficult to prove a negative in this way because you can never be sure that you have covered every eventuality. It can take a great deal of tweaking, both of assumptions and variables, before researchers arrive at the 'correct' outcome.

Modeling the initial planet-forming stages of the solar system is hard enough. Astrophysicists also have to explain later events which are now known to have taken place in the solar system. The theory about the formation of the solar system becomes increasingly convoluted to successfully provide accurate predictions for events such as these, especially when they have emerged from prolonged periods of calm. Theory is required to not only explain the water cycle, but also why you can have calm before a storm.

This brings us neatly to the "Nice model". But, first of all, let us consider how Planet Nine might fit into the Milky Way galaxy's menagerie of worlds.

Super-Earths and Observational Bias

Konstantin Batygin and Mike Brown consider Planet Nine to be a super-Earth object, of about 10-20 Earth masses (2016). Super-Earth exoplanets are now considered commonplace. Indeed, they appear to be the most common form of planet found

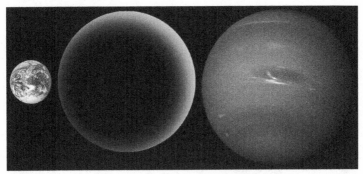

Fig. 8-3 Planet Nine Comparison image

orbiting other stars. This could be a finding which results from observational bias: Common sense alone indicates that the methods of detecting planets will promote the discovery of larger planets over smaller ones. Astronomers often look for tiny wobbles in the position of stars which imply the presence of orbiting planets. This method clearly favors the discovery of large planets located close to the star itself, where the impact upon its motion will be maximized.

Other exoplanets are discovered through the observation of a regular dimming of a star's light. This periodic variation occurs when a planet passes in front of the star (the plane of the planet's orbit must line up with the line of sight from the Earth for this to work). Again, this tends to favor bulky planets which will block out more starlight as they transit across the fact of the star.

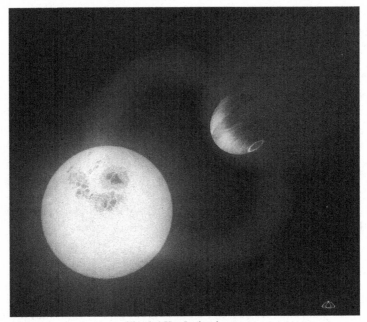

Fig. 8-4 Hot Jupiter image

These biases towards the discovery of massive, intimately-located worlds may explain why "hot Jupiters" have been observed so regularly, despite seeming quite exotic to our minds (after all, these big balls of gas seem to defy the odds by orbiting so close to their parent stars). By their very nature, hot Jupiters optimally affect the variation in their parent star's observed light and movement.

In the same way, super-Earths will be easier to find than Earth-like planets. Nonetheless, the prevalence of super-Earths out there contrasts markedly with their absence in our own planetary system. The more super-Earths that are discovered, the weirder the absence of them seems in the solar system. Perhaps, as has been suggested, Planet Nine is our missing super-Earth (Brennan 2017)? In a way, the discovery of a super-Earth beyond Neptune would satisfy the statisticians, as well as get the bubbly flowing at Caltech.

It raises another question, however: How would a super-Earth beyond Neptune fit into theories of how the planets of the solar system formed?

The Nice Model

Konstantin Batygin teamed up with Alessandro Morbidelli, one of the architects of the "Nice model" of solar system formation, to test whether the model could support Planet Nine's place in the Cosmos on a theoretical level (Batygin & Morbidelli 2017).

Morbidelli is an Italian astrophysicist, working in the south of France, who is a proponent of the Nice model for solar system evolution (named after the rather wonderful French city where he works). This model arises from a comparison between our solar system's dynamics, and those of the many other planetary systems now known to us, many of which seem bizarre and chaotic in comparison to our own. Thus, the Nice model seeks to blend the kinds of dynamical fluctuations which might occur during the evolution of a star's planetary system with both the outcomes witnessed in our own solar system, and the more extreme exoplanets observed elsewhere (Morbidelli 2010).

To do so, the Nice model invokes significant changes in the positions of the major planets during the history of the solar system. The planets form out of the spinning disk of dust, gas and accumulating rock, and begin competing for space and resources as they build. The leftover remnants of the protoplanetary disk are substantial enough to cause their own gravitational effects upon the newly forming planets in their midst. Within this chaos, there can be shifts in gravitational power, which can invoke the sudden large-scale migration of even the most massive of the planets.

Planets are constantly in motion, moving around the Sun. When astrophysicists discuss migration, they mean a shift of that orbital path. This may be inwards, contracting the orbit and speeding up the planet's orbital period. Or it may be outwards, making the planet take much longer to circumnavigate the Sun and increasing its average distance from the Sun.

Modeling Chaos

The growing complexities of solar system formation have led to a great many hypotheses about migrating gas giants. One example is the "grand tack hypothesis". This hypothesis sees Jupiter shifting its position to move into the inner solar system, dispersing asteroids and planetessimals, and limiting the growth of the planet Mars (Walsh et al. 2011). It builds upon the Nice model, which seeks to comprehensively explain the evolution of the giant planets in the solar system (Tsiganis et al. 2005). The migration of these outer ice giants results in a re-shaping of the Kuiper belt while at the same time increasing the eccentricity of Saturn and Jupiter during this migratory phase. Finally, it disrupts Uranus, Neptune, and probably Planet Nine(?). According to Levinson the migration of these outer ice giants results in a re-shaping of the Kuiper belt (Levison et al. 2007).

Migrations of Jupiter, Saturn, Uranus and Neptune across the solar system would clearly have had knock-on effects upon other, smaller bodies in the early solar system, as well as upon each other and the leftover primordial gas clouds. The Nice model thus proved capable of predicting a cascade of disturbances which disrupted the status quo of the early solar system. Furthermore, it went on to simulate conditions which might explain the cataclysm which arose over half a billion years later. The ensuing chaos of all of this migration activity can be used to explain the destructive wave of later cataclysmic collisions (Gomes et al. 2005).

By far the greatest catastrophe to befall the solar system took place about 3.9 billion years ago – an event known as the "Late, Heavy Bombardment".The evidence for this devastating event, or series of events, is etched into the surfaces of asteroids, moons and terrestrial planets across the solar system. They are pock-marked with craters aged back to this time period, some 600 million years after the formation of the planets. The reason for this cataclysm is not known, but advocates of the Nice model argue that it was the inevitable consequence of a sudden alteration in the solar system's architecture.

Something tipped the balance of forces in the solar system. The gas giants and ice giants shifted positions, and these substantial alterations in the dynamics of the solar system caused the smaller bodies, like asteroids and comets, to change their orbital trajectories. Chaos ensued. The result was a massive uptick in collisions, lasting for perhaps hundreds of millions of years, and laying waste to the early terrestrial worlds.

Then, order was returned to the solar system. The planets and minor bodies fell back in love with each other, migrating back into the harmonious arrangement set out by

Fig. 8-5 Asteroid impact image

Bode's Law. So, although the present solar system seems to be a nicely ordered collection of objects which seem to respect each other's personal space, at other points in the past things were a great deal more chaotic.

In this "having your cake and eating it too" model, stability and chaos are both affirmed through dynamical processes. Complex computer simulations of the early solar system, using multiple random variables, might prove capable of generating the current solar system architecture. But, if they do not, then the chaos generated within the programs succeeds in simulating other very different patterns of planetary distribution observed elsewhere, e.g. hot Jupiters.

Advocates of the Nice model, which is now generally accepted among astrophysicists, are drawn to this catch-all approach. And who can blame them? A relatively small set of assumptions about how planets behave dynamically can capture a wide variety of diverse outcomes (Morbidelli 2010).

A Closed System

However, let's say you were inclined to approach this skeptically, you could argue that if researchers were to carry out enough simulations, one of them is bound to strike gold eventually. This is the old "monkeys with typewriters" argument. However improbable it might be to randomly generate high complexity, given an infinite series of circumstances it will be achievable eventually.

I think the main problem with the Nice model is that planetary systems are not closed. One might reasonably argue that catastrophic events like the **Late, Heavy Bombardment** (LHB) could equally have been triggered by the sudden presence of an object visiting from outside the solar system. This visitor might be a passing star, or a brown dwarf, or a substantial free-floating planetary mass object. Catastrophe coming from outside might even be triggered by the influx and disruptive effect of a nebula, or a particularly dense molecular cloud. A massive, long-lasting influx of comets, disturbed from their lofty perches in the Oort cloud, could have caused the LHB, particularly if a loosely-bound planet or brown dwarf were to itself migrate inwards, and go on the rampage through the outer solar system.

There is no particular need to keep the reason for the cataclysm "in-house". But the Nice model has the advantage of explaining matters without having to resort to such an explanation. Everything you need is already on the table, and if your simulation comes to a different conclusion then (a) it successfully models other, more bizarre planetary systems and, (b) you just need to tweak your underlying values until you get the answer you want.

Fig. 8-6 Protoplanetary disk image

Astronomers bear witness to similar chaotic effects written into the planetary distributions of other young star systems, as well as within the visible patterns that elegantly swirl within observed protoplanetary disks. Theoretical astrophysicists have emulated the structure of our own solar system from similar processes by working backwards to explain how that structure first came to be.

The resulting theory presents an increasingly complex model. After all, it attempts to simultaneously predict long periods of quiet, interspersed with catastrophic events which happened long after things in the solar system should have settled down nicely. Where one might have expected a linear evolution of planets accreting from dust and gas, and sorting themselves out into some kind of stable order, the neatness of linear evolution has been disrupted by the awkward fact of catastrophism. The Nice model is an attempt to square this circle, where astrophysicists shape the theoretical conditions deemed capable of generating intermittent bouts of turmoil seemingly from nowhere.

Fig. 8-7 Observatoire de la Côte d'Azur, Nice image

To achieve this, the Nice model invokes Byzantine complexity. There is a sense of chaos theory about it, where, to take the familiar path through the theoretical rain forest, the flap of a butterfly's wings eventually triggers a distant hurricane. In the case of the planets, chaos becomes the result of order when certain conditions are met.

Further problems then materialize when you have to incorporate new variables. As our known (or theorized) understanding of these complex systems builds, the circumstances under which massive lurches in conditions like this are triggered become more extreme.

Enter Planet Nine

This brings us back to Planet Nine. Can its presence in the outer solar system fit within the Nice model? Or would the presence of Planet Nine derail this model by adding a variable-too-far? Given the seeming capacity of the Nice model to explain pretty much any scenario, one would imagine it should have no problem with Planet Nine. Particularly when both the authors of the scientific paper we are about to consider are advocates of the existence of Planet X.

Alessandro Morbidelli has had more than a passing interest in Planet X for some time. I first wrote about his interest in the subject when the discovery of the enigmatic scattered disk object now known as Sedna began to rock the astrophysics world (Lloyd 2004 & 2005). There are a number of scenarios which might explain how Sedna ended up in its long, dislocated orbit. Among other ideas, these include the prior passage of a brown dwarf through the solar system, or the existence of an as-yet undiscovered massive planet beyond Neptune (Morbidelli & Levison 2004). Computer simulations carried out to examine these various hypotheses were coincident with an article in a French magazine which featured an interview with Morbidelli. In the article, he appears to have advocated the possible existence of an undiscovered Mars-sized planet in the outer solar system (Sitchin, 2004). A diagram of the orbit of a possible Planet X body which featured in the article (Greffos, 2003) bore a strong resemblance to the Planet X body advocated by the late Zecharia Sitchin.

The dynamicist, Konstantin Batygin has more recent form on the topic with his collaborator, astronomer Mike Brown on Planet Nine, but had also contemplated a similar idea before (Batygin 2012). Batygin & Morbidelli's theoretical paper discussed the dynamic relationships between anomalous bodies in the outer solar system and Planet Nine, particularly with respect to the kind of resonance relationships capable of maintaining the long-term stability of these patterns (2017). They demonstrated mathematically how the dynamical evolution of the trajectories of these anomalous objects over the lifetime of the solar system may have been driven by the presence of an undiscovered super-Earth, i.e. Planet Nine.

This theoretical analysis covered the following three areas with respect to the Nice model: (1) Orbital clustering of long-period KBOs; (2) The dynamical detachment of KBO orbits from Neptune; (3) The generation of highly inclined/retrograde bodies within the solar system.

All of these points were parts of the original Planet Nine paper, which attempted to model these observed effects using computer simulations (Batygin & Brown 2016). The general thrust of the answers provided in this new theoretical paper can be succinctly summed up by the phrase "It's complicated..." That is very much in keeping with the general gist of the Nice model. It seems to wrap itself up in knots in an effort to keep all of the factors in-house. So, although the end-point is the desired complexity, the starting conditions to achieve it are too simplistic. The solar system is not a closed system, as we shall see.

Skeptics would no doubt note that both of these theorists have considered – even advocated – the presence of undiscovered planets in the solar system. They might argue that a degree of wishful thinking is driving these theoretical efforts. But, despite these reservations, it is clear that there is no theoretical difficulty arising from the presence of Planet Nine – as far as the Nice model goes, anyway. Both of these scientists are highly respected astrophysicists and their interest in this, or similar, solutions should in no way undermine their technical abilities to explore the topic in depth.

An Outside Bet

The question I would pose is rather different: Does the presence of a substantial Planet X body change the working assumptions of the Nice model itself? Adding this planet into the mix provides an opportunity to explain the LHB in an entirely different way. Planet Nine may have originally been captured by the Sun from interstellar space, or at least from one of the Sun's sister stars from the primordial birth pool. In that case, it is an added extra which need not have featured in the early solar system. That potentially removes a huge layer of complexity from the Nice model, as a massive Planet X body may have indirectly brought about the serial bombardment of the inner solar system 3.9 billion years ago.

At least two possibilities could work here:

1. The maverick planet from interstellar space may have catastrophically interacted with another planet in the solar system as it passed the Sun. It may then have subsequently been captured into a comet-like orbit before eventually migrating outwards.

2. Or it might have originally been an Oort cloud planetary denizen which was perturbed into the planetary zone of the solar system, causing widespread chaos.

In both of these cases, there is no need to invoke the complex shuffle of major planets advocated by the Nice model. The historical arrival of an outsider is quite sufficient to generate long-term chaos sometime after the solar system had settled into shape, while at the same time explaining the continuing presence of a substantial planet beyond Neptune. For such an event to become an accepted part of the model for solar system formation, the case must first be made for its existence. It may have been ejected from the solar system, in which case it generated the effects discussed prior to leaving the scene of the crime. Or it may still be hanging around, waiting for us to find it.

Further Planet X Evidence Among The Minor Bodies

Part of the problem with this subject is that the term "Planet X" can be, and often is used to describe a wide variety of different objects. Partly, this is because of the sheer spread of possible objects that might still be in the outer reaches of the solar system: Massive comets, dwarf planets, terrestrial worlds, super-Earths, ice giants, gas giants, sub-brown dwarfs, and perhaps other types of exotic planetary objects we have not even dreamed of yet. After all, the solar system is a vast place, the tiniest fraction of which we can claim with any certainty to know much about.

However, it is not just the type of world we might be seeking out when we discuss the subject of Planet X. There is also the issue of how to define what a planet actually is, and whether a given object may even be classified as such. As time has gone by, and a greater variety of bodies have been discovered in the solar system, the classification of objects has changed. Sometimes, these new definitions create controversy, like the row over whether Pluto is really a planet or not.

Shape-Shifting Worlds

On the face of it, Pluto seems to have all the hallmarks of a planet. First of all, it orbits the Sun, rather than another body in the solar system. We can compare Pluto to a number of planet-sized moons in the solar system. The biggest example is Saturn's moon Titan, which has its own substantial atmosphere. Despite these obvious planet-like properties, Titan is classified as a moon because it orbits Saturn, not the Sun.

Pluto has a round shape, and even a substantial moon, Charon. But its planetary status was revoked in 2006 by the International Astronomical Union, despite 76 years proud years of planetary status. The arguments about this demotion of Pluto continue to play out. Some astronomers argue for the reinstatement of Pluto's prior status (Boyle 2010), still others are firmly against making such a volte-face (Brown 2010).

Fig. 9-1 Various Moons, Pluto image

Anyone paying attention to the growing family of known objects in the solar system will recognize the expanding list of "dwarf planets", to which Pluto now belongs. These are planets whose mass has reached the point where their internal gravity pulls the body into a spheroidal shape, essentially allowing it to take the form of a planet. The gravitation of the body reaches the point where powerful internal forces render the insides of the planet plastic, allowing for the filling in of cavities, and the pulling down of protrusions. Sufficient gravity smooths out geography over time. Contrast this with asteroids, the shapes of which can vary significantly. Small asteroids are irregularly shaped, while larger examples become potato shaped.

So, the bigger the body, the more dominant are the internal gravitational forces on its internal structure. The shape of an asteroid tends towards a smooth sphere as you reach a certain threshold of mass. Thus, a planet is rounded by its own gravity. When it becomes as round as it can be then it may be defined as a dwarf planet. There is no exact definition of when this internal drive towards a spheroidal shape should take place: There are a number of factors to take into account, like the object's composition, and its thermal history. A diameter of 800km seems an acceptable size as a rough guide, but opinions among planetary scientists on this topic vary, and there is no established convention as yet.

It is interesting to note that many major planets are also not exact spheres, which is why the term spheroidal is used. Many planets bulge in the middle, but are symmetrical around their axis of rotation. The gas giants Jupiter and Saturn each have pronounced equatorial bulges, and take the shape of "oblate spheroids". Their equatorial bulges are created by forces exerted by the planet's rotation. The Earth also has such an equatorial bulge, although not nearly as pronounced as the gas giants.

The Earth is far from being "flat"!

An interesting result of our world's misshapen nature is that if you measure the highest point on the Earth from the planet's physical center, then you would find it to be the peak of Mount Chimborazo in Ecuador, rather than that of Mount Everest. Being closer to the equator, Mount Chimborazo enjoys the added advantage of the Earth's central bulge, which extends the distance of the oceanic surface from the center of the planet.

Odd shapes occur at the lower end of the dwarf planet spectrum, too. Some of these bodies are "ellipsoidal", rather than spheroidal. An extreme example is the dwarf planet Haumea which resides in the Kuiper belt. Haumea is shaped like a rugby ball. Its major axis is twice the length of its minor one, making it an extremely squashed oblate spheroid. This is due, at least in part, to its rapid rotational period of four hours. Although its spin appears to draw this object into an ellipse, it has nonetheless undergone the internal stresses driving its reshaping.

A Clean Sweep

A spheroidal, or ellipsoidal, shape defines the lower limit for a dwarf planet. The upper limit for a dwarf planet is determined by whether it is capable of sweeping clean its orbital environment of other space materials. If a body's gravitation is sufficiently great for it to clear out its own orbital domain – either by pulling smaller objects towards it, or flinging them out of its path – then it earns the right to be called a planet. Dwarf planets are not sufficiently massive for their gravitational fields to affect nearby objects in this way.

Although Pluto orbits the Sun, and is round, it fails to achieve the modern definition of a proper planet because it does not clear other bodies out of its way. Where there should be no other bodies of comparable size in its region of space, apart from its own satellites or other objects held under its gravitational influence, Pluto instead shares its orbital neighborhood with other KBOs.

In a way, though, one might argue that Pluto is exempt from this issue, because its orbit is inclined and eccentric. It does not occupy the same band as the main disk of TNOs. Instead, it mostly travels through zones which are off the plane of the ecliptic. These zones are less populated than the main planar disk, and so Pluto's ability to sweep out these zones is not tested in practice. However, Pluto shares an orbital relationship with Neptune. Its 2:3 mean-motion resonance with the ice giant is also shared with a number of other objects, and these collectively have become known as the "Plutinos". Pluto has failed to establish dominance over these other related objects, curtailing any dodging of the non-planet bullet.

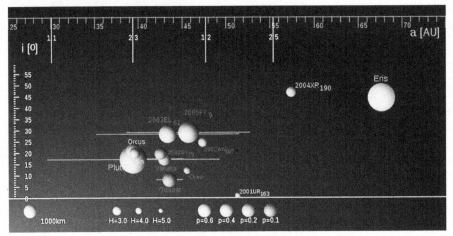

Fig. 9-2 Inclined KBOs image

But is the application of this definition of a planet any clearer in other cases? Some astronomers, like Alan Stern, argue that other planets also fail to clear their neighborhoods of neighboring objects (Rincon 2006). The Earth, for instance, shares its neighborhood with thousands of Near-Earth Objects. So does Mars. Jupiter has a mighty collection of followers (and fore-runners) in the form of tens of thousands of Trojans (these objects sit in Jupiter's stable L5 and L4 Lagrangian points, respectively). Neptune's own intimate relationship with the Kuiper belt beyond indicates that the ice giant has done everything but sweep its zone clean. Pluto even crosses its orbit.

However, one might argue that in the cases of Jupiter and Neptune, the related objects are held under their gravitational influences, having been shepherded into position through celestial mechanics. The case of the inner solar system worlds is harder to argue away, though. Under the IAU definition, should Earth and Mars really be designated as *planets*? Furthermore, the Sun's powerful influence was significant in clearing out Mercury's territory, so how can we firmly establish tiny Mercury's planetary status?

Those astronomers who are more inclined to accept the IAU definition of a planet offer the counter-argument that the accumulated mass of the debris in the inner solar system (NEOs, cometary fragments, etc.) does not manage to reach even one percent of Mercury's mass. So, the dominance of the inner solar system planets over their orbital territories is sufficient for the definition to apply. Therefore, there needs to be the presence of a more substantial object with a planet-crossing orbit before the planetary ship gets rocked down to a mere boat.

Applied to Planet X

Let us take this sweeping argument to a more extreme conclusion. Let us say that there exists a massive Planet X body demonstrating an eccentric orbit that periodically brings it close to the rest of the planets during its perihelion passage. Could we still define it as a planet if it fails to cause the kind of collateral damage in its flight path that might be expected of a similarly-sized object in a more standard orbit on the ecliptic? What if Planet X's orbit is highly inclined, so that the only contact it potentially has with bodies along the ecliptic is during its two crossing points, or nodes?

How would one be able to test its eligibility for planetary status under these circumstances? As you can see, the potential inclusion of a Planet X body capable of periodically crossing swords with other planets may raise questions about how a planet is actually defined.

Too Big for its Boots

Skeptics might argue that a substantial planet should, indeed, cause a tremendous amount of scattering in the solar system over its lifetime, during these short, periodic incursions. They could argue that the apparent lack of chaos in the solar system argues against the presence of a massive Planet X body describing an eccentric, inclined orbit.

Perhaps surprisingly, early calculations carried out to determine the feasibility of a Nemesis object passing several times through the planetary zone of the solar system showed that an object of less than about ten Jupiter masses need not have a significant effect upon the planetary orbits (Hills 1985). This seemed to provide a maximum mass for any Planet X object capable of getting anywhere near the planetary zone during its perihelion passage. Below that mass (equivalent to a sub-brown dwarf), anything goes, it seems, at least theoretically.

One might point to the scattering of extreme KBOs as indicative of this very effect taking place, albeit at a more distant point than the classical planetary zone. Massive distant objects can affect the orbits of smaller objects orbiting within them, causing periodic exchanges between their eccentricities and inclinations. This exchange brings about an oscillating movement known as "libration"(an apparent or real oscillation ...by which parts near the edge of the disc that are often not visible from the earth sometimes come into view). The effect caused by this mechanism is likely to be a factor at play with Planet X: It will have a measurable perturbing effect upon other objects located within its own orbit. As we have seen in the case of the ETNOs, working backwards in this way may provide information about the whereabouts of Planet X (or, possibly, more than one undiscovered planetary objects beyond the Kuiper belt (de la Fuente Marcos & de la Fuente Marcos 2014).

This is all well and good, and needs to be taken into account by any serious researcher investigating the Planet X puzzle. But I am just as interested in Planet X's ability to clear out space around it – as all 'real' planets should. There is, after all, a substantial gap of about 2,000AU between the Kuiper belt and the torus-shaped inner Oort cloud beyond. We now know that this zone contains a scattered disk of objects, and there may well be many more. Nonetheless, there remains a significant deficit of objects lying between the Kuiper belt and inner Oort cloud, and I would suggest that this is because this 'zone' is part of the domain of Planet X.

Over time, the eccentric movements of this substantial planet have re-shaped the population of this zone, and largely cleared it out. The extended scattered disk may therefore mark the limit of Planet X's inner influence. At the far end of its orbit, the torus-shaped inner Oort cloud incorporates the outer extent of Planet X's gravitational influence. In this sense, at least, the missing world really would be a "planet".

Dwarven Worlds

Coming back to little old Pluto, this distant but highly celebrated world was perceived to be a sufficiently light-weight entity to qualify for a demotion to dwarf planet status. This was despite the fallout from dissenting astronomers, and the American public. (Pluto was, after all, an American discovery).

In a similar way that Pluto has not swept clean its neighborhood of other Plutinos, the very existence of the asteroid belt between Mars and Jupiter precludes the existence of a planet within it. The only dwarf planet in the asteroid belt is "Ceres". Its diameter is approximately 945 kilometers, taking it over the 800km rule-of-thumb for a dwarf planet. As a member of the asteroid belt, Ceres is clearly incapable of clearing out its orbital domain. If it was, it would have disrupted and absorbed the rotating debris field of neighboring planetessimals over time. So, therefore, Ceres is not a planet.

If we compare Ceres to the second largest member of the asteroid belt, "Vesta", which has a diameter of 525 kilometers, we will immediately notice that Vesta is potato shaped. Vesta is therefore not a dwarf planet.

This definition of terms is important for the concept of Planet X. If one accepts these terms, then one cannot talk about a Planet X body embedded within the solar system's second asteroid belt, the Kuiper belt. Why? Because, by definition, if it is a planet, it should be capable of sweeping the zone clear of other materials. Given the self-evident existence of a disk of asteroids from Neptune's orbit at 30AU, out to about 48AU, then it is not possible for a planet to be found in this zone.

There could be any number of dwarf planets in this zone, of course. Pluto is one. Three of its Kuiper belt neighbors, Eris, Haumea, and Makemake, are also officially designated by the International Astronomical Union (IAU) as dwarf planets. At the time of writing, there are three other contenders for dwarf planet status, too: Sedna, Orcus, and Quaoar (Gibbs 2017). But a dwarf planet is as good as an object can get in this zone – by definition. Any Mars-sized 'planet' found in or very near to the Kuiper belt should properly be categorized as "Dwarf Planet X". This does not have really the same ring to it!

If Planet X is a real 'planet', as opposed to dwarf planets such as the examples above, then it cannot regularly pass through the Kuiper belt.

Ice Queens

The composition of objects plays a part in how they are shaped. Ice is more malleable than rock, and it is thought that TNOs have icy cores. This predominantly icy composition may lower the threshold where internal plasticity starts to occur, allowing for the achievement of hydrostatic equilibrium and leading to a spheroidal shape. KBOs are therefore more easily shaped by their own gravity than the more rocky asteroids closer to the Sun, like Vesta and Ceres. It has been suggested that TNOs with a diameter of just 400km may gain a spheroidal shape through these internal processes (Brown 2008). If this turns out to be correct, then many of the TNOs whose diameters lie between 400km and 800km should qualify as dwarf planets.

That having been said, how can we be so sure about the internal composition of TNOs? Astronomers have conducted studies of their surface chemistry from observational spectroscopy, and found a variable mixture of surface compositions among them (Barucci et al. 2008). However, finding out the interior composition of these objects is far more difficult. These objects are not comets, and therefore there is no out-gassing to observe so we cannot study their composition.

One method would be to study materials strewn around TNOs by impacts which have penetrated deep into their mantles. In other words, study the dusty debris clouds which may accompany these distant objects. (There will be much more on this dusty theme later in the book). A prime candidate for such a study is the rugby ball-shaped dwarf planet Haumea, whom we have already met. It proudly sports a high albedo (reflectivity) which is suggestive of a prior impact that has, at least partly exposed Haumea's bright, icy interior. Indeed, Haumea may be just one of a broader collisional family of TNOs, all of which appear as bright, icy objects (Rabinowitz et al. 2008).

A major cosmic impact may have been a causal factor in molding Haumea's odd shape. Alternatively, some kind of natural resurfacing may have taken place – the mechanism for that remains unclear. Either way, the youthful crystalline surface composition of Haumea seems to suggest an icy interior, potentially lowering the threshold for gravitational molding for this class of object.

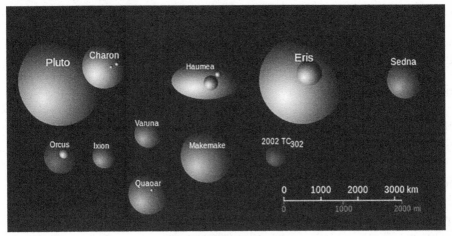

Fig. 9-3 KBOs sizes image

Factors such as these will have a bearing upon the possible planetary status of many of these newly discovered objects. The properties of these objects are clearly difficult to establish with any degree of certainty, due to the distances involved, and the tiny amount of light being reflected back from them across the solar system. Astronomers have to painstakingly piece together clues to determine these properties, and sometimes make precarious assumptions. Given the controversy around the IAU definition of a planet – and, more relevant in these cases, that of a dwarf planet – it is easy to see why many of these objects have yet to be properly classified. Perhaps the New Horizons probe will provide some of the answers as it gets closer to Kuiper Belt Objects on its extended mission beyond Pluto.

A Level Playing Field

There seems to be an additional unexpected correlation between the ETNOs. In 2017, another anomaly emerged about the invariable plane of the solar system. This also seemed to point towards the influence of an undiscovered planet.

The solar system is like a sheet, the uniformity of which provides an invariable plane where most of the action takes place. As already noted, the ecliptic is twisted with respect to the Sun's own tilt axis, but it is at least consistently twisted. Individual solar system objects, like asteroids, Centaurs and KBOs, may deviate from this plane, but taken overall they tend to average each other out.

This maintains an overall angular momentum for the solar system which has been determined to a high degree of accuracy. The main planets, and in particular Jupiter, dominate this value. If a massive Planet X object exists, then its massive presence would add to the solar system's overall angular momentum. If it already lies along the ecliptic, then its overall effect upon the system may be minimal. If, on the other hand, Planet X is inclined to the ecliptic, then it creates an additional vector for the solar system's total angular momentum which none of the other planets currently provides.

A comparison of the total angular momentum of the planets in the solar system with exoplanets elsewhere has shown that there is a great deal of missing angular momentum in the solar system. It has been calculated that the solar system is missing about 2.5 Jupiter masses (Gurumath, Hiremath & Ramasubramanian 2016). The Sun appears to have transferred much of its angular momentum to the planets early on. A massive Planet X body, which is inclined to the ecliptic, might help to explain this contentious paradox.

Several studies have been carried out over the years to measure the mean plane of the Kuiper belt, which have generally arrived at the rather uneventful conclusion that it is generally in alignment with the rest of the solar system. KBOs which lie within the main section of the Kuiper belt have orbital tilt inclinations which pretty much average out. These objects are shepherded by Neptune, so perhaps we should not be surprised. They seem to behave themselves pretty well.

However, since those previous studies were conducted, there has been a rash of discoveries of new TNOs lying beyond the Kuiper Cliff – the point where the main belt abruptly comes to an end. Astronomers use the word **truncated** to describe this cutoff phenomenon at the edge of the Kuiper belt. Values vary as to the exact distance of the Kuiper Cliff, ranging from 48AU up to about 55AU. For simplicity's sake, we'll go with 48AU for the main belt. The fact that the Kuiper belt is cut off at all is suggestive that something odd is going on out there.

However, there are some objects located well beyond this outer boundary – namely, the scattered disk objects (or ETNOs), and their more extreme cousins beyond (the Sednoids). Nonetheless, despite the presence of these populations of objects, a distinct truncation in the Kuiper belt is apparent.

Warping the Scattered Disk

The question then arises: Do these scattered objects average out in the same way as the KBOs? Or, does their mean deviation from the ecliptic plane represent a break from the established norm?

Given the seeming erratic nature of the extreme scattered disk objects, which seem to revel in their anomalous nature, we might be forgiven for expecting the unexpected. However, theory suggests that these objects should also fall into line with the mean plane of the solar system. After all, if Neptune is to blame for scattering the objects out, then it should have done so under a common gravitational pretext. With that consideration in mind, the mean plane of the scattered objects should not become unduly distorted, even at these distances.

Kathryn Volk and Renu Malhotra, astrophysicists at the University of Arizona's Lunar and Planetary Laboratory, carried out a study to answer this very question. They analyzed the distributions of main TNOs (adding in a multitude of new objects that had been discovered over the last decade or so), as well as the objects in the scattered fringe beyond.

They found that the scattered objects beyond the main population of the belt showed a collective deviation for "the value of their average spin axes" (Volk & Malhotra 2017). For objects lying between 50AU and 80AU, the mean plane for their extended disk is found to be pitched at about 9° to the ecliptic.

This was an unexpected finding, and the researchers argued that 9° was too large a value to be attributable to a statistical blip. This warped distribution has important implications for the dynamical history of the outer solar system. It calls for the presence of another planet: An unseen mass located beyond the Kuiper Cliff would be capable of affecting the mean plane of the solar system, as long as it is moving along an orbit which is sufficiently inclined to the ecliptic to pull objects out of the invariant plane en masse.

The best fit, according to the University of Arizona team, would be an object with the mass of Mars, orbiting roughly 60 AU from the Sun, and on an orbit inclined by about eight degrees. This would have sufficient gravitational influence to warp the orbital plane of the distant TNOs within about 10 AU to either side.

However, as we have noted previously, such an object could hardly be said to have "cleared out its zone". It would be sharing its space with the objects in the scattered disk. As a result, a Mars-sized planet should qualify only as a dwarf planet. In response, Volk & Malhotra argue that Mars itself does not really clear out its zone. The classification of a planet over a dwarf planet is problematic in this regard. A terrestrial world embedded in the scattered disk beyond the Kuiper belt of slightly greater than two Mars masses should be sufficiently massive to clear out the ETNOs (Volk & Malhotra 2017).

This scattered disk band from the Kuiper Cliff outwards is about 30AU deep. The University of Arizona team thinks that an area this size could have been warped by a single planetary mass object in its midst, or just beyond (Stolte 2017). Their solution involves a relatively small Planet X object located less than 100AU away.

It is interesting to note that their conclusion is in keeping with a previous Planet X proposal made 15 years ago, this time by Adrián Brunini and Mario Melita. They also proposed an embedded Mars-sized object at around 60AU, arguing that this could explain the observed presence of the Kuiper Cliff. They have also concluded that a Planet X body would need to have quite an eccentric, inclined orbit in order to account for the observed anomalies (Brunini & Melita 2002).

The University of Arizona team does not think it is likely that the warping effect they noticed is due to a super-Earth like Planet Nine. Neither, do they think, there are other, more distant gravitational factors likely – even at this distance, the bodies are too close to the Sun to be tempted into deviation by galactic tidal effects, for instance. An alternative possibility is the transient gravitational nudge provided by a passing star, or rogue planet.

However, there is a self-correcting principle at work with these collective anomalous motions. Following such a passage, things should settle down again within about 10 million years. This limits the date of the flyby to a relatively recent point in time. There do not seem to be any obvious candidates in the stellar neighborhood capable of having glanced by the Kuiper belt in the last several million years.

I would argue that there is another important possibility to consider. Let us say that there is a more substantial Planet X body moving along a highly eccentric trajectory which periodically brings it within range of the Kuiper belt. Although it approaches the no-man's land of the extended scattered disk upon achieving perihelion at, say, 100AU, Planet X reaches out as far as the distant inner Oort cloud at its furthest point, some thousands of astronomical units away.

That would explain, quite neatly really, why there is such a significant gap in bodies between the Kuiper Cliff (~48AU) and the inner Oort cloud (~2000AU). Essentially, this area has been swept out over time by the highly eccentric orbit of a massive Planet X body. The planet would periodically recreate the observed anomalies in the extended scattered disk with each perihelion passage, thus preventing any self-correction over time. If the Planet X object is inclined to the ecliptic – as seems likely – then it would also generate the warping effect noted in the extended scattered disk. Its last perihelion incursion could easily have taken place within the last 10 million years, preventing the self-correcting adjustment of the mean plane from yet having taken place.

If it is near aphelion, Planet X could be thousands of astronomical units away at the current time. This would go some way towards explaining why it has been so difficult to detect.

A Resonance Relationship with Planet X?

The year before, in 2016, the University of Arizona team looked into a possible resonance relationship between the orbits of the most extended scattered disk objects, and Planet Nine (Malhotra, Volk & Wang 2016). Objects in the solar system very often fall into resonance relationships with one another. Neptune, in particular, enjoys a great many such relationships with the KBOs it shepherds through the belt beyond. The planet's name is apt. You might consider Neptune to be the Lord of the Deep. He commands a great shoal of objects in the dark, swirling cosmic seas beyond.

Fig. 9-4 Neptune's Fountain image

The UA team determined that the orbits of the ETNOs they studied were inter-related by integer ratios. This is interesting in itself because, unlike the TNOs, these extended scattered disk objects are scattered away from Neptune's influence. So why would they show such patterning? Are these extreme objects locked into a mean motion resonance with a massive planet beyond?

The team duly crunched the numbers. By taking the four most distant ETNOs known at that time, and assuming that a relationship exists with an unseen perturber, the ratio relationships provide some numbers for the putative Planet X object. The planet, of about ten Earth-masses, would have an orbital period of 17,117 years, and a semi-major axis ~ 665 AU (Malhotra, Volk & Wang 2016). This is fairly similar to the scenario for Planet Nine, arrived at through other means (Batygin & Brown 2016).

These figures were offered as tentative support for the Planet Nine hypothesis. However, because there are so few objects within this study, the resonance relationships between them may simply turn out to be merely coincidence. However, if the observed pattern were to continue with the discovery of further ETNOs, these relationships could help flesh out Planet Nine's orbital path.

I was intrigued by why the same team would be, on the one hand, providing supplementary evidence for a distant super-Earth Planet Nine while, on the other hand, attributing the warping effect to a much closer object. I wrote to one of the UA scientists, Renu Malhotra, to ask whether there was any correlation between their Mars-sized Planet X, and this possible resonance relationship with Planet Nine. In her reply, she indicated that these were largely unrelated issues, and explained the thinking behind each piece of research:

"Our paper from last year was on possible orbital resonances of distant KBOs with an unseen planet (so-called Planet Nine) of about 10 times Earth's mass orbiting at several hundred astronomical units distance from the Sun."

"The new paper is about the mid-plane of the Kuiper belt and its measured warp at distances of 50-80 AU. We think that this warp is unrelated to the so-called Planet Nine – that hypothetical planet would be too weak to influence the Kuiper belt's mid-plane at 50-80 AU."

"In the resonance work last year, the useful sample of KBOs was limited to just the 4 most distant KBOs (whose mean distance from the Sun exceeds ~200 AU and whose perihelion distance exceeds ~40 AU); this was because we wished to probe the effects of a very distant planet, some 10-20 times more distant than Neptune. We used the resonance work to compute the possible planet mass range and the possible planet sky location."

"In the new paper, we are probing a closer range of heliocentric distances, so the KBO working sample is much larger – known KBOs whose mean distance from the Sun is between 30 and 150 AU (a data sample size of ~600). We find that the KBOs whose mean distance from the Sun is between 50 and 150 AU (a sample of more than 150 objects) have a mid-plane that does not follow expectations (it deviates greatly from the solar system's invariable plane)."

"Our inventory of the outer solar system remains highly incomplete, but as we discover more KBOs we can use their orbital distribution to probe the structure of the solar system at larger distances, in a kind of bootstrapping way! But we have to be very cognizant of observational biases and small number statistics." (Malhotra 2017)

So, we have two quite different scenarios. They need not be mutually exclusive, however. This may present us with a two-body solution:

- The first planet is the Planet Nine Super-Earth, located in an inclined orbit well beyond the Kuiper belt. This body may be occasionally drawing up

KBOs into the anomalous extended scattered disk cluster, creating resonance patterns with the distant shepherding body.

- The second planet is a much closer, Mars+ sized object whose orbit largely lies within that of the ETNO cluster, but which is capable of distorting the outer sections of the main Kuiper belt.

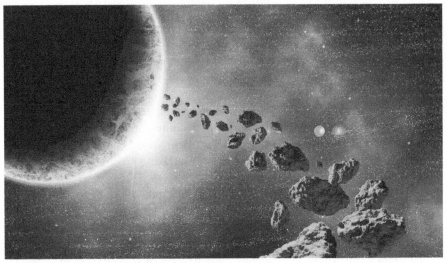

Fig. 9-5 Planet-asteroids image

The Mars-sized object located at 60AU may then warp the shape of that disk out of line with the invariant plane. But, can it explain the other anomalies attributed to the cluster of scattered disk objects? This seems unlikely. If you exclude Planet Nine, it may be that these extended scattered disk objects originated elsewhere. As previously described, it has been suggested that the retrograde KBOs, like Niku & Drac, may have been inner Oort cloud objects originally, occasionally nudged towards the planets in much the same way as long-period comets are perturbed from their more distant locations towards the Sun.

But if the extended scattered disk objects, which cause astrophysicists and dynamicists so many sleepless nights, have also originated from a more distant location, then how did they pick up the claimed-for relationships between them – like their similar arguments of perihelion, and other contentious patterns of behavior like resonance? Something still needs to shepherd them into their related organized orbits.

It is not clear to me how a Planet X object embedded within the scattered disk could achieve this. One would expect, instead, that its influence would create a more homogeneous arrangement, in the same way that Neptune shepherds the main belt of objects beyond it.

In 2005, I argued that a substantial PX body, whose inclined and eccentric orbit creates a broad range of orbital values (perihelion ~80AU to aphelion ~2000AU), might explain why there is such a striking absence of objects between the Kuiper belt and inner Oort cloud (Lloyd 2005). It still seems to me that a very eccentric orbit, allowing Planet X to periodically gently perturb the outer zone of the Kuiper belt, as well as graze the distant inner Oort cloud, is the best bet to help explain all of these varied issues. It is effectively a two-body solution rolled into one.

In 2005, this was highly contentious. Skeptics argued that such an object would be dynamically unstable, and have a provenance that is difficult to explain. Now, however, this subject has been blown wide open. Similar ideas about this perturbing body circulate among astrophysicists as a result of new evidence. How it might have gotten there in the first place remains a sore point. We shall look at the possible origins of this object in detail in the next chapter.

The Origins of Planet X

At the time of writing, there appears to be two major unsolved problems with the various Planet X theories on offer.

First, if the indirect evidence pointing to the existence of a substantial Planet X body is correct, then how has it evaded detection for so long? There is no shortage of groups looking for it, and the various constraints applied to its position have served to cut down the search area considerably. Sky searches are increasingly powerful, and the thresholds of their ranges are steadily lowering. Yet, Planet X continues to evade detection. Of course, an announcement of its discovery could be made at any time. Nonetheless, at the time of writing, two years on from Caltech's original press release, there is still no sign of it. I think that this is itself something of a mystery – although skeptics may say that this merely confirms that there is "nothing to see here".

The second problem is; how did it get there? The known planets formed out of a protoplanetary nebula, clearing out wide swathes of space around them. They then migrated radially to take up stable positions in the solar system. By contrast, a very distant Planet X body seems cut off from this family. In the case of the proposed Planet Nine body, it is located well beyond the presumed limit of the Sun's primordial protoplanetary disk.

So, what did Planet Nine form out of exactly? The Kuiper belt appears to truncate around 48AU, with a disparate band of freakish objects located beyond. This Kuiper Cliff may signal the edge of Planet X's clearing influence, but the generally low population of the disk seems at odds with the kind of density which would be expected if a substantial planet had formed beyond it early in the life of the solar system (Lawler et al. 2016).

Planetary Origins Beyond the Kuiper Belt

Having established the potential for one or more planetary objects to exist way out beyond the Kuiper belt, the question that desperately requires answering is

… how is this even possible? Not so long ago, the whole concept of Planet X was derailed within the astronomy community by the simple argument that massive planets could not form beyond Neptune. Even if they could, they would be so loosely bound to the Sun that they would be at risk of being pulled away from the solar system by the gravitational attraction of passing stars or galactic tidal influences. In other words, there would be a significant instability inherent with such worlds, effectively ruling out their existence.

To some extent, these arguments continue to dominate discussions about Planet X within the scientific blogosphere. This is despite an increased understanding of the chaos of young planetary systems, the seeming necessity to call upon complex patterns of planetary migration to explain various solar system characteristics, and a range of bizarre and unaccountable exoplanets, many of which call into question previously commonly-held assumptions.

The new evidence for the existence for Planet X has triggered a renewed examination of how a loosely bound planet might be possible way beyond Neptune and the Kuiper belt. In this chapter, we will look at some of these ideas, which include migrating gas giant remnants; a captured exoplanet from a passing star; and the gradual accretion of diffuse material in the outer solar system.

Ejected Worlds

A flurry of research papers emerged after the 2016 Planet Nine paper by Brown and Batygin. Many of them focused upon the tricky question of the planet's origin. As we have noted previously, current theoretical models of planet formation struggle with the accretion of major planets beyond the Kuiper belt. According to the prevailing theories, there simply isn't enough material available out there to build large planets. Many astrophysicists regard this as a strong argument against the possible existence of Planet X: It shouldn't be there…therefore, it can't be there.

However, if Brown and Batygin (and Scott and Trujillo, and several other groups of scientists) are proven correct, then that argument fails. I think that the confirmed presence of a massive Planet X would create an existential crisis for current theories of solar system formation – including the Nice model, described in Chapter 8. The existence of such a far-flung world would need to be robustly accounted for. This predicament has become a hotbed of theoretical speculation and computer modeling.

Brown and Batygin themselves think that Planet Nine may have started out as a gas giant core among the ice giants Uranus and Neptune. This additional object was then scattered from this outer zone by the gaseous component of the primordial nebula. This may have taken place during a turbulent period in the early history of the solar system when planets interacted gravitationally with the large number of remaining planetessimals, or by exchanging angular momentum with the rotating disks of gas still remaining (Batygin & Brown 2016).

Fig. 10-1 Early system image

It is widely accepted that many primordial planets were kicked out of the nest during these early periods of turbulence. Many would have been flung into the Sun by these encounters. Many would have escaped the solar system entirely, becoming free-floating orphans in interstellar space. An exception may have been an ejected planet that didn't quite make it out of the door, ending up in a loosely bound, wide orbit. In other words: A world like Planet Nine.

So, within the context of the Nice model, Planet Nine may have been a sibling of the ice giants Uranus and Neptune, becoming the unfortunate victim of the fluctuations in the orbits of Jupiter and Saturn. In this case, astronomers would expect Planet Nine to look a lot like the ice giants – or, at least, how they would have looked during this early phase of the solar system. Like the comets, Planet Nine might be a planet lost in time – a remnant of the primordial solar system, forever frozen in its early compositional state. The super-Earth identity attributed to it may reflect that early primordial state; as a core planetesimal under construction at the point of its ejection.

Planet Nine's Size

In their original paper, the Caltech team placed the range of masses for Planet Nine between five and twenty times that of the Earth – although they left some wriggle room for it being more massive (Lloyd 2016).

If Planet Nine is 'only' about 10 Earth masses, Mike Brown considers it likely that it is an ejected ice giant core, of about two to four times the radius of the Earth. This would be consistent with observations of exoplanets of similar mass, which are often dubbed super-Earths. Far from being large terrestrial worlds, though, Brown's working assumption is that Planet Nine looks much like a mini-Neptune (Brown 2016).

Exoplanet experts from the University of Bern, Switzerland modeled Planet Nine to provide some additional insight into what kind of world it might resemble. They were keen to explore the possible properties of the undiscovered planet, given what is now more generally known about the diverse wealth of different exoplanets orbiting alien suns. This context might be important when trying to work out how the planet took shape in the outer reaches of the solar system, or where it might have first originated from. Their modeling was closely based upon Caltech's projections of mass, at about ten Earth masses. This provided them with an extremely cold, gaseous planet which has a radius 3.7 times that of the Earth (Linder & Mordasini 2016). Given the continuing paucity of observational data, this remains a speculative punt into the dark.

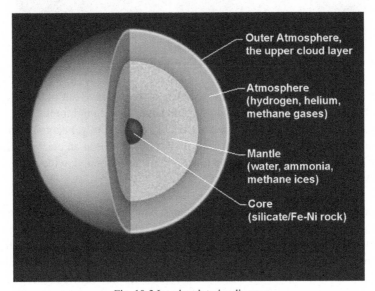

Fig. 10-2 Ice giant interior diagram

At ten Earth masses, this analysis simply reflects conventional planetary modeling of the 'ice giants' Uranus and Neptune, providing us with a Planet X object which began life as the runt of 3-piece ice giant litter. At around 20 Earth masses, Planet Nine would resemble a true ice giant, like its cousins.

Could it be any bigger? Taking into account its projected influence upon Saturn's orbit, the Italian physicist Lorenzo Iorio, of the Ministry of Education, University and Research in Rome, carried out some further work on the range of values for Planet Nine. Looking at the best available data for Saturn's position and trajectory, Iorio determined that the upper limit for the mass of Planet Nine could be considerably greater than the 20 Earth masses cited by Brown. Instead, masses up to 30+ times that of the Earth often worked in his calculations. This could rise to an extreme value of over 90 Earth masses if invoking a particular set of orbital values (Iorio 2017).

This projected upper mass for Planet X is close to that of Saturn! Could Planet Nine really be a small gas giant – a "gas dwarf"? Saturn has 95 Earth masses, but appears as an enormous gas giant because of its very low density (Saturn would float, if you could find an ocean big enough). Its remarkable rings make it more spectacular still. Another planet of the same mass could be denser, and so smaller in size, and therefore comparatively more difficult to observe.

Using their planet evolution model, the University of Bern team calculated how parameters like Planet Nine's planetary radius and reflectivity evolved over the age of the solar system (Rincon, 2016). This provided a view on how its relative magnitude might vary with size and distance from the Sun. Based upon this modeling, they argued that an infrared search would be far more likely to spot Planet Nine than telescopes detecting light from the visible part of the spectrum. They also commented that, at the distances involved, an upper limit for Planet Nine would likely be around 50 Earth masses before it should have been picked up by NASA's Wide-field Infrared Survey Explorer (Starr 2016).

How such an object might have evaded detection by an infrared sky survey as powerful as WISE seems to be at the forefront of most planetary scientists' minds. I cannot help but think that the variable parameters have been carefully selected to provide the scientific community with a best-fit scenario for Planet Nine. Suggestions that it might be a larger body become problematic, because a more massive object should have been observed by now. However, in my opinion, a larger, more distant body remains a distinct possibility.

Swapped at Birth

There are alternatives to the idea that Planet Nine was scattered out of the realm of its alleged siblings, Uranus and Neptune. Scott Sheppard's working hypothesis is that the planet was a surrogate world obtained from the planetary litter of a sister star within the Sun's early birth cluster (Kaufman 2016).

Let us rewind here a little. Binary star systems are very common, which is one of the reasons why there have been suspicions that out there, somewhere, there may yet lurk an unseen companion to the Sun. As we have discussed previously, this hypothetical body was initially given the name Nemesis. This ominous name reflected the popular theory in the mid-1980s that Nemesis periodically sent comet swarms our way from its distant abode in the outer Oort cloud (Davis, Hut, & Muller 1984; Whitmire & Jackson 1984).

Nemesis (the mass of which was gradually revised downwards to a sub-brown dwarf of several Jupiter masses) has not yet been discovered, despite sky searches using infra-red telescopes like IRAS and, more recently, WISE. However, as the saying goes, absence of evidence is not evidence of absence. It is difficult to prove a negative in astronomy, but skeptics of Planet X argue that the better the sky searches get – and the longer a planet remains undiscovered despite these advances – the lower the expectation should be that Planet X can exist.

A planet of this nature could be considered to be a binary object. We are quite accustomed to the notion that the Sun is a sole entity. To challenge such an assumption may seem foolhardy, however the more we learn about how stars form elsewhere, the more reasonable such a notion becomes.

Although the numbers are vague and subject to change, perhaps about half of the Sun-like stars in the Milky Way are in systems with multiple stars. Stars are born in star clusters, within nebulae. The varied distribution of the stellar birth clusters could well affect the solitude or otherwise of stars. Just because our Sun is considered a solitary star, we cannot assume that things were always that way. Rather astoundingly, it looks increasingly like stars are rarely, if ever, born alone.

Dense Cores

It is becoming clear that stars like the Sun tend to form alongside companion objects (Sanders 2017). Stars are born inside egg-shaped cocoons called "dense cores". These Cosmic Eggs are found within stellar nurseries – immense clouds of cold, molecular hydrogen which are often dark and opaque to visual telescopes. Astronomers have to use radio telescopes to study their contents. These observational studies show that the younger the stars within these nebulae, the more likely they are to be binary stellar objects. This has been predicted by some theoretical simulations showing that stars like the Sun tend to form as binary pairs (Kroup, 2011).

Observations of star formation in the Perseus molecular cloud, using the Very Large Array of radio telescopes in New Mexico and the SCUBA-2 survey in Hawaii, have confirmed the binary nature of initial star formation within the dense cores. The positions and alignments of the components of the systems vary within these cores, depending upon whether they are tightly knit together, or exist as loosely bound, wide binaries.

Fig. 10-3 Binary star image

The model which manages to explain this observed diversity the best requires that all stars initially form as wide binaries. They then either draw together over time to strengthen their binary nature, or they move further apart, leading to the break-up of the binary completely into solitary stars (Sadavoy & Stahler 2017).

Perhaps a third possibility should be added: That some stars retain these wide binary objects over the long-term, at varying distances. In the case of formation of brown dwarf companions, theoretical work has shown that these can quite readily migrate into highly eccentric orbits around the parent star. Alternatively, they may face possible ejection from the young system (Thies et al. 2010).

These observations of extremely young planetary systems imply that most, if not all, star systems start out as binaries, or multiple-star systems. That has implications for the concept of a solar companion for the Sun. According to this model, a brown dwarf or sub-brown dwarf companion may well have existed – at least during the early solar system. Of course, the model allows for its subsequent loss to the depths of interstellar space, but according to this new thinking the companion object might have formed about 500AU away from the Sun, before being ejected from the solar system in the early chaos (Sanders 2017).

That kind of distance is in the same ballpark as a possible Planet X object. So, one might speculate that a Planet Nine object was once a part of a brown dwarf system at this distance, and remained tethered to the Sun when the brown dwarf companion was ejected. Perhaps Planet Nine started life as a world orbiting that early companion, and became separated from it during the companion's eventual ejection. If that was the case, then the Sun grabbed one of its early companion's planets as a parting gift to itself: Something for the Sun to remember its lost partner by? This, then, may have been the origin of Planet Nine alluded to by Sheppard (Kaufman 2016). Although born of a birth sister to the Sun, the planet was swapped at birth. It therefore counts as a captured exoplanet.

Alternatively, one might speculate that the observed effects that are attributed to Planet Nine instead rely upon the looming presence of a more distant companion object from the Sun's original birth litter. An even more tantalizing possibility is that the original stellar sister from the dense core is still out there. In this case, the effect created by the proposed Planet Nine is a misinterpretation of the perturbing influence of a larger, more distant object: Essentially, that of the Sun's sub-brown dwarf companion. This was the basis of my own **Dark Star** Theory (Lloyd 2005).

The emerging realization that stars all start out with binary companions opens up possibilities about the nature of the primordial interstellar realm beyond the heliopause. It confounds the assumption that there was just empty space beyond the Sun's protoplanetary disk/accretion cloud. Instead, the Sun's dense core contained another major condensing zone.

Fig.10-4

The entire evolution of the solar system may have been fundamentally affected by one or more early solar companions, perhaps now gone (or perhaps still around in the depths of the solar system)! The construction and dissolution of the birth cluster, lasting up to perhaps 100 million years, could be crucial to understanding Planet Nine's origins (Li & Adams 2016).

Planet Nine from Another Place

In 2016, the concept of a binary capture of Planet Nine was explored using computer modeling. The research team who carried out this work factored in a number of constraints developed from observations of the outer solar system (Mustill, Raymond & Davies 2016). As with all simulations, to a certain extent you will get out of it what you put into it. A number of assumptions were made. The Sun should capture the planet from a sibling star early in its lifetime when the chaotic dynamics of the early planetary systems propelled planets, or planetary embryos, into wide, highly eccentric orbits where they would be ripe for picking.

The lead author, Alexander Mustill from the Lund Observatory, Sweden, argued that the Sun likely had a number of close encounters with other stars present within the extended birth cluster. There is far greater potential for planet swaps during this early, chaotic period than after the primordial nest has dissipated (Hall 2016a).

To fulfill the requirements of the present solar system, the stellar flyby should take place more than 150AU away, and the captured exoplanet should be very loosely bound to its own star. It should then be captured into the kind of orbit which is capable of sculpting the extended scattered disk.

One might immediately contend that if a young sister star can have a planet in such a loosely bound orbit (>100AU), then why can't the Sun? In which case, one removes the necessity for cradle-snatching in the first place...

That objection aside, the results of the modeling indicate that a large population of TNOs should align with the captured planet, and the nature of that alignment might provide clues as to the origins of the planet itself. The problem is the probability of this occurring. Although the Sun's ability to grab a passing planet is reasonably good, the chances of it slotting neatly into the Planet Nine configuration is far more remote, perhaps only as high as 2%.

There is a further downside. If the Sun can steal a planet from another star so easily, then it follows that another large passing star might similarly be able to snatch a loosely bound world like Planet Nine away from the Sun. Over the lifetime of the solar system, Planet Nine's projected orbit is not considered to be particularly stable, although once the Sun had emerged from its birth cluster, the incidence of such flybys decreases dramatically. The time spent within the birth cluster is therefore important, and further computer simulations have shown that any time spent in the cosmic nest beyond 100 million years significantly increases the chances of Planet Nine being stolen (Li & Adams 2016).

All in all, Planet Nine appears to be a low probability world. Whichever way you look at it, each of the formation scenarios seems as unlikely as the next. The formation of Planet Nine, and its insertion into the Sun's family, appears to have the same probability as placing a winning bet on number 9 at one of Monte Carlo's Casino roulette tables.

However, these odds do not always give the right impression. Say there were 100 Suns like ours, each similarly set up with a competing array of protoplanets, and jostling for position in a busy star cluster. Maybe only two or three manage to form a Planet Nine-like body, and hold onto it. There is no reason why our Sun could not be one of those. There are more than enough stars out there for these kinds of statistical games to play out in our favor.

Adopted Cousins

Another factor to take into account here is the origin of members of the extended scattered disk. As previously noted, some of these objects are referred to as the "Sednoids", after the first of their kind to be discovered; Sedna. Sedna is a curious reddish object, a rare and rather strange occupant of the extended scattered disk beyond the Kuiper belt. Its discovery in 2003 sparked great controversy at the time because of the object's highly unusual orbit. As we have seen, that controversy continues to rage on, with competing theories attempting to explain the elongated orbits of this and other extended scattered disk objects.

Sedna's strange properties sparked off a wealth of Planet X speculation, both in academic circles and in popular Internet forums. As far as the former goes, calculations using computer simulations attempted to work out how massive a planet would need to be to set Sedna's bizarre orbital trajectory into motion. The mass of the perturbing influence concerned goes up rapidly with its distance from the Sun. It was calculated that the unseen planetary companion would need to be a Jupiter-mass planet at more than 5000AU, a Neptune-mass planet within 2000AU, or an Earth-mass planet anywhere within a few hundred astronomical units (Schwamb 2007). Caltech's Planet Nine super-Earth sits neatly within this spectrum of Planet X possibility. Planet Nine therefore remains consistent with the 2007 projections for Sedna's perturber.

After Sedna's discovery – and the subsequent contemplation of the huge can of space worms that had been opened up – at least two groups of researchers proposed that Sedna had been captured by the Sun from a passing star, or brown dwarf (Kenyon & Bromley 2004; Morbidelli & Levison 2004). This new alien world did not seem to fit in the solar system. The potential for it being a captured object created a blueprint for the capture of other objects, too – including, perhaps, a whole planet.

A decade later, computer simulations were carried out to see if the Sun might have been capable of snatching Sednoids from a larger passing star (Jilkova et al. 2015). It provided a positive result, showing that Sedna might be one of hundreds of such captured objects, the majority of which would be located well beyond the range of current observational technologies (specifically, in this case, beyond about 75AU). By extension, the Sun should be equally capable of capturing larger objects from passing stars, including, perhaps, an exoplanet.

I would add a further possibility to this suite of capture scenarios. If there are more free-floating planets moving through interstellar space than is currently accepted, then there is an increased chance of one entering the planetary zone of the solar system as an independent, rogue planet, and being captured by the Sun. The probability of such an event occurring hinges on how much planetary material is ejected from young star systems, and, more contentiously, whether planets can form independently in interstellar space. We shall explore these ideas further when we discuss interstellar asteroids later in the book.

Is Planet Nine Nibiru?

Seriously pursuing the hypothesis that Planet Nine is a captured world from another star brings with it inherent dangers for scientists. Some of the vocabulary may have changed since the late Zecharia Sitchin wrote his first book about the mythical planet Nibiru (1976), but there's no escaping the fact that he claimed that this was a rogue planet captured by the Sun. If he was still alive today, Sitchin might now be exploring questions like "was Nibiru an exoplanet captured from one of the Sun's siblings within their mutual birth cluster"? Back in 1976, he was content to write about a rogue planet from interstellar space passing through the Sun's planetary system. His proposed interjection of this free-floating planet not only caused chaos to the early solar system, but also resulted in its capture into an inclined and eccentric orbit around the Sun.

At the time, this seemed like wild conjecture. It was vehemently dismissed on a number of levels. Indeed, despite many of the initial theoretical objections having been overtaken by new scientific findings, skeptics continue to pour scorn over Sitchin's "12th Planet Theory". To illustrate this point, consider these advances in our scientific knowledge over recent decades, and how they might support Sitchin's original scenario:

- Around 700 million years after the birth of the solar system, the planets were bombarded with a colossal stream of cosmic debris, lasting a few hundred million years. This period is known as the Late, Heavy Bombardment, the cause of which remains unknown, but seems to reflect a huge upheaval in the solar system.

- Stars are born in clusters (and start life as binaries within dense cores). During their early lives, young stars are much closer together, and encounter each other far more readily than the Sun does now. As a result, stellar flybys are more likely early on, with an enhanced potential for exoplanetary capture.

- We now know that planets and brown dwarfs move through interstellar space independently of stars. Whether they were ejected from chaotic star systems, or formed in tiny, fragmenting nebulae – or both – remains an open question. So, rogue planets can drift through space, just like stars, creating interstellar hazards.

- The chances of the Sun capturing a planet during any given transition may be statistically small, but if enough events like this occur during the early life of the solar system, then the overall chance of a capture event having successfully taken place during the lifetime of the solar system begins to look realistic. In trying to 'make Planet Nine work', astrophysicists have modeled scenarios where the capture of an exoplanet by the Sun becomes feasible.

- Astronomers have identified a new classification of planets: The super-Earths.

- Astronomers have observed wide binary objects in young star systems, showing that Planet X scenarios are possible.

- Planets likely migrated from one orbital position to another during the early stages of the solar system, rather than remaining fixed in place for the duration. The now broadly accepted Nice model positively embraces chaos and catastrophe, rather than the calm, orderly development of planets. That is not to say that the Nice model is correct in its assertions, but the very fact it has grown in acceptance by the astrophysics community illustrates the central importance of a catastrophic and chaotic nature for the early solar system. Complexity is king.

There are other reasons to draw a parallel between the hypothetical Planet Nine body, and the mythical planet Nibiru:

- Both have eccentric, highly elliptical orbits.

- Astronomers have observed wide binary objects in young star systems, showing that Planet X scenarios are possible.

- Both share orbits that are acutely inclined to the plane of the planets, by 30° or more.

- Sitchin argued that Nibiru was a sizable world, perhaps ten Earth masses, or more. This correlates with Caltech's estimates for Planet Nine, making it a super-Earth, or mini-Neptune.

- There are mutual constellations of interest in the path of these objects, particularly at aphelion and perihelion (Lloyd 2016a).

However, there are also important discrepancies between Nibiru and Planet Nine:

- Planet Nine stays well clear of the planetary zone of the solar system, while Sitchin claimed that Nibiru periodically returns to the inner solar system, like a comet.

- Sitchin claimed that Nibiru is a terrestrial world capable of supporting life in the outer solar system. Whatever exotic self-heating mechanism may have allowed this to be the case (an issue I tackled in **Dark Star** (Lloyd 2005)), this would ipso facto be a warm world to allow for life. In comparison, Planet Nine is thought to be an intensely cold, gaseous world.

- The orbital periods of these objects are vastly different. Sitchin claimed 3,600 years for Nibiru's orbit. Planet Nine's own circuit around the Sun is likely to be measured in tens of thousands of years.

It seems that I am more inclined to see the potential of Planet Nine than many of Sitchin's advocates. The journey this planet takes remains a big hurdle for this kind of eclectic synthesis, and may indicate a more complex solution than a simple 1:1 correlation between them.

Comparing Orbital Periods

If Sitchin's planet Nibiru is a traditional terrestrial world, with an orbital period of 3,600 years, then it would have a semi-major axis of about 235AU (this represents its average distance from the Sun). This scenario is in keeping with the kind of terrestrial Planet X body proposed by several astronomers studying the outer solar system's mysteries.

By contrast, if Planet Nine has an average distance of about 600AU, then its orbital period is just shy of 15,000 years. This is more than four times as long as Sitchin's Nibiru.

I have long argued that a more massive and more distant object is required to explain various aspects of the mystery (Lloyd 2005). The scenario I have depicted, where Planet X moves through the gap between the scattered disk and the inner Oort cloud, would give it a semi-major axis of about 1,000AU. This would give us a value for its orbital period of about 30,000 years. However, if it moves through aphelion beyond the inner extent of the inner Oort cloud, then those numbers would quickly go up. A semi-major axis of roughly 1500AU provides us with an orbital period of about 60,000 years (Shipway & Shipway 2008). Stretching the mid-point of Planet X's orbit to 2,200AU (now within the inner Oort cloud) gives it an orbit of over 100,000 years, which is equivalent to the length of time modern humans have been around!

A Ferry Across the No-Man's Land

The solution to the outer solar system's mysteries may involve multiple objects. This notion has been suggested by the de la Fuente Marcos brothers, who are astrophysicists working in Spain. Their computer simulations, carried out in the wake of the Planet Nine proposal, suggested the potential for more than one massive Planet X body beyond Pluto (de la Fuente Marcos & de la Fuente Marcos 2016a). A similar implication arises from Volk and Malhotra's work, through a side-by-side comparison of two of their Planet X papers (Malhotra, Volk & Wang 2016; Volk & Malhotra 2017). Note that none of these researchers is advocating the existence of Nibiru!

I think it would be reasonable to speculate that there may be a small planet that acts like a comet, and that such an eccentric world could be Sitchin's Nibiru. It is interesting to note that the Sumerian word "Nibiru" can mean "a ferry". A Planet Nine or Dark Star object could then be this ferry's other location of interest.

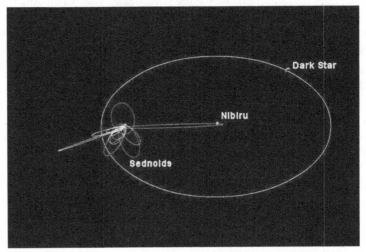

Fig. 10-5 Dark Star ellipses diagram

In the same way that some of the clustered objects in the extended scattered disk align themselves with Planet Nine, so too could this comet-like planet, Nibiru. By being directly aligned to the larger, more distant Planet X body, it could point the way towards it. However, for Sitchin's ideas to work, this object would need to act more like a comet, and move through perihelion within the inner solar system. This makes it very different from the objects in the extended scattered disk. ETNOs and comets may be related, however, particularly if the ETNOs originated from the inner Oort cloud. In which case, these objects are simply cousins at varying points along a spectrum of transition.

Planets IX, X, XI, XII, *etc.*

When you take into account the ever-broadening exoplanet-scape, the odds against a captured exoplanet seem to be steadily shortening by the year. Let us take this statistical argument in another direction. Supposing Planet X is out there, defying the odds. According to theoretical computer simulations of the solar system, the chances of this body having been originally captured by the Sun from another star are just a few percent at best. Having found a precarious home in the outer solar system, by whatever means, the odds of managing to stay out there for the duration are similarly long.

So, what if Planet Nine was just the lucky survivor? What if it was one of perhaps hundreds of similarly jilted planets, most of which were ejected into interstellar space? Then, the game of percentages takes on a new emphasis. Perhaps there are a number of 'survivors' out there in the far reaches of the outer solar system which managed to avoid the fate of their kith and kin earlier in the lifetime of the solar system. For every surviving Planet X body, then, perhaps fifty other planets were completely ejected.

This concept picks up on the ideas put forward by Eugene Chiang, an astronomer at the University of California, Berkeley back in 2005. His hypothesis was in response to the discoveries of various KBOs (Clark 2005). Based upon the oligarch theory of planet formation, where substantial numbers of similarly-sized planetary embryos are produced en masse, Chiang argued that many of the primordial planetessimals would have been flung out of the emerging planetary zone into the far reaches of the solar system. Many would have been thrown out of the solar system altogether, becoming free-floating planets. Others may still be orbiting around the solar system at a great distance, among the comets.

Chiang's proposition envisioned large-scale planetary migration early in the solar system's history. It suggests not just the presence of a Planet IX, a Planet X, but also Planets XI, XII, XIII and so on. Many of these would be Mars-sized or greater, reflecting the early planetesimal admixture (Clark 2005). This notion is not particularly new, especially with regard to what are now termed dwarf planets (Couper & Henbest 2002). If the comets could have migrated outwards, forming the Oort cloud, then there is no reason to doubt that a selection of primordial terrestrial planets might have also migrated outwards alongside them.

TXOs

Yet, there does not appear to be a continuum of objects from the Kuiper belt out towards the inner Oort cloud. Instead, there is a long, long gap, punctuated only by the extended scattered disk objects, which are themselves so tantalizingly suggestive of a distant Planet X object. How is that consistent with the prospect of multiple Planet X bodies?

Perhaps this extensive zone of relative emptiness is the domain of more than one additional planet. Relatively small, icy objects at the edge of the solar system become increasingly difficult to find with distance, and so the discoveries of objects beyond the Kuiper belt tend to involve those which are currently at their closest approaches to the Sun. There may be larger populations beyond, scattered out towards the inner Oort cloud, which do not approach the Kuiper belt at all, and therefore evade detection.

Perhaps further belts of objects exist in this extensive 'gap', shepherded together by their proximity to one or more farther planets. If these belts are stable enough, how

Fig.10-6 TXOs image

would we ever know whether they are there? One can imagine a series of extra planets beyond Neptune, rolling back through the solar system, each shepherding distinct belts of objects. In that case, we might have trans-Planet Nine Objects, trans-Planet Ten Objects, and so on, right out to the Oort cloud. Let us call these "Trans-planet X Objects", or TXOs. In which case, one might consider the disk-shaped inner Oort cloud to consist of a population of TXOs, rather than comets. Effectively, then, the Hills cloud is the solar system's third asteroid belt, this time shepherded by a substantial Planet X object.

We can only speculate at this time. But the clues in the outer solar system provide evidence for at least one planetary mass object out there... and perhaps two. As we have seen in this chapter, the origins of such an object may have been from within the solar system during its formative stages, or by the capture of an exoplanet.

There is also another possibility to consider; that planets actually accrete, or form, in the dark, outer reaches of the solar system. If this is possible, then the concept of multiple extra planets really opens up. We will explore this in some depth in the next chapter.

Building Planets in Interstellar Space

In the last chapter, we began considering the origins of Planet X. We looked at the favored hypothesis of the Caltech team that Planet Nine is an embryonic world akin to early versions of Uranus and Neptune, and that it had been ejected from the planetary zone as a result of chaotic migrations of the major planets, and planet-to-planet interactions. Their solution is in keeping with the Nice model of solar system evolution where planets migrate radially due to interactions with the extant gaseous disks (Batygin & Brown 2016; Batygin & Morbidelli 2017).

An alternative possibility is that Planet Nine is a captured exoplanet, ripped away from a passing star along with a massive retinue of other scattered objects, like Sedna (Mustill, Raymond & Davies 2016). This concept of a rogue planet being drawn into the Sun's family of planets bears a similarity with Zecharia Sitchin's theory about Planet X, which he argued had been captured from interstellar space (1976). In his case, the rogue planet was likely a free-floating interstellar planet which wandered unwittingly into the solar system, with catastrophic consequences.

Both of these ideas are challenging mathematically, because, rather like attempting to put a satellite into a stable orbit around the Earth, the trajectory of the capture needs to be finely-tuned. If the ejection incidence of planetessimals from the young solar system is high, or flybys of stars and planets early on is commonplace, then it is possible that the presence of Planet Nine represents the 'lucky' occasion when the math worked out just right.

In this chapter and the next, we will explore an entirely different possibility. Instead of a mathematically difficult ejection or capture, what if Planet X simply formed where it is, far beyond both the Kuiper belt and the heliopause? If so, then would the nature of that world be somehow different to the ones we know about?

From a classical perspective, there is simply no way that this could happen. According to accepted theories of planet formation, the protoplanetary disk of the early solar system does not extend far enough out to provide sufficient material to form substantial planets. However, among other things, the diversity of known exoplanets has shaken some long-held assumptions about planetary formation, opening the doors to new possibilities.

Nebulae

Stars form out of condensed nebulae of dust and gas, often left over from ex-stars which have disintegrated and/or exploded in their dotage. During their death throes, immense amounts of material are ejected into interstellar space. As a result, beautiful clouds of interstellar dust and gas are formed by the gravitational collapse of interstellar gas. These are known as nebulae. They become the building yards from which new stars are born. As a nebula swells, and its overall mass increases, its gravitational pull upon nearby interstellar gases generates a cascading momentum.

Fig. 11-1 Carina nebula image

Weighed down by their own gravity, nebulae can condense back down into clumps, or dense cores, which then give birth to new stars. The young stars that are born within the dense cores spin like swirling dervishes. The light from these stars illuminates the remaining disk of rotating gas and dust left over from the primordial nebula.

Planet formation takes place within these rotating protoplanetary disks. Astronomers are able to study the disks directly, and then try to simulate the accretion of planets within them using computer programs. Eliciting the exact mechanisms has proven to be a surprisingly difficult task, but planets clearly do form out of these protoplanetary disks.

Much work has been done on accretion mechanisms, where rocky and icy materials clump together in space. Processes like "collisional cascades" and "collisional damping" can affect outcomes, as well as factors like the size of the colliding materials and how well they are knitted together. The processes of the universe seem like a constant flux of recycling, breaking things up and putting them back together. Within the protoplanetary disk, the creative flow seems clearly established. But it is less clear whether such a creative drive can also take place beyond the main disk, and whether the balance shifts towards a more destructive pattern of behavior in the outer solar system.

Planet-Forming Beyond the Disk

A year before Brown and Batygin put forward their Planet Nine hypothesis, the question of whether a terrestrial super-Earth planet could form at these kinds of distances had been discussed in a paper by Scott Kenyon, of the Smithsonian Astrophysical Observatory, and Benjamin Bromley of the University of Utah. They wondered whether planetessimals ejected from the solar system might become lodged in the outer solar system, and then mop up remaining gaseous material around them to grow in size.

Fig. 11-2 Rotating disk image

During this early period, the abundance of gas in the solar environment would create a drag effect upon these dislodged worlds that would reduce their orbital eccentricity over time. As a result, these refugee planetessimals from the heart of the solar system's protoplanetary disk could settle down nicely in the outer solar system to become the focal points for a continuing process of accretion into more substantial planets (Kenyon & Bromley 2015).

Forming solid structures in space involves a number of factors. Gravity will attract new materials to a massive enough body, causing it to grow in size over time. With gaseous objects, like gas giant planets, materials are absorbed relatively easily – as observed during the spectacular impacts of the Shoemaker-Levy comet fragments into Jupiter. However, if a comet strikes another comet head on, the results are likely to be destructive, rather than creative. The collision will disperse materials, rather than allow them to merge together. The force of the impact is weighed up against the binding forces of the materials involved.

It seems likely that the Moon formed out of an orbiting stream of debris resulting from a planetary impact between a Mars-sized body and the Earth. Within the debris cloud of material swirling around the battered Earth, impacts between materials are more cohesive than destructive. Over time, the cloud of debris draws together to form a larger body, in this case the Moon. Similarly, then, if ejected materials from the protoplanetary disk can crowd together in the outer solar system, then their encounters with each other are likely to be relatively more constructive than destructive.

Surprisingly, even way beyond the outer edge of the classic primordial protoplanetary disk, an exploration of these mechanisms by computer simulation can bring about positive outcomes. Kenyon and Bromley's work showed that, under certain circumstances, super-Earths might form from relatively small planetesimal seeds over the course of billions of years (2015).

The most distant limit for this kind of mechanism to work seems to be about 250AU. Much beyond that zone, a super-Earth planet forming in situ would require the Sun to have had a much more extensive presolar nebula than is currently accepted. Planet Nine is considered to lie beyond this limit, making this an implausible scenario. That is not to say that it could not have formed around 200AU out, and then been subject to further outward migration. However, that creates an additional layer of complexity. This explains why Brown and Batygin consider Planet Nine to be an ejected planetary embryo, rather than a full-grown ice giant (2016).

Where Are the Warm Disks?

In their work, Kenyon and Bromley considered the life-cycle of pebble-sized objects in the outer solar system. Balancing fragmentation through collisions with mergers due to gas pressures within the disk, they argue that pebble-like building materials can survive the duration of the protoplanetary disk. The pressures of the gas in the protoplanetary disk can help accrete these materials together to form larger bodies.

This is fine for the early stages of the solar system, when there is still a protoplanetary disk available to create planets from, and provide the external pressure to bind materials together. Observational evidence suggests that planetessimals are assembled from these pebbles relatively quickly, in the order of about a million years, and then quickly merge into protoplanets. The next stage seems to be the sticking point. Sometime over the course of the next 100 million years, the protoplanets become terrestrial planets, like the Earth, through a series of cosmic collisions. These impacts should provide a massive dispersal of collisional debris across young planetary systems, which should be observable as warm dusty disks.

Fig. 11-3 Protoplanetary disk image

As more rocky exoplanets are discovered orbiting neighboring stars, questions have been raised about whether there are sufficient numbers of warm debris disks observed around young stars to statistically match the extrapolated populations of planetary products. One solution to this glaring discrepancy, again advocated by Kenyon and Bromley, is to argue that the speed with which planets form takes place more rapidly (Kenyon, Najita & Bromley 2016).

Bear in mind that the number of rocky planets being discovered is likely to be the tip of the proverbial iceberg, as the technology to detect small planets continues to improve year on year. Observational bias towards larger planets implies a future bonanza of terrestrial worlds as observational technologies improve. Either the low incidence of warm dust disks needs to be explained, or the increasingly large quantity of rocky planets must be accounted for in other ways. Perhaps planets do form more quickly than thought, within the short lifespans of the original protoplanetary disks. Or perhaps the formation processes creating terrestrial worlds involve less cosmic collisions, thus accounting for the lack of warm debris disks.

Dust in the Wind

Let us return to Kenyon and Bromley's idea about long-term planet building in the outer solar system. I wonder whether there are further on-going processes assisting planet-formation. If so, where is all the material to come from to carry on building planets? Must it all be leftover material from the early solar system, or could the Sun be mopping up interstellar material as it bobs along through the solar system, in the same way that it might occasionally steal objects from passing star systems?

Passing stars are a rare event in comparison to the Sun's movement through vast interstellar clouds of materials. It seems to me that there is plenty of scope here to mop up extra materials along the Sun's journey, especially when it encounters dusty and gaseous cosmic environments. This happens when the Sun travels through gigantic molecular clouds and nebulae. These may provide the right conditions to continue the building process – at least for planets in the outer solar system which also reside in interstellar space. I believe that this distinction is an important one: It potentially allows planets in interstellar space to accumulate small nebulae of materials around themselves.

If we are to consider the possibility that interstellar dust and gas plays a part in shaping objects in the solar system – and in particular planets well beyond the Kuiper belt – then we should first determine whether there is sufficient exposure to interstellar dust, and penetration of that dust into the solar system, to make an on-going building mechanism possible.

The Interstellar Medium (ISM) is composed largely of molecular hydrogen, as well as helium and small quantities of carbon and other elements. 1% of the ISM consists of sub-micrometer-sized dust, coated with frozen carbon monoxide and nitrogen, which serves to darken the clouds. Molecular hydrogen is hard to detect in infrared, and the carbon monoxide component of the clouds is used for detection purposes instead. Interstellar dust contains silicates, soot and greasy aliphatic hydrocarbons. Laboratory experiments have shown that the amount of the latter, greasy component has been significantly underestimated. There may be about 10 billion, trillion, trillion tonnes of aliphatic hydrocarbons in the Milky Way, created in stars, and distributed across space (Günay et al. 2018). This 'space grease' likely plays a role in sticking dust grains together in interstellar space. Perhaps this may even aid planetesimal formation as larger and larger pebbles clump together. An analogy might be cement holding bricks in a wall together.

The solar system currently resides within a Local Interstellar Cloud (LIC), of a lower density compared to the galactic interstellar medium. However, over its lifetime, the solar system has passed through many interstellar gigantic molecular clouds. It has been estimated that the solar system passes within 16 light years of star-forming nebulae every ~50–100 million years (Napier 2007). This has intermittently exposed solar system bodies to a ready source of interstellar gases, dust and complex organic molecules.

High Speed Chemistry

Like all materials in the galaxy, ISM is in motion, and is light enough to be blown around by the cosmic wind. The cosmic wind can easily push low-density clouds of interstellar gas and dust around, but not high-density clouds (Kenney, Abramson & Bravo-Alfaro 2015). I would argue, then, that a dust cloud or tiny nebula bound to an interstellar planetary body is similarly undeterred by the action of the cosmic wind. So, although the dissipative effect of the cosmic wind may prevent new stars and planets forming from scratch in interstellar space, that effect is overcome by the protective gravitational field of a planetesimal. As a result, free-floating planets should continue to accumulate interstellar materials around themselves.

Of course, materials in interstellar space are untethered, as far as parent stars go. They can meet each other at very high velocities. These small scale cosmic collisions are considered generally to be more destructive than creative. In other words, the impact of ISM dust grains against interstellar comets would erode them, rather than build them up further. However, given that much of the content of these interstellar materials is 'greasy' (Günay et al. 2018) there is the potential for a 'gnats-on-the-windscreen' effect, too.

Laboratory experiments carried out to investigate how some comets throw out oxygen into their comae suggests that such small-scale bombardments can drive up an atmosphere from rocky silicate surfaces. If the bombarding molecules are ionized, then the bombarded silicates on the comet's surface can undergo what is termed an "Eley–Rideal reaction". This releases molecular oxygen. In the case of a comet during its perihelion passage around the Sun, the bombarding material is ionized water, already driven off by out-gassing, and then falling back onto the cometary surface (Yao & Giapis 2017).

However, this mechanism could equally be applied to interstellar comets subject to high-speed collisions with ionized plasma in the ISM. What we are seeing here is high speed interstellar chemistry. The presence of molecular oxygen enveloping an interstellar comet can then drive other chemical processes, particularly oxidation. Far from having a purely erosive effect, the surfaces of interstellar comets should become increasingly complex as a result of ISM bombardment.

The Fate of Interplanetary Dust

A quite different set of circumstances applies within the heliosphere of a star, compared with interstellar space. The Sun creates within its complex magnetic environment an outward movement of plasma and gas. This is best seen as coronal mass ejections from the surface of the Sun during a total solar eclipse, but is a constantly on-going process. This solar wind is a powerful force and helps to 'clean' the space between the planets.

Many of the early building blocks of the solar system, which might have gone on to become parts of planets, were obliterated by collisions with their peers. They were pulverized into dusty grains. The resultant fine dust was gradually removed from the planetary zone of the solar system by the action of the solar wind. The Sun acts upon interplanetary dust and gas in two ways:

- Through the outward pressure of the solar wind, the Sun can blow dust and gas away from itself. This effect can be seen in out-gassing comets, whose tails point away from the Sun, despite the comet's actual motion through the solar system.

- The Sun can drag larger particles back towards itself to be absorbed. Most interplanetary dust spirals down into the Sun as a result of a combination of "Poynting-Robertson drag" and solar wind drag (Klačka 2013).

Through these damping effects, much of the 'captured' interplanetary dust spirals down towards the Sun over time. The rest of the dust is destroyed during collisions, ejected from the solar system, or gradually accreted by the planets (Messenger 2014; Grazier 2015).

Fig. 11-4 Zodiacal dust image

The dust which remains, or, rather, which is in the process of removal when observed, can be seen just after sunset as the phenomenon known as "Zodiacal light". This visible interplanetary dust, which is concentrated in the plane of the ecliptic, is thought to be continuously replenished by fragmenting Jupiter-family comets (Schulz 2015). The Sun plays a predominant role in systematically clearing out this dust. It does this through the influence of solar radiation forces upon small particles, which creates drag. As particles lose momentum, they gradually fall towards the Sun. As a result, the Sun is an efficient housekeeper. t continuously clears its domain clean of the interplanetary dust which arises from cosmic collisions in the solar system.

If this process is interrupted for some reason, then the consequences for life on Earth could be dire. The mechanism for dust removal has been linked by some with periodic catastrophism on Earth, specifically when connected with the periodic movement of the solar system through the galactic plane (Thaddeus 1986). But that is not a situation anyone has experienced directly. Fortunately, our planetary zone is not full of dust, blocking out the heat and light from the Sun.

This established mechanism of gradual dust removal is not available beyond the influence of the solar wind. The effect of the solar wind peters out eventually. Indeed, there is a boundary condition where the lowering pressures of the dissipating solar wind match the pressures of the cosmic windswept ISM beyond. This fluctuating border is currently encountered around 123AU and is known as the heliopause. It is clear that the solar wind in the planetary zone is significantly stronger than the galactic winds beyond.

Most of the dust orbiting the Sun beyond the heliopause is thought to reside within the inner Oort cloud, at about 3,000 AU. This is in keeping with the zone likely to contain the highest distribution of comets (Kaib & Quinn 2008) – or TXOs, if you wish to link this torus-shaped belt with a massive neighboring Planet X object. Mechanisms for the removal of this dust have also been proposed. These involve various scenarios dependent upon whether the solar system happens to be in and out of a gigantic molecular cloud (Belyaev & Rafikov 2010).

It should be noted, however, that the presence of foreground zodiacal dust obscures the view of infrared emissions from dust contained within the Kuiper belt (Vitense et al. 2012). This makes it difficult to determine what the levels of dust are in the interstellar zones of the solar system beyond the heliopause.

At odds with expectations about the paucity of dust in the outer solar system, an anomalously high level of externally-sourced, large dust grains entering the planetary zone has been observed by the Ulysses, Cassini and Galileo satellites. These large dust grains move with the flow of interstellar medium towards the inner solar system. This effect has yet to be satisfactorily explained, and implies that there is an unexpected source for these particles beyond the heliosphere (Draine 2009). Calculations indicate that this source of large dust grains is ranged some 500 AU from the center of the solar system (Belyaev & Rafikov 2010). Curiously, then, this anomalous incursion of large dust grains may provide a breadcrumb trail back to an unseen celestial object located beyond the heliosphere.

The Heliopause

Let us take a moment to consider this boundary between the Sun's domain, the heliosphere, and interstellar space beyond. The boundary is not set in stone, but instead represents a balancing point between two forces, internal and external. The inner force, flowing outwards from the Sun, is the solar wind. The outer force, pressing inwards, is the galactic stream of interstellar medium.

Note that the heliopause is not a tangible, solid surface. Nonetheless, it may help to think of it as the surface of a magnetized bubble blown out from the Sun, whose pressure has equalized with the galactic atmosphere beyond.

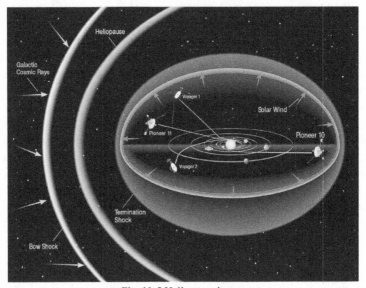

Fig. 11-5 Heliopause image

However, just to make things slightly more complicated, there is more than one boundary condition at play here. The outward movement of solar plasma and gas flows towards two boundary layers beyond the heliosphere, interfacing with both the galactic magnetic field, as well as the stream of Interstellar Medium (ISM) of dust and gas:

- The ionic density of the solar wind quickly falls away with distance from the Sun, to the point where the solar wind velocity drops below sub-sonic values. This creates an initial, roughly spherical boundary region, which is known as the "termination shock".

- Beyond this zone, within a second, tear-shaped boundary region known as the heliopause, the slowed wind is swept back by its interaction with external forces.

Until recently, it was thought that a third boundary layer would be created beyond the heliopause as the heliosphere moved through the interstellar medium beyond, creating a bow shock, in a similar way to a boat through water. However, data from NASA's IBEX has ruled out the existence of this third boundary feature (McComas et al. 2012).

Even before the realization that there is no bow shock, it was understood that only a small fraction of interstellar ions are actually diverted by the termination shock. The vast majority of molecular material penetrates into the solar system, although these boundary conditions will cause the invading material to slow and heat up. These complex boundary conditions can lead to instabilities at the heliopause, allowing ISM to intersperse with solar wind in complex finger-like patterns, similar to those observed in some nebulae.

The position of the termination shock is itself variable. It is dependent upon the relative outward pressures within the heliosphere, compared with the plasma and neutral gas density in the local interstellar medium beyond (Cooper 2003). One of the scientific findings which came out of Voyager 1's transit through the heliosheath was that the strong magnetic field of the interstellar medium is either turbulent or distorted in the solar vicinity (Opher et al. 2009).

Planet X and the Local Fluff

NASA's Interstellar Boundary Explorer spacecraft, often referred to as IBEX, studies the magnetic conditions beyond the heliosphere. Over several decades, the IBEX spacecraft has observed changes in the direction of travel of the interstellar medium encasing the solar system. Unexpectedly, it detected a roughly circular ribbon of energetic material circumscribing the heliosphere, involving areas with the lowest pressures of local ISM. There appear to be localized effects taking place beyond the heliopause showing an unexpected degree of complexity and interaction (Grzedzielski et al. 2010). The heliosphere is now thought to be asymmetrical in shape. This is considered to be the result of interaction with ISM in the Local Interstellar Cloud – known colloquially by scientists as the "Local Fluff".

Perhaps, like the large grains of material unexpectedly seeping into the planetary zone from a source some 500AU away, these localized interactions may indicate the additional presence of substantial Planet X body beyond. It is probably too distant for its own magnetic field to directly create such a discernible effect at the heliopause. However, if Planet X, is surrounded by an extensive local nebula, then in-falling materials from that nebula towards the Sun could be the source of this observed 'Local Fluff' interaction. These observations add to the lengthening list of anomalies which could readily be explained by the presence of a distant world located in the interstellar space beyond the heliopause.

A Flexible Boundary

The size and shape of the heliosphere is not set in stone. The inner pressure created by the outward push of solar plasma is balanced against the external forces exerted by interstellar gases and other material. If the Sun were to flare up, then the internal pressure within the heliosphere would rise, pushing the heliopause outwards. Similarly, if the Sun moves through a rich region of space containing dust and gas, then this external 'dust storm' can push the heliopause back, effectively shrinking the heliosphere.

Such an effect could, potentially, be significant. Modeling shows that the shock position should fall back deep into the planetary zone when the solar system moves through a particularly dense interstellar cloud (Zank & Frisch 1999; Yeghikyan & Fahr 2004). Indeed, some of the modeling shows a complete collapse of the heliosphere under external pressure, albeit temporarily. The Sun's power is such that it can regain its composure and reform its field strength, pushing the heliospheric bubble back out through the solar system. This has much to do with the finger-like instabilities woven through the 'surface' of the heliosphere, where charged particles meet neutral materials, interacting chaotically. This fluctuating effect is akin to a sailboat making its way through blustery weather.

The implication of this is that the whole solar system can occasionally become immersed within dense gigantic molecular clouds. During this immersion, the outer solar system experiences significant penetration of neutral interstellar dust, gas and other greasier materials associated with these immense clouds (Frisch et al. 2002).

Where the heliosphere boundary generally provides a defensive shield to repel immersion by the continuous stream of fine interstellar medium, significant external pressures can quickly break down these stellar defenses, allowing a flood of materials into the planetary zone. However, even when this happens, the inner solar system generally remains protected by the solar wind.

Building Planets in the Beyond

There is an obvious implication here about the outer solar system beyond the heliopause. If dust in the interstellar medium can penetrate into the planetary zone, covering worlds with a fine layering of extra materials, then it should be clear that planets located outside the heliosphere are actually constantly exposed to this kind of immersion. I propose that planets outside the heliosphere are, in effect, always growing, however grindingly slowly that process may actually be (Lloyd 2016b).

Over the lifetime of the solar system, that slow accretion should add up to a considerable dusting. This effect should be boosted during the times when the Sun moves through dense patches of interstellar cloud, localized nebulae, or when bobbing through the galactic plane. The conditions for renewed accretion should improve significantly during these times. As a result, planet growth in the outer solar system should be intermittent in character. It would very much depend upon the particular galactic environment the Sun happens to be moving through at any given time.

Right now, there's probably not much happening out there, relatively speaking. The local interstellar cloud offers thin gruel (although the IBEX results, and the trail of granular materials penetrating the outer solar system hint at something more). The process might be minimal, providing the false impression that this is always how things are. But the current doldrums may be nothing compared to the times when the Sun traverses dense clouds of interstellar gas and dust. If I'm right, then these blustery journeys would trigger heydays of rigorous planetary accretion.

This is why I believe that there is significant potential for planet-formation beyond the heliopause. It is not just because planetessimals get ejected into interstellar space from the chaos of early planetary systems and absorb what is left of the outer edge of the solar nebula. It is that these seeds, or kernels, can become the core for further planetary accretion over the lifetime of the solar system. Compare this with the planets left behind within the heliopause, which are effectively protected from such growth by the clearing action of the Sun.

It is important to point out here that dust can also be removed from the outer solar beyond the turbulent termination shock. However, the mechanism for ejecting granular material of various sizes may be orders of magnitude slower than that for removing similar material in the inner solar system (Belyaev & Rafikov 2010). Interstellar material gathered beyond the heliopause therefore has greater time to coalesce around a planet or planetesimal in that zone. Studies of the lifetime of dusty grains within the Kuiper belt show removal rates several orders of magnitude slower still (Scherer 2000).

Dust which falls into a resonant orbit with Neptune is capable of clumping, forming larger particles of debris matter which can statistically withstand the opposing mechanisms of collisional cascades (Vitense et al. 2014). Beyond the heliopause, the greater the mass of an interstellar planet, the more difficult it would be for the cosmic wind to disperse its attendant nebula of material. The conditions at the boundary of the heliosphere may provide an opportunity for perturbed interstellar material to accrete and stabilize when in the presence of a substantial gravitational core, like a planet.

While the planets in the planetary zone of the solar system stop growing (particularly in the well-protected inner regions), planets beyond the heliosphere might just keep on getting bigger, aided by these processes.

Amassing Outer Solar Dust

Such a scenario depends upon the pooling of usable interstellar dust in the outer solar system beyond the heliopause. While findings from various spacecraft allude to an anomalously high stream of large granular particles from this zone into the inner solar system (Draine 2009; Belyaev & Rafikov 2010), most work about the dynamics of dust in the outer solar system concentrates upon its ejection from the solar system by the action of the galactic stream of ISM. Similarly, interactions with GMCs and thick interstellar nebula material focus upon the erosion of outer solar system, creating an expulsion of cometary debris (Stern 1990).

What is generally not described is the effect of the solar system's passage upon the medium through which it is traveling. Could it be that this is surely subject to drag, inelastic collisions and field interaction on a scale where at least some of that medium becomes bound to the solar system? Given the vast scale of these clouds, if only a tiny proportion of the medium interacting with the solar system were to be captured in this way, that would still amount to a great deal of bound material to consider.

In effect, the solar system might act like a sponge subjected to a constant stream of fluid. Charged plasma may be diverted around the heliosheath. Other materials move seamlessly through the solar system. Within this flux, some material may be slowed and absorbed by the solar system. For instance, heated and decelerated neutral hydrogen which has been diverted in this way, piles up within the heliosheath (Zank & Frisch 1999). As we have seen, ejection of that material from the outer solar system is very much slower than its commensurate removal from the inner solar system, providing opportunities for new phases of accretion and activity each time the Sun traverses such thick fields of material.

There has been speculation that layered deposits of cosmic materials in the Earth's geological record may have been caused by deep immersions of the solar system within interstellar clouds, which pushed the heliopause right back to the zone of the inner planets. Modeling of such interactions has been carried out in an attempt to explain relatively recent catastrophic events on Earth, as well as anomalously high concentrations of 10Be in Antarctic ice cores (Zank & Frisch 1999). Other researchers have attributed this observed anomaly in terms of increased cosmic ray flux resulting from ancient supernovae shock waves (Sonett et al. 1987).

If such processes have caused the Earth to accrete interplanetary dust originating from interstellar space to this extent, then presumably the same can be inferred for other planets in the solar system. Indeed, one might reasonably expect a greater level of accretion around outer solar system bodies during these periods of termination shock collapse and dust/ISM inundation. The further away from the Sun a planet is, the less well protected it is from inundation when the heliosphere collapses inwards. If the solar system happens to be moving through a particularly thick zone of nebula material, then the accumulation of decelerated, heated granular material entering the solar system would be that much greater, and so the potential for capture by outer solar system planets much more promising. Furthermore, once the shock-wave has subsided and the Sun has recovered its usual heliosphere, it will be the inner solar system that is cleared out first. The dust-removal processes in the outer solar system will be slower, and less efficient, allowing more time for accretion effects to take place.

Outer Solar System Complexity

This argument envisages zones of effect within the solar system during times of interstellar medium inundation. The inner solar system remains relatively dust-free and shielded from the worst of the periodic inundations. The outer planetary zones, home to the gas and ice giants, are more exposed, leading to a higher potential for re-shaping and reforming. The greater propensity for a refashioning of this outer zone would then provide the potential for more complexity – rings, moons, reformed surface features.

Fig. 11-6 Chariklo with rings

The periodic infiltration of high levels of interstellar medium might play a major role in shaping what we are increasingly observing in the outer solar system, i.e. unexpectedly higher levels of geophysical activity, and greater variation of components within the planetary systems as one proceeds into the darker recesses of the outer solar system. I would argue that the unexpectedly diverse compositions of objects, prevalence of complex rings and satellite systems around planets, as well as bilobate (two-lobed) and contact binary objects, may point to a degree of accretion and clustering that has been an on-going process through the lifetime of the solar system.

These accretion periods may occur intermittently throughout the lifetime of the solar system in a quasi-periodic way, i.e. when the Sun encounters gigantic molecular clouds as it bobs through the galactic plane every 35 million years or so, or during its rarer travels through dense galactic spiral arms. The intervals may also be random and unpredictable, as the Sun moves through extensive clouds of interstellar gas and dust. The outer solar system is affected by the alteration in the galactic tide it experiences during these slow carousel movements. This can trigger the nudging of Oort cloud comets over time (Matese et al. 1995) and so, by extension, the shifting around of smaller, granular materials, too.

The further away from the Sun the solar system object is, the more likely it will be affected by periodic immersion in interstellar medium. This is because the protective action of the Sun will be weaker. We could extend this speculative thinking still further, concluding that the complexity of outer solar system objects should increase with an object's distance from the Sun.

This complexity (the number of moons, the extent of an object's ring system, the variance in its geophysical composition), may result from the occasional heavy in-wash of dust and gas when the solar system moves through particularly dense zones of galactic molecular clouds or nebulae, re-shaping its overall structure. The multitude of collisional cascades involved with all these myriad components, creates more dust.

Building is like that: It generates a mess! In the case of recently inundated planets in the outer solar system, the dust will take a while to disperse. It will form a foggy haze around the planets. While the pressurizing interstellar gas is in situ, that dust will press down and accrete into bigger lumps. But once the gas disperses again, due to the renewed action of the solar wind and other drag effects, the haze will gradually disperse.

This, however, is not what will happen beyond the heliopause. Beyond that point, the Sun's clearing effect will not occur. The processes for gas and dust removal out there are much, much slower, and rely upon the gradual erosion of debris by the streaming interstellar medium. This leads me to put forward a further argument about the hazy environments around planets outside the heliosphere. An argument that may just solve one of the greatest problems in the history of the hunt for Planet X: If there is so much evidence for another planet in the solar system, then why haven't we spotted it yet?

C H A P T E R

12

The Shroud Hypothesis

In this chapter, I will continue to build on a hypothesis which could explain how Planet X has evaded detection so far. My thesis is based on the idea that interstellar space provides a different environment for planets compared to a classical planetary zone. Although astrophysicists recognize the very different properties of interstellar space, it would not appear that anyone has considered how these variable conditions might apply to free-floating planets.

Using the available evidence, some logic, and a little conjecture – I hope to present a thesis which will stimulate new lines of scientific inquiry while providing Planet X with a bit more breathing space. That breathing space may prove to be rather dusty…

Dust Exospheres

As discussed in the previous chapter, space dust is not a major feature in the known solar system. We can look out into the cosmic void and, on a clear night, we can see a myriad of twinkling stars. The only localized indications of dust are from zodiacal light, and the lunar dust kicked up by micrometeorite impacts. The solar wind acts to clear the dust away. This action is most efficient as we get closer to the Sun. Light atomic gases, like hydrogen, can become charged, or "ionized". These are blown away from the Sun in a stream of discharged solar plasma. Heavier materials, like dusty grains, lose momentum over time and are dragged down into the Sun.

Yet, despite these powerful ongoing processes, permanent dust clouds orbiting solar system objects have been observed and studied. For instance, a dust disk has been detected around Saturn, and is thought to originate from Phoebe, one of Saturn's irregular satellites. Similarly, the irregular satellites of Uranus have created collisional dust debris which is gently falling down onto the ice giant (Tamayo & Burns 2013). Collisions with meteorites have created observable dust clouds, or "dust exospheres", around the Galilean moons of Jupiter, and around the Moon (Horányi et al. 2015).

All of these processes are dynamic: They are off-set by continued collisions, erosion, or disintegration. In the case of the Moon, for instance, the dusty ejecta clouds thicken up after the annual meteor showers. It seems highly likely that all bodies in the solar system which lack atmospheres have these thin dusty exospheres. They are continuously replenished by cosmic micro-impacts.

I suggest that these effects are magnified significantly around worlds located beyond the heliosphere. Here, the clearing processes are far less efficient. It seems logical, therefore, that sizable dust clouds can accumulate around planets in interstellar space. Dust exospheres could be routine features of these distant worlds, where the dissipating action of the solar wind is not felt. Arguably, then, Planet Nine, or its equivalent, may be surrounded by a significant envelope of dust.

To a lesser extent, the same may be true of distant comets which do not routinely cross back into the heliosphere during perihelion. At the moment, we have no way of knowing either way: The out-gassing effects during perihelion would drive these atmospheric features away. What we do see, however, is that the surfaces of comets are not the pristine, icy environments once envisioned. Instead, their surfaces look more like the bottom of a quarry. This, presumably, is due to the accumulation of in-falling interstellar dust over the course of billions of years.

Complex Collisions

Each of the known gas giants has a ring, a feature lacking among the inner planets. They also each have a series of irregular satellites. These appear to have been depleted significantly over the age of the solar system, releasing huge amounts of interplanetary dust into the vicinity of their parent planet.

During early epochs of our own solar system, the gas giants probably had similar prominent circumplanetary disk clouds. This has been implied from unusually shallow size distributions of the gas giants' current moon systems, and the violent collisional history that likely brought that distribution about (Tamayo & Burns 2013). This dust is eventually removed by ejection further into space, or by the action of drag, pulling debris material down towards the main planet.

Collisional activity involving irregular satellite systems can create a swarm of debris dust around a major planet (Kennedy et al. 2010). If the giant planet is located out beyond the heliopause, then the external clearing processes in the interstellar environment are significantly slower.

Spinning clouds of interstellar gas orbiting around stars tend to flatten into disks. Although these dusty disks are not readily apparent around the known planets, rings are. However, we cannot automatically assume that dust clouds around free-floating planets will get fashioned down into neat rings.

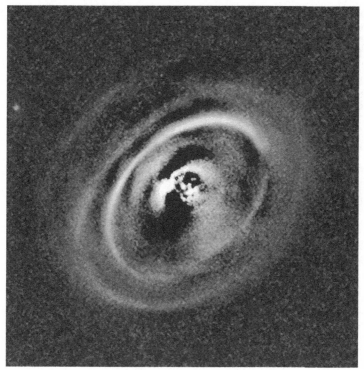

Fig. 12-1 Planetary disk image

After all, nebulae are often highly irregular in shape, despite the shepherding effects of stars within them. The mechanisms at work beyond the heliosphere may result in different shaping processes, and so different visual features.

Lessons from Dagon

Fomalhaut b is a possible example of this principle at work. It is an exoplanet candidate orbits its parent star, Fomalhaut A, the brightest star in the constellation Pisces. The discovery of this exoplanet was made from analysis of optical images taken in 2004 and 2006 by the Hubble Space Telescope (Kalas et al. 2008). This discovery was unusual because of the difficulty in directly imaging planets. The dimly reflected light from planets tends to get washed out in the glare of its parent star. Fomalhaut b was the first exoplanet to be directly observed in visible light.

Or, at least, that appears to be what has been observed. The light from this object is fuzzy. Potentially, this implies that another phenomenon is also present. The two most popular theories among astronomers appear to be an extensive, spherical dust cloud, or an enormous circumplanetary ring system.

The mystery provided by this quirky orbiting patch of light has been ongoing for almost a decade, providing sufficient notoriety for Fomalhaut b to be given the name

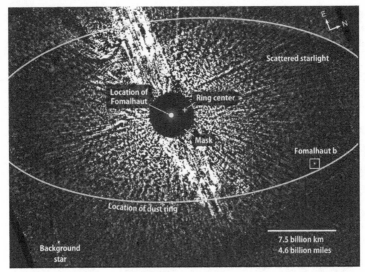

Fig. 12-2 Fomalhaut b image

Dagon, after the Semitic half-man, half-fish deity. Dagon is equated with the Babylonian deity Oannes/Adapa, the first of the seven antediluvian sages known as the Apkallu.

Carl Sagan and the *Apkallu*

As a brief aside, I find it a rather curious coincidence that this fuzzy exoplanet should be given an epithet that is so obviously tangled up with ancient Mesopotamian mythology. The Apkallu were once cited by the eminent scientist and broadcaster Carl Sagan as the most likely examples of extra-terrestrial contact in historical times. Contrary to the skeptical characteristics often attributed to him, Sagan thought that numerous advanced lifeforms existed in the cosmos (Zygutis 2017).

Fig. 12-3 Nimrud Apkallu

Over the galactic lifetime, many interstellar-faring civilizations must have emerged, Sagan argued, based upon the statistical probabilities emerging from the famous Drake Equation. Writing a report for NASA during his early career, Sagan advocated that NASA, and other interested organizations, conduct an in-depth study of ancient myth to trace early contact with alien visitors. Predating Zecharia Sitchin's work by at least a decade, Sagan proposed that these might have included the Mesopotamian mythological creatures, the Apkallu (Sagan 1963).

Fomalhaut b's Dusty Shroud

The nature of the exoplanet named after this deity remains controversial, as the defining feature of the object in question is a large optical flux of what is widely considered to be a circumplanetary dust cloud. Fomalhaut b is most likely a sizable gas giant surrounded by a spherical dust cloud (Currie 2012), generated and maintained by ongoing collisional cascades with associated planetessimals (Kenyon et al. 2014), such as a related system of irregular moons (Kennedy et al. 2010).

In order to be capable of sculpting Fomalhaut's extensive debris disks in the way observed, it was initially thought that Fomalhaut b was between one and three Jupiter masses. However, the Spitzer Space Telescope failed to be detect the planet Dagon in infrared (Janson 2012). This constrained Dagon's upper mass to twice that of Jupiter.

Because no heat-signature has been detected for this enshrouded planet, it has not been possible to confirm the presence of a planet of any size within the cloud that envelopes it. As a result, Fomalhaut b stands as an exoplanet discovered by direct imaging, even though no thermal image of this world has yet been obtained (Currie 2012). The thermal data is at odds with the sculpting pattern carved out by this gas giant through the star's extensive debris disks, creating something of a headache for astrophysicists. A consensus seems to have built up around a mass of half that of Jupiter for the enigmatic exoplanet. But other theories abound, including a cluster of terrestrial worlds, a swirling swarm of planetessimals, or a pair of coupling planets.

Perhaps the presence of the enveloping dust cloud blocks out the heat signature of the enshrouded planet. The planet may be hidden from view in infrared. If true, I suggest that this could have a significant implication for Planet X. If Planet X is similarly enshrouded, then this might explain why it has not been found by infrared sky surveys.

Does Dagon's distance from its star have any bearing upon its fuzzy appearance? Fomalhaut b, which is currently off-set from the disk, has a highly elliptical orbit, thought to extend from 50AU to 300AU (Harrington & Villard 2013). These are Planet X-like distances! Fomalhaut A is twice as massive as the Sun, so a larger protoplanetary disk might be expected for this young star. Even so, the debris disk around the star Fomalhaut A is truly enormous. It is thought to span out over a range from about 150AU to over 200AU (compare this to the Kuiper belt's meager 30 – 48AU).

It is possible that Dagon periodically plows through this debris disk. In that case, the collisional maelstrom that would take place during that violent crossing may explain the planet's shrouded appearance. We will find out sometime after 2030, when it is projected to encounter the disk – assuming the orbit of the planet and the disk share the same plane.

Wings of Dust

Dagon may take up a cloud of debris from the disk every 1,700 years or so. Or, it may miss the disk completely. In that case, further explanation for the planetary nebula is needed. I suggest that this is simply what loosely-bound planets actually look like. I think that planets beyond the heliopause struggle to shake off an accumulating sheath of materials. These local nebulae are buffeted about by the galactic winds, but not actually dislodged from the planet's environment. Like the larger, more familiar nebulae, these planetary nebulae take on an indeterminate shape with unclear boundaries. I suggest that loosely bound planets in interstellar space are surrounded by billowing 'wings' of dusty debris.

Fig. 12-4 Winged disk image

Where our Sun may be symbolized by a shimmering circle, and our Moon by a crescent, Planet X might be symbolized by a winged disk. Again, this logical argument sits well with Zecharia Sitchin's claims for ancient Mesopotamian and Egyptian Winged Disk symbols (1998, p247). Rather than symbolizing the Sun, these esoteric motifs symbolized another celestial object: Planet X.

Dagon seems to be an example of this mechanism. It is not as distant from its star as Planet Nine, and moves within the heliosphere of Fomalhaut A during perihelion, so its dusty shroud is at risk. However, Fomalhaut A is a young star, and the system's extensive dusty debris disk indicates that the interplanetary 'clearing' processes are far from complete. Fomalhaut b may be an extreme example of a generalized principle – exacerbated by its periodic interactions with the broader debris disk around its star: The dust clouds may well be periodically replenished by collisions between materials in the cloud and the debris ring (Kennedy et al. 2010).

The enhanced magnitude of a diffuse dust cloud over a lone planet in the outer solar system would provide a subtler effect. In the case of our solar system, a Planet X body enshrouded in dust might become a faint, fuzzy object for astronomers, making it indistinguishable from a distant galaxy or nebula. Even so, shouldn't a brighter dust cloud be easier to spot than the relatively small planet enclosed within it? By extension, wouldn't Planet X be easier to spot if it is wrapped up in a dusty shroud?

There is a problem with observing dust in the outer solar system: The view can be affected by the interplanetary dust, or Zodiacal Dust. The light reflected back by this dust obscures dusty signatures beyond it – particularly within the plane of the planets. As a result, it is difficult to assess dust levels in the outer solar system from the vantage point of Earth. There is too much dusty white noise in between. Even so, we can see most nebulae easily enough, despite this issue. These nebulae are internally lit by birthing stars, or are super-heated remnants of exploded stars. By contrast, far removed from its parent star, Planet X is enshrouded in relative darkness.

If the enshrouded planet approaches its star, then these dusty features would strongly increase the planet's overall luminosity. That is the case here with Dagon, which currently lies near perihelion along its highly elliptical orbit. Saturn also provides a good example of how the magnitude of light from a planet can be substantially increased by a set of rings. However, Saturn's rings are a sharp, well-defined feature, and it retains its planetary appearance. A third example of the increased visibility of an object surrounded by gas is a comet approaching the Sun. Without their immense tail, comets would be difficult objects to detect.

Misidentification

Because a dust cloud can increase the visibility of objects, the Sun's light should illuminate the dusty clouds around a Planet X object, reflecting back dim, diffuse, dusky light. But the dimmer these objects get, the more this fuzziness works against detection. How so? If you are specifically seeking out a planet, you will be looking for a pinpoint of light slowly moving across the background field of stars. You seek out candidate objects for further study over a period of time. To refine your search, you will be actively eliminating other features from your line of inquiry. If you are not expecting to find a planet within a fuzzy blotch, you will disregard such a feature. After all, it might be a tiny nebula or dust cloud, or one of billions of distant galaxies. Either way, it is not what you are looking for.

If Planet X lies in the galactic plane, or even in front of the central mass of the Milky Way's galactic core, then the bright, busy background makes it a difficult object to identify at the best of times. Add into this frame a dusty, obscuring shroud and it makes matters more complicated still. The light from this localized nebula would be difficult to identify because of the diffuse foreground light of the solar system's interplanetary dust. The dusty shroud would also partly conceal the planet within, turning a sharp object into a nebulous one in various wavelengths of the electromagnetic spectrum, including infrared.

In the noise-ridden search for outer solar system objects at the very edge of modern computerized capability, that obscuring effect may well make a significant difference to the probability of the Planet X needle emerging from the stellar haystack. It could explain why this object continues to evade detection, against all odds.

Dark Nebulae

In 2015, the European Southern Observatory published a remarkable image of a star field being obscured by a dense, dark nebula. This amorphous black spot in the southern constellation Serpens Cauda is named LDN 483 (Hook 2015). It is a striking image, for sure. At first glance, it appears as if there is a dark hole in the starlit sky. This is something of an optical illusion – the stars are still there, but the light from them has been blotted out by a particularly dense, dark cloud, full of gas and dust, which lies about 700 million light years away.

Fig. 12-5 Dark Nebula LDN 483 image

This 'dark nebula' is the blackened birthing ground of new stars in the fetal stage of their development, just prior to their lights gradually turning on. The protostars within the dark nebula LDN 483 are currently condensing out of the dense interior of the molecular cloud, under the force of gravity. This striking example of a dark nebula is one of many examples cataloged by the American astronomer Beverly Turner Lynds, back in 1962.They were discovered during a painstaking visual inspection of the Palomar Sky Survey photographic plates. The name of this particular dark nebula, LDN 483, stands for Lynds Dark Nebula 483.

The smallest dark nebulae found by this method was about 2 square minutes of arc (Lynds 1962). Finding dark nebulae relies upon a sufficient density of background stars for them to emerge visually. This helpful background light varies significantly with distance from the galactic plane. Star fields beyond the Milky Way are much less dense with stars that those within it. Self-evidently, it is far more difficult to identify dark nebulae in dark parts of the sky.

When molecular clouds become dense enough to block out the light from behind them (essentially eclipsing background stars) they qualify as dark nebulae. They vary in their darkness, or opacity, scaled from one (the least dense) to six (the darkest).

Could LDN 483 itself be a densely obscuring smog surrounding Planet X? How do we really know that LDN 483 is a vast molecular cloud 700 million light years away? Could it not be a tiny, smog-veiled object in the outer solar system?

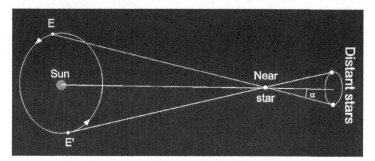

Fig. 12-6 Annual parallax image

There are a number of ways of determining an object's distance in space, most of which involve the measurement of properties of its light. However, by definition a Dark Nebula does not emit light. Alternatively, astrophysicists calculate distance based upon parallax. This is how an object's position varies against background features as a result of the seasonal movement of the Earth around the Sun.

There are thousands of dark nebulae in the Lynds catalogue. What is more, these are just the clearest examples of dark spots in the sky – there are potentially millions more whose size and location against a dark sky make them too insignificant visually to make it into the catalogue.

Other Dark Blobs

Fragmented, wispy examples of dark nebulae have been observed in the open star cluster IC 2944.These appear to be leftover remnants of these larger star-forming globules. These diffuse dark blobs are called "Thackeray's globules", and are strongly silhouetted against the glowing gas of the star-forming nebula beyond (Thackeray & Wesselink 1965).

Fig. 12-7 Thackeray's globules image

These globules are not thought to contain protostars, but are instead the dying wisps of the dense dust clouds. They are being blasted away by the intense ultraviolet radiation from the emerging stars within the main birthing cluster.

In the battle of darkness over light, objects which match up with the background will be obscured. A fuzzy Planet X 'cloud' could be easily camouflaged in visible light if set against a similarly bright cosmic background. Contrast is everything, and without it, objects disappear from view.

How would a dusty shroud affect a search for Planet X in infrared? I suggest that the heat signature of the enshrouded planet would be severely curtailed by a localized nebulous cloud. Dark nebulae tend to be extremely cold, because of their capacity to quickly lose their heat to space. The inner regions of these freezing fog patches in space are still, dank and cold. Not much gets out of them beyond microwave radiation. Therefore, the denser the dark nebula cloud surrounding a planetary body in interstellar space, the more difficult it would be to detect in infrared. This would readily explain why infrared searches like WISE have failed to detect Planet X, even though astrophysicists have calculated that it should have been detectable.

Birthing Interstellar Planets

Can planets form within dark nebulae? It is now recognized that some free-floating planets in interstellar space may have started their lives within relatively small nebulae, completely independent of star-forming processes.

Nebulae come in all shapes and sizes. Nebulae are not just vast, protostar-bearing entities light years across. For instance, the dark Bok Globules are relatively small, at perhaps a light year across. Yet, they are still capable of forming stars within them, especially when associated with larger complexes (Yun & Clemens 1990).

Fig. 12-8 OTS 44 image

Smaller, darker nebulae are capable of building brown dwarfs. Usually, planets form from protoplanetary disks rotating around nascent stars. The dust and gas contained within these vast clouds whirl and spiral as the disk rotates, creating eddies, then clumps, which then accrete into protoplanets. However, some free-floating interstellar planets and brown dwarfs are thought to form within "Gahm globulettes" (Gahm et al. 2013). An example is OTS 44, a very young free-floating planet of 12 Jupiter masses observed to have an accretion disk of gas and dust around it (Joergens et al. 2013). Its youth provides some light, but, in time, it will burn up the available fuels and sink into the darkness. OTS 44 will probably become a dark, rogue world wandering through interstellar space.

Nebulae range in size from the truly colossal, down to the kind of neatly wrapped bundles which might provide us with free-floating sub-brown dwarfs, or even gas giants. Perhaps they can be smaller still, wispy remnants of dust and gas embroiling protoplanets. These would be extremely difficult to detect in interstellar space. Not only would they be dark and cold, but extremely small relative to other galactic features.

Extrapolating to our own solar system, I suggest that breakaway clouds of the presolar nebula may have formed loosely-bound planets in wide orbits around the Sun. Essentially, these small breakaway clouds would be examples of Thackeray's globules on a tiny scale. It is possible that Fomalhaut b represents an example of this phenomenon. This mechanism would provide a simple means to create one or more surviving Planet X objects which now skirt the peripheries of the solar system. More generally, this hypothesis would greatly expand the theoretical population of interstellar worlds in the galaxy.

Magnetized Fluff

The above examples largely relate to nebulae in young star-forming regions of space. The Sun is in the middle of its life, and its protoplanetary disk is long-gone. One might argue that the forces that drove away that disk over time would have removed localized clouds around Planet X objects, too.

However diminutive they might be, these hypothesized localized nebulae are still subject to galactic tidal forces, and the streaming currents of interstellar medium (ISM) surging through the space between the stars. The solar system exists within a 300-light year wide zone known as the hot local bubble. This immense cavity has been largely cleared of ISM. The reason for this clearing is thought to be due to the forces created by a number of supernovae about ten million years or so.

While recent experiments appear to confirm this hot local bubble theory (Galeazzi et al. 2014), the efficiency of this ISM wipe-out is seemingly at odds with the presence of the Local Interstellar Cloud (LIC), or "local fluff", which spans some thirty light years across (Phillips 2009). If the local interstellar medium has been so comprehensively driven away by multiple supernova events 10 million years ago, then how did this 30 light year diameter bit of fluff manage to survive? A possible reason for the LIC oasis is that the ISM within the local cloud appears to be magnetized to such an extent that it was capable of withstanding the forces unleashed by the supernovae (Opher et al. 2009).

Magnetic fields seem to be the factors preventing the destruction of clouds of interstellar materials by supernova hot gas ejecta. If these tiny globulettes were wrapped around gas giants, or larger failed stars, then they might enjoy the protection of the vast magnetic fields that these planetary behemoths generate around themselves.

The Sun can provide such a field within its heliosphere. Beyond this, external planets with a sufficiently strong magnetic field prevent the erosion of dust fields around them. A body as large as Planet Nine would generate a sizable magnetic field environment. This effect would be even stronger for a sub-brown dwarf. These localized protective environments would magnetize and thus shield dust clouds wrapped around such worlds.

Dust in the Outer Solar System

But other forces are at work, too. The gravitational wells of external worlds would still drive a downfall of materials, and planetary accretion. So, if I am right to hypothesize about dusty shrouds obscuring hidden planets in interstellar space, then there would need to be a balance of forces at work. There would need to be a dynamic equilibrium between the build-up of dust and gas materials around the planet, and its losses to interstellar space or downfall onto the planet itself.

It is recognized that debris fields of dust can be created through collisional cascades following cosmic impacts (Kenyon et al. 2014). Chipping surface materials off rocky bodies can create a mess of debris. This can be subject to a further series of impacts, generating ever finer rounds of grainy materials. Although the Kuiper belt is a highly dispersed asteroid belt, spanning a disk thickness of some 20AU, there will inevitably be the odd collision between KBOs, or between KBOS and comets. At these distances, where the action of the solar wind is significantly less than the inner solar system, KBOs probably come with dust exospheres of varying intensity.

But this represents thin gruel. The Kuiper belt may well have dusty environments, particularly around the KBOs themselves. The New Horizons probe is set up to investigate this very issue. However, the kind of environments around interstellar planets that I am arguing for here require a great deal more material than just the debris strewn off from a few colliding rocky bodies. Where does it come from?

Space Weather

I suggest that there are times when the dust exospheres of Planet X objects are enhanced by massive infusions of gas and dust. This is a very difficult thing to observe happening to dark objects in interstellar space, and so would be difficult to prove through direct observation. However, there is evidence of this kind of complexity occurring during the formation of stars of various sizes. Birthing stars can be fed by huge gas clouds. This resembles logs being thrown onto a fire.

The so-called 'FU Ori phenomenon' relates to the occasional dumping of vast clumps of material from an associated nebula onto accreting dust disks. This has been observed to happen around certain pre-main sequence stars and results in flaring in the inner portion of the disk. Such young protostars (colloquially known as FUors among astronomers) are "eruptive variables". They display extreme changes in magnitude and spectral type. They are all associated with reflective nebulae which glow in visible and infrared wavelengths (Tsehmeystrenko 2013).

There is also evidence of massive stars being similarly aided in their development by the dumping of immense quantities of material from the protoplanetary disk onto them. This creates strong variability during the star's accretion. This kind of episodic accretion creates something of a light show, with spectacular outbursts as material is dumped onto the young stellar fire.

These mechanisms appear to be played out in the formation of truly massive stars, too, creating instabilities in the accretion disks which provide variable photometric bursts, and some truly bizarre distortions. Computer simulations carried out by a research group based in Tübingen University have modeled how high-density clumps within a gravitationally unstable circumstellar disk form and then migrate through the disk. These clumps finally fall spectacularly into the central star (Meyer et al. 2017).

There may be universal principles involved here. The monumental turbulence associated with stellar birth may strongly influence these processes. Might this also be true in dwarf star formation? Might this principle apply to the formation of planets in interstellar space, themselves formed from tiny nebulae, or globulettes?

Fig. 12-9 Dust disk image

I wonder whether this intermittent accretion could be happening in interstellar space at the planetary level. This would be based upon encounters with globular nebulae.

172

As with clouds in our own atmosphere, the turbulence created by external and internal conditions shapes and buffets nebulae. In a way, nebulae have their own weather. Continuing this analogy, rain and hail stones can form within the clouds. These patterns of accretion provide dense pockets of interstellar materials which defy the ideal of uniformity.

When the solar system moves through a denser region of space, then there will be the opportunity for it to encounter this "space weather". I suggest that the gigantic molecular clouds effectively rain on the solar system. This bombardment of ISM would push the heliopause inwards. It would expose the outer regions of the solar system to this cosmic weather.

The same accretion process may happen to planets within a star's heliosphere, too, if the protective heliosheath retreats under significant pressure from a particularly dense external cloud. But it's beyond these stellar environments where a significant new mechanism might be going on, in the darkness, potentially feeding an immense population of free-floating planets. Once made wet by the clumps of ISM within these clouds, Planet X-like objects would struggle to become dry again. They are not subject to hair-dryer action of the solar wind in the same way as their inner neighbors.

Accretion, then, could be an ongoing process in interstellar space, heavily dependent upon the changes of environment through which the free-floating planet moves. This is the conclusion my logical sequence of arguments leads me to. But what is feeding this process? Where is the fuel for intermittent accretion coming from?

Spiral Arm Catastrophism

As our spiral galaxy, the Milky Way, rotates around its core, the Sun slowly bobs up and down, like a carousel horse. It takes about 230 million years to complete one rotation. The Sun has completed about 20 circuits during its 4.56 billion year lifetime. Its carousel motion causes it to move through the galactic plane every 35 million years.

During its long circuitous journey around the galactic core, the Sun traverses the galaxy's spiral arms once every hundred million years or so. This has occurred about fifty times during the Sun's lifetime. It takes about ten million years to cross through the spiral arms. The Sun is exposed to a higher number of environmental dangers during these spiral arm transits. There is a greater stellar density at the heart of the spiral arms, so there are likely to be a greater number of local supernovae affecting the Sun's local environment.

However, in contrast to the bright star-fields within them, the spiral arms are bordered by "dusty lanes". These outer zones pose an entirely different hazard to the solar system. Due to the potential for a catastrophic influx of dust into the inner solar system, this periodic dusty lane transit has been cited as a possible trigger for Ice Epochs on Earth (McCrea 1975; Shaviv 2003; Gies & Helsel 2005).

The (disputed) idea that the Sun's movement through spiral arms might be coincident with the onset of Ice Epochs is known as the "Galaxy-Ice Age Hypothesis". Various mechanisms have been offered to explain the cooling of the Earth including cosmic radiation upon the Earth's atmosphere ionizing gas into cloud-forming plasma (Duhamel 2013); and the absorption of the Earth's atmospheric oxygen by the interstellar clouds, dramatically reducing the ozone layer (Davidson 2005).

A third possibility is that the Earth's immediate space environment could be altered by the intrusion of dense interstellar matter into the solar system.

Dust Inhalation

In this scenario, the Sun's heliosphere would be pushed inwards by the external pressure of the ISM in the dusty lanes, potentially shrinking as far in as the Earth's orbit. For a time, this concentrated ISM would envelope the Earth, and most of the other planets, within a particularly cloudy interstellar environment (Begelman & Rees 1976). In the absence of the solar wind, this interstellar dust would then penetrate the Earth's atmosphere, bringing about what we might now call a "nuclear winter" effect.

Recent work on the iridium layer associated with the extinction event, thought to have triggered the extinction of the dinosaurs 66 million years ago, has shown that the Earth was exposed to this cosmic material over a period perhaps as long as eight million years. The implication of this may be that the iridium was deposited during prolonged exposure to a gigantic molecular cloud (Nimura et al. 2016). The encounter with this vast dark cloud would have led, over time, to climate change and mass extinctions.

Naturally, the Earth would not be the only planet in the solar system so affected. If the Earth had been inundated with dust in this way, then the solar wind had already retreated from the outer solar system, exposing all of the planets beyond ours. Once the Sun has moved on through whichever dust cloud has immersed the solar system – whether a gigantic molecular cloud, or the dusty lanes bordering galactic spiral arms – the heliopause would have been restored to its 'usual' place, and the action of the solar wind could once again clear the Sun's planetary domain of the accumulated dust.

The dust and materials accumulating around the Sun's traditional set of planets is then removed through the usual mechanisms. But this same mechanism does not apply beyond the heliopause. I suggest that a massive Planet X body would maintain a great deal of dusty material around it, forming a long-standing local nebula environment. There is no solar wind at that distance to clear it all up after the Sun has emerged from the dust fields. Driven on by a series of collisional cascades, the enshrouded Planet X body might then form this renewed material into complex rotating rings, irregular moons, and so on.

This on-going accretion mechanism would be possible for all interstellar, free-floating planets so exposed. It would be repeated many times over the lifetime of the galaxy, as these objects are intermittently exposed to such dusty environments.

We are currently located within a minor spiral arm known as the Orion arm, which connects the nearest two major spiral arms of the Milky Way. As such, during the Sun's lengthy journey through this part of the galaxy, the solar system may have been exposed to effects like this. One might expect, then, that a Planet X body would still have wrapped around it the accumulated nebula from relatively recent transits through the minor spiral arm's dusty peripheries. That claim somewhat depends upon whether the planet's magnetic field is sufficiently strong to hold on to its dusty cloud: As mentioned, the Sun was exposed to a string of 'local' supernovae events within the spiral arm around ten million years ago. The magnetic and gravitational fields enveloping a massive planet provide the optimal opportunity for retaining a dusty nebula in the face of this onslaught.

Galactic Gas Transfer

You might imagine that galactic dust should be steadily removed by supernovae, and that the overall stock of interstellar medium in the galaxy has been gradually eradicated by such events. However, the Milky Way's supplies of dust and gas are being constantly replenished from beyond the galaxy.

Computer simulations have shown that about half of the material in the Milky Way may have originated from other smaller galaxies (Anglés-Alcázar et al. 2017). This startling find opens up a whole new debate about the movement of materials within the universe, and provides evidence that we are all at least partly composed of matter which originated from distant galaxies (Carlisle 2017).

It has long been understood that although galaxies are generally shifting away from one another within an expanding universe, they sometimes end up on a collision course with each other, merging and exchanging materials. We are not immune: The spiral shape of the Milky Way results from a long-term interaction with (and cannibalization of) the Sagittarius Dwarf galaxy. This small galaxy has been getting shredded by the powerful gravitational tug of the Milky Way, leaving streams of debris swirling around our galaxy in its wake (Choi 2011).

You might imagine that this 'big-fish-eats-little-fish' scenario may be sufficient to explain inflows of materials from other galaxies. It had been thought that perhaps just 20% of matter in our galaxy was derived from external sources. However, the new research indicates that on-going movements of gas flowing between galaxies adds a significant new dimension to this picture, bringing that figure up to 60% for a galaxy the size of the Milky Way (Anglés-Alcázar et al. 2017).

This matter, moving unseen in the dark space between galaxies, is swept along on galactic winds generated by supernovae. As galaxies age, the effect of this inflow of intergalactic dust begins to dominate the accretion mechanisms within larger galaxies over the internal recycling of in situ materials. If in-coming materials from other galactic sources are simply composed of the lightest elemental gases, like hydrogen and helium, then their impact will be confined to the creation of new stars of low "metallicity". However, if these transferred materials are richer in complexity (as one might expect from older local galaxies), then that inflow might include more dusty materials capable of clumping together to form sub-stellar structures like free-floating planets.

The implication is that our galaxy is subject to a steady inflow of a great deal of gas, and quite possibly immense quantities of extra-galactic dust, too. This creates more opportunities for the Sun to encounter immersive fields of gas and dust during its long travels through the galaxy.

Fig. 12-10 Asteroid in nebula image

This further increases the potential for intermittent dust accumulation around planet-sized objects beyond a star's heliopause, in turn allowing continued accretion throughout their lifetimes. This could be true as much for free-floating planets in interstellar space, as it is for loosely bound planets orbiting their parent stars in wide, Planet X-like orbits.

An Abundance of Dark Stars

In the last chapter, I suggested that massive, free-floating planets offered the perfect environment for the intermittent clumping of interstellar dust to take place. These interstellar planets would effectively act as seeds for growing dark nebulae, which, in turn, would hide the host planets away within their dark shroud. Because these objects do not emit a stellar wind, like the Solar Wind, they do not blast the dust away. However, their massive magnetic and gravitational fields provide a haven for a localized nebula around them.

For instance, Jupiter's magnetic field is immense. If you could see it at night, it would stretch across the sky about four times the diameter of the Moon. A visible magnetic field would make Jupiter look like an Aurora Borealis comet. Imagine moving Jupiter outside the heliosphere, beyond the clearing influence of the Solar Wind. Then imagine the solar system moving through a dense molecular cloud in space. Beyond the heliopause, Jupiter would become immersed within this dust cloud. It would magnetize the fog which penetrates its extensive magnetosphere, and hold on to at least some of it. This effect would be an ongoing process: An interstellar Jupiter would then really look like a comet, as the magnetized shroud is buffeted and molded by the galactic tide and ISM.

I suggest that this mechanism provides the basis for on-going planet-building process in interstellar space, including within the peripheries of star systems. It opens up the potential for one or more **Dark Star** neighbors within the solar system, hidden away within their dark, dusty cocoons. Although I consider it likely that Dark Stars are ideal masters of such dark shrouds, little is known about them. That is a shame, because I think they offer great potential to solve a great many mysteries, including a partial answer to the Milky Way's missing mass. Another potential solution, of course, is a massive Planet X body.

If one exists within the comet clouds of our solar system, then I suggest that it would have built its own planetary system around it through this intermittent inundation of dust and gas. Within a Dark Star's dark shroud, there is the potential for habitable environments to thrive in interstellar space. Think again about the enshrouded Jupiter and its moons, stuck out in space beyond the heliopause. Its dark nebula would not only cocoon and hide the gas giant system, but also insulate it. The infrared emission from Jupiter, its warm glow, would stay in-house.

It has been over a decade since I wrote **Dark Star** (2005), and more Dark Stars have been discovered in that time – although these are usually the young, bright examples of low mass sub-brown dwarfs which emit their own light. Given that brown dwarfs can essentially live for ever, this early 'shiny' phase represents only the tiniest fraction of their lifespans. After they have burnt their available nuclear fuels up, they turn dark. This much we know. The rest is often tainted by ambiguity: Even their classification remains problematic.

Astronomers use several different names for these objects: sub-brown dwarfs, cool Y Dwarfs, and even "planemos". I use the term Dark Star, usually referring to planetary-mass objects several times more massive than Jupiter. I will continue to use the term here, interchangeably with sub-brown dwarf. You may start to see why as this chapter proceeds.

Observing Interstellar Planets

It is tricky enough to spot exoplanets around other stars. To detect exoplanets, astronomers often rely upon the gravitational tugs they exert on the parent stars. It is far more difficult to find similar objects floating through the void, disconnected. Population counts of planets are subject to considerable observational bias, which plays out in several ways:

- First, it is clearly more difficult to find dark planets in interstellar space compared to exoplanets cuddled up to their parent star.

- Second, there is a natural assumption that exoplanets are going to be similar to those we already know about: Planets furrowing neatly-defined zones close to their parent stars. Recognized planetary types are sought in preference to possibilities unthought of. The bizarre characteristics of many exoplanets have opened the door to a more varied universe, but it is still not easy to predict the nature of these "known unknowns" in advance.

- Third, currently understood mechanisms for planet-formation favor accretion within the protoplanetary disks, over, say, continued accretion of planets moving independently through interstellar space.

- Fourth, there is a clear observational bias towards sought objects which can be readily identified, over those which cannot.

These assumptions and observational constraints add up to a significant bias in favor of the planetary status quo. I would argue that interstellar space is littered with planets of all varieties. The chaos inherent in early planetary systems causes planetessimals, planets and other debris to get flung all over the place. Smaller objects, like comets, may not develop much beyond that, and remain in their primordial state for all eternity. More massive objects, however, may draw material towards themselves, becoming the catalyst of further accretion in the dusty darkness of interstellar space. Indeed, there may be many planet formation processes going on out in the inky darkness.

There are good reasons to think that there may be vast quantities of planets waiting to be found in interstellar space, perhaps even outnumbering those found around stars.

Appearance is Everything

Very young brown dwarfs light up quite brightly, and are readily detectable. But brown dwarfs use up their available fuel in just a few million years. From that point on, their light output wanes significantly. This preserves their integrity as stellar objects: Theoretically, brown dwarfs could outlive the cooling and expanding universe.

However, brown dwarfs become much more difficult to detect once they have lost their youthful glow. The vast majority of these stellar objects become essentially dark. Their detectability often relies upon occasions when their gravity influences the movement of their parent star, betraying their dark presence. If older *Dark Stars* float freely through the void of interstellar space, then they are almost impossible to detect directly. Their population levels can only be extrapolated from what is known about their birth numbers.

As an analogy, imagine a sub-species of chameleon that has evolved such remarkable camouflaging skills that they become almost invisible when fully-grown. To determine the populations of these super-chameleons, biologists would become reliant upon counting their still-visible young. Continuing this analogy, you would have no problem detecting well-camouflaged chameleons if you looked for their heat signature at night.

This is just as true for the spectral classes of objects that drop below the deuterium-burning level. Infrared is the way to go. Sub-brown dwarfs may never emit the kind of light provided by their bigger brown dwarf cousins, but they are still relatively warm objects compared to the background heat-signature of space. Their massive planetary bulk enables them to retain a significant fraction of the heat generated during their formation.

Helpfully, infrared telescopes, like WISE, have filters set at certain wavelengths in order to discriminate the variation in light from these different classes of ultra-cool Dark Stars. These variations identify the presence of different molecules, like methane, in the planetary atmospheres, which alter their spectrograph in infrared (Kuchner 2017). This allows astronomers to make distinctions between these categories of object. But one must be careful to distinguish warm sub-brown dwarfs from hot Jupiters!

Super Jupiters

A further complicating factor is the burgeoning schema of exoplanets, which create their own planetary menagerie. Included in this exoplanet zoo is the "hot-Jupiter". Hot-Jupiters whiz around their parent stars like hyperactive toddlers around their mother's legs. Although the name 'hot-Jupiter' implies a warm gas giant, it is not the same as a sub-brown dwarf, even though the latter is essentially a warmer, denser version of Jupiter.

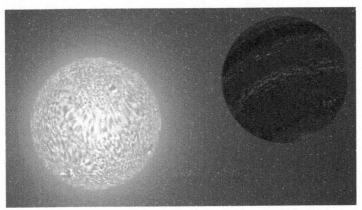

Fig. 13-1 Hot brown dwarf

Terrestrial planets of about 10 Earth masses are sometimes known as "super-Earths". In the same way, then, we might describe sub-brown dwarfs as "super-Jupiter s". This term is sometimes used for exoplanets straddling the communal planet/brown dwarf hinterland.

If a super-Jupiter has, say, thirteen Jupiter masses (13Mj), then is it still a planetary mass object – or is it now a brown dwarf 'star'? If we call an 8Mj object a "super-Jupiter", then it sounds more like it is a planet. Compare that terminology to an 8Mj "Y dwarf": Now this sounds distinctly like a star – albeit an ultra-cool one. The terms themselves contain a prejudice towards either stars, or planets.

You may then counter that making a distinction between the two rests upon where they are found. Hot-Jupiters are located very close to their stars. Regular gas giants, and the more massive Super Jupiters, are those exoplanets that orbit their stars like our own do. Sub-brown dwarfs, by comparison, might be more loosely bound objects, which take on a greater degree of independence. If only it were that simple!

Brown dwarfs versus Sub-Brown Dwarfs

It is a curious quirk of nature that sub-brown dwarfs are actually smaller in size than Jupiter, despite being heavier. No one could accuse mighty Jupiter of being a brown dwarf, but if Jupiter were to start piling on the pounds in a big way, then after a while it would start to look more and more like its darker, denser cousin – the slightly diminutive sub-brown dwarf. At what point does this object become a brown dwarf 'star'?

It is generally accepted that sub-brown dwarfs have an upper mass limit of about 13 times Jupiter's mass. Objects with masses above 13 Jupiter masses (>13Mj) are considered to be brown dwarfs, and therefore fall into the category of 'star'.

This weighing-in limit marks the point where the planet's internal pressure-cooker begins to "burn" deuterium. In this particular nuclear fusion process, a deuterium nucleus and a proton combine to form a helium-3 nucleus. This mass limit is known as the "deuterium burning minimum mass" (DBMM) and marks the lowest theoretical mass for a brown dwarf. This marks the point at which a planet lights up to become a low mass star (Martin 2002). As previously mentioned, the bright flames of young brown dwarfs quickly die out, having quickly burned up all their available nuclear fuel. They are the James Deans of the stellar community.

However, after a while, the charred, darkened remains of these once bright worlds become quite generic in their appearance. Objects of a certain mass (between 10-13Mj) may form in different ways, depending upon their initial conditions. But, after their early shiny period, there would be on way of telling whether they began life as a planet, or a star.

Brown dwarfs seem to form much like stars do. They condense under their own weight from dense dust and gas nebulae. They may do this completely on their own, or within striking distance of a companion star. They seem to spontaneously and rapidly condense from a dynamical unstable nebula of gas and dust.

Giant planets, on the other hand, accrete out of a star's rotating disk of cosmic after-birth; a process known as "core accretion". It is thought that planets like Jupiter begin life with a substantial rocky core. This terrestrial core accumulates from a multitude of collisions between grains of dust and ice. This substantial core then attracts to it a very significant shroud of gas. The result is a coalescing gas giant planet.

However, if these planets really pile the weight on, and go over the 13Mj deuterium-burning limit, they will ignite in much the same way as their much larger 'stellar' brown dwarf cousins. The problem, then, is distinguishing between these two sets of objects. Compared to the condensing stellar gas ball route, the rocky core component of a burgeoning super-Jupiter planet may affect its deuterium-burning limit, or DBBM.

Fig. 13-2 Brown dwarf, Sub Brown-Dwarf image

However, even if this was a significant effect during a given brown dwarf's early life, it becomes impossible to tell the difference after a couple of hundred million years (Mollière & Mordasini 2012). As these objects age, it is no longer possible to determine by which method they formed.

A Star is Born

So, the defining point distinguishing planets from brown dwarfs remains rather fuzzy. Essentially, mass is not sufficient on its own to tell the difference between brown dwarfs that formed by the planetary route, and the brown dwarfs that formed by the stellar route.

A new definition has been offered by Kevin Schlaufman of John Hopkins University (Schlaufman 2018). Rather than focusing solely upon the deuterium-burning limit, this instead addresses the issue of where the brown dwarf or super-Jupiter planet formed. In particular, it looks at whether the object formed around an iron-rich star.

The thinking goes that if the parent star is iron-rich, like our Sun, then there will be a greater potential for the formation of rocky worlds in its midst, and therefore a greater potential for super-Jupiter planets. This stems from the observation that gas giant worlds are usually found orbiting iron-rich stars: Condensing iron and rock within the protoplanetary nebula disk supplies the initial seed for the formation of planets, including the mighty gas giants.

This is not the case with brown dwarfs, which form in much the same way as stars – by condensing directly from collapsing clouds of gas and dust. The formation of brown dwarfs, therefore, is more indiscriminate. It does not rely upon the chemical make-up of the primordial protoplanetary disk around a young star. There is a crossover between these two mechanisms of formation that appears to occur around 10Mj (Sneiderman 2018). Schlaufman argues that the lower limit for brown dwarfs needs to shift downwards from 13Mj to 10Mj.

The implication of this is that a Nemesis-like object in our own solar system either formed from the presolar nebula, alongside all the other planets, or, formed independently as a >10Mj brown dwarf.

I am not entirely convinced by this, for reasons which will become clear later in the chapter. This analysis rests upon standard models of planet and star formation. If a Dark Star was born within the same stellar nursery as the Sun, but at a considerable distance from it, then it would have formed in the same way as a star does: In this case a tiny, dense nebula collapsing under its own weight. According to Schlaufman's analysis, for that to happen, the collapsing 'star' would have to be at least 10Mj. In other words, that particular mechanism of formation should produce a fully-fledged brown dwarf.

In this book, I have advocated an on-going mechanism of planet-building within interstellar space. Massive worlds can continue to pile on the pounds as they move through dusty regions of space. Over the course of billions of years, this can make a substantial difference to their mass. As a result, planetessimals which escaped the early chaos of the protoplanetary disk, and migrated out towards the comets, can continue to build over time. They may still have an iron core, as per the planet-forming route, but they may go on to become Dark Stars independently.

Dark Star Populations

It has been my contention for some time that the galactic population of sub-brown dwarfs has been significantly underestimated. It is recognized generally that ultra-cool dwarf stars and planetary mass objects may be encountered as free-floating objects in interstellar space.

There are two ways such objects might have made themselves a home in interstellar space. They may be orphans that have been forcibly ejected from young star systems, as young planets boisterously jostled for position. Alternatively, they may have formed independently from relatively small nebulae.

Opinions about the numbers of Dark Stars vary greatly among astrophysicists. Some think that there may be twice as many of these objects as stars. This statistic emerges from studies involving gravitational micro-lensing surveys of the galactic bulge (Sumi et al. 2011). Other studies produce results which conflict with this conclusion. They imply that there may be as few as one Dark Star object (defined here as 5-15Mj objects) per 20-50 stars in a cluster (Scholz et al 2012).

Fig. 13-3 Dark Star image

This glaring discrepancy is important: The difference is perhaps as high as two orders of magnitude. This significantly affects our expectation of how many free-floating Dark Stars are out there. In turn, this has implications for the feasibility of a Dark Star lurking unseen in our own backyard.

As previously discussed, the formation of sub-brown dwarfs is likely to be different from brown dwarfs and stars. As a result, any free-floating sub-brown dwarfs moving independently through interstellar space – and good luck with finding them – are likely to have been ejected from their original star systems. Similarly, many might end up occupying loosely bound orbits around their parent stars, having not quite managed to escape the nest.

If the number of free-floating sub-brown dwarfs is on the high end of expectation, then it follows that there are likely to be more of these objects in wide, distant orbits around their parent stars. This, in turn, increases the likelihood of there being a Dark Star object in our own immediate solar neighborhood.

Free-floating Dark Stars are extremely difficult to locate – certainly in visible light. Within our own galactic neighborhood, astronomers stand a better chance of imaging them using near-infrared sky searches, like WISE and 2MASS. The sub-brown dwarfs emit sufficient heat to help them to stand out against the frigid background of space. Even so, they are still relatively cold compared with their more massive brown dwarf cousins. So, just because infrared sky searches have not had great success at finding local sub-brown dwarfs (Clavin 2012), it does not mean that they are not out there.

Herd Instinct

Studies of open clusters and associations of young stars have received much attention in recent years. Young clusters may host bright brown dwarfs, along with their darker planetary-mass cousins, the sub-brown dwarfs. The youngest of these 'local' star clusters, is the TW Hya Association (TWA). This star cluster, located about 100 light years away, contains a few dozen 10-million-year-old stars, all moving together through space (Carnegie Institution for Science 2017). It is an important cluster to study because the young age of its stars is helpful for studying the late stages of star and planet formation (Donaldson 2016).

It is known that many of TWA's low-mass stellar objects are missing. This is despite careful observation of the cluster by astronomers for over a decade: Unfortunately, these objects were too faint to have been detectable by ESA's Hipparcos satellite. Hipparcos has been replaced by the Gaia space observatory, which is mapping the skies during its nine year mission. Gaia is creating accurate astrometry measurements for about one percent of the visible population of the Milky Way – no small task! As these new catalogs are made available, then clusters like TWA may give up some of their hidden ultra-cool dwarf stars for further study.

In the meantime, a census of TWA has made use of computer simulations to try to fill in some of the gaps in the sub-stellar populations of this open cluster of stars. This is a bit like working out how many minnows there might be lurking within a large school of fish. The cluster might contain unseen objects in the 5–7Mj range – classic sub-brown dwarfs. Obtaining an accurate census of these objects within the cluster would help astrophysicists gauge the extent of the missing populations of sub-brown dwarfs in the galaxy.

Astrophysicists have attempted to calculate the missing population of sub-brown dwarfs in TWA. To do this, they looked at the "Initial Mass Function" [IMF] of the cluster, to work out the distribution of mass within the TWA Hya Association. From there, inferences were made about the number of planetary-mass objects that might remain undiscovered within its midst (particularly towards the far side of this extensive association). The results demonstrated that there may be10-20 unseen sub-brown dwarfs in the TWA cluster (Gagné et al. 2017).

This large number of potential objects has surprised astrophysicists. It significantly surpasses the expected numbers of low-mass objects compared to the number of stars in the current TWA census.

More Dark Stars than Stars

The new estimate is in keeping with other studies of galactic interstellar populations of sub-stellar objects of this size (Gagné et al. 2015), and results from micro-lensing surveys (Sumi et al. 2011). A higher galactic population of sub-brown dwarfs can also be implied from the relatively recent discovery of a free-floating planetary-mass object in the Sun's own stellar neighborhood (Luhman 2014). The difficulty in detecting sub-brown dwarfs contrasts strongly with the growing numbers of brown dwarfs observed in several moving groups of stellar objects (Schneider et al. 2017). This is perhaps to be expected, given the detection difficulties regarding interstellar sub-brown dwarfs.

Fig. 13-4 NGC 3590 open cluster image

So, it looks like the number of free-floating sub-brown dwarfs may well be more than the number of actual stars in our galaxy after all. If they are so numerous, and our near-infrared sky surveys have become ever more powerful, then why have we not been detecting more of them? The observations of young, free-floating brown dwarfs indicate the potential for a large population of their older, darker objects. The same seems to be true for the sub-brown dwarfs, too. Nevertheless, they continue to evade detection.

This highlights the deficit between theory and observation regarding the galactic population of dark sub-brown dwarfs. Arguments against such an object existing in our own solar system can therefore be challenged: The lack of discovery locally may simply reflect general failure elsewhere.

Binary Dark Stars

Two candidate brown dwarfs have been observed within TWA, by the near-infra-red Two-Micron All Sky Survey (2MASS). Incomprehensively, they are known as 2MASS J11193254-1137466 and 2MASS J11472421-2040204 (Gagné et al. 2017).

These two objects reside at the near-side of the moving cluster, like calves at the edge of a herd. They may be paired up with parent stars, or they may be independent, planetary mass objects moving with the cluster. Working out which could provide valuable information about how objects of this size end up becoming free-floating objects in interstellar space (if that is the case here).

It turns out that the former may actually be a binary pair of smaller sub-brown dwarfs (Kohler 2017). Astronomers studying the Dark Star pair think that the distance between the two sub-brown dwarfs is about 4AU – an equivalent distance from the Sun to the outer asteroid belt. They are of roughly equal mass (~3.7Mj each), and rotate around one another (Best et al. 2017). They are remarkable both for their low masses, and their identity as a binary coupling, floating freely through interstellar space.

Other research work has revealed that stars may all begin life as binary objects, which emerge from dense cores within birth clusters (Sanders 2017). As the core continues to develop, the binaries may, or may not, split apart. In that sense, then, it may not be too much of a surprise that a relatively young duo of sub-brown dwarfs (a mere 10 million years old, it is estimated) would be found revolving around one another. But that would be assuming that they have formed within their own segregated dense core, away from a larger parent star, rather than having been ejected very early on from a larger system.

At this mass, one would expect the ejection route, according to Schlaufman's definition of brown dwarfs. Again, though, we know so little about planet formation. Perhaps accreting as a binary out of a dense core allows planet-mass sub-brown dwarfs to form independently after all. This could provide a sensible means of creating sub-brown dwarfs by the stellar method.

It is not firmly established at this stage whether the brown dwarf binary twins that make up 2MASS J11193254–1137466 belong to TWA. If they do, then the relative proximity of the rest of the grouping of stars would provide a potential basis for their ejection from one of their larger neighbors. One would then have to imagine this duo being flung off from a parent star, spinning off into interstellar space like a planetary sized bolas.

On the other hand, given how dispersed this cluster is, and how young all of its stars are, it is possible that the binary evolved independently, like stars do, within their own dense core. If so, then the population of dark, free-floating sub-brown dwarfs may be substantial across the galactic piste. Their independent star-like formation in interstellar space adds a whole category of ultra-cool star systems, which go largely unseen in the Milky Way. If these systems have their own collections of planets, then it is also conceivable that they themselves may eject asteroids and comets, and even planets, out into interstellar space.

This potentially increases the amount of interstellar debris significantly. It provides new sources of material that, by definition, is loosely bound to dark, ultra-low mass dwarf stars. There may be other reasons to think that interstellar space is littered with a great deal more material than previously thought, routinely expelled from young star systems (Laughlin & Batygin 2017). We will explore this in the next chapter.

Without the light of young stars to illuminate these mechanisms for us, planetary materials may be forming in dark nebulae and gigantic molecular clouds unseen. This logical argument builds the case for a much greater quantity of materials moving through interstellar space than previously considered.

Observing Dark Stars

Brown dwarfs come into the Universe with a blast, shedding light and heat in an infantile display of vigor. But within just a few million years, they have burned their available nuclear fuels, and settled down to consume their leaner elemental pickings. Their visible light dims considerably with time to perhaps just a magenta shimmer. Nonetheless, they still produce heat. The older they get, the greater reliance on detection using near-infrared.

Making a search for these objects in the infra-red spectrum is fraught with difficulties, particularly for ground-based infrared telescopes. To catch these faint heat signatures in the night sky, you first need to have a cold night sky: A very cold night sky. Worse, water vapor in the atmosphere absorbs infrared light along multiple stretches of the spectrum.

The presence of atmospheric water vapor severely affects the capability of ground-based infrared telescopes to pick up water signatures of brown dwarfs within dusty nebulae. How do you know whether that elusive signature is not simply due to water in the Earth's own atmosphere? The warmth and humidity of the Earth's atmosphere obscures infrared searches, even in frigid climates or atop mountains.

Detection of water vapor in space is best achieved using space-based infrared telescopes. The downside of space-based platforms is that they rely upon coolants to bring the operating temperature of their infrared detectors close to absolute zero. Volatile coolants, like liquid helium, can get lost to space quite quickly. This shortens the system's operating lifespan considerably.

Fig. 13-5 Orion nebula image

Despite these issues, astronomers have tracked down a new set of brown dwarfs in the Orion Nebula, by searching for the tell-tale signs of water vapor in their atmospheres. The Hubble Space Telescope is equipped with a near-infrared detector and this has been used to identify over a thousand candidate objects in the nebula.

Following detailed analysis, many of these objects turned out to be background stars, whose light had been turned reddish by the intervening nebula dust. This is similar to the effect that causes atmospheric light to turn red during sunsets. Differentiating background stars masked by intervening dust from bona fide young brown dwarfs embedded in the nebula is difficult. It was the presence of water vapor in the atmospheres of the brown dwarfs that allowed for a breakthrough in identification. Sub-brown dwarfs were also identified using this method (Jenkins & Villard 2018).

The astronomers conducting this work used similar detection techniques used for searching for exoplanets orbiting around stars. They adapted the techniques for use in high-quality snapshots of an entire region rich with new stars and planets (Strampelli et al. 2018). This refined technique could be used to trawl through Hubble's entire near-infrared archive, opening the possibility of many more discoveries in the future.

This is exciting in itself, but perhaps more important to our inquiry is the method used to distinguish sub-brown dwarfs from the nebula materials enshrouding them. The implication is that before this technique was made use of, it would have been particularly difficult to find sub-brown dwarfs 'hidden' within nebulae. They were effectively indistinguishable from the dust-reddened near-infrared light from background stars occasioned behind the nebula. I would argue that this problem of misidentification could also apply to a candidate Planet X object wrapped up in its own localized nebula.

Bigger, Better Infrared Searches

The first major sky search using a space telescope was IRAS, back in the 1980s. After a pause in space-based observation, astronomers were able to use the Spitzer platform at the turn of this century, followed by Herschel, and then WISE. Some infrared telescopes conduct broad searches across the sky for heat traces, while others zoom in on candidate objects for closer inspection. Usually, each new telescope exceeds its predecessors in performance, sometimes by orders of magnitude. This means that faint objects that might have been missed by early searches stand more of a chance of being picked up in the newer searches.

The next big thing in infrared astronomy will be the James Webb Space Telescope (JSWT), which is due for launch in early 2019. The JSWT should provide the kind of observational power provided by the Hubble Space telescope, but this time in infrared.

The reason why astronomers want to view the universe in detail using infrared wavelengths is that very distant objects are red-shifted to such a degree that their light tends to be found in the infrared spectrum, generally outside Hubble's operational parameters (Masetti 2018). Essentially, the JWST will be able to see deeper into space, and, therefore, look for objects sending their red-shifted light to us from further back in time when the first stars and galaxies emerged.

One of the advantages of this shift into the red is that this extremely powerful telescope will also have the capability to seek out new brown dwarfs and sub-brown dwarfs. Precious telescope time using the JWST is currently being allocated to various teams of astronomers. Two announcements about these allotted slots involve brown dwarfs.

The first is a closer look at a young low-mass brown dwarf located in the Sun's stellar neighborhood. This particular object, known as SIMP0136, is a member of a 200 million-year-old group of stars called Carina-Near (Ramsay 2018). SIMP0136 is a free-floating, low-mass brown dwarf. Its light is not obscured by the presence of a parent star. This makes it a lot easier to study, providing the opportunity to analyze the components of its cloudy atmosphere. SIMP0136 lies at the low end of the range of masses for brown dwarfs, at about 13Mj (Gagné et al. 2017a). This study could provide some critical information about the boundary conditions between planets and ultra-cool stars.

Fig. 13-6 NGC 1333 image

The second JSWT project announcement aims to study the NGC 1333 stellar nursery, located in the constellation of Perseus (Ramsay 2018). This nebula contains a great many birthing stars and young brown dwarfs. It is thought to also contain an abundance of sub-brown dwarfs. It is difficult to spot sub-brown dwarfs at the best of times, but the search for them is all the more difficult here because of the obscuring clouds of gas and dust within the nebula. A powerful infrared search should allow them to pop out from their cosmic hiding place, especially as the space-based platform will enable scientists to pick out the nuanced signatures of brown dwarf and sub-brown dwarf atmospheres – particularly the presence of water.

This issue is at the heart of the shroud hypothesis developed in this book. Obscuring micro-nebulae form around planetary mass objects in interstellar space, particularly when they are exposed to dense patches of interstellar medium (Lloyd 2016b). Without the wafting action of the solar wind, and other dynamical processes driven by a mainstream star's proximity, interstellar dust and debris may intermittently accumulate around any massive planet located in interstellar space. I suspect that planetary objects orbiting beyond the heliopause become enshrouded within a dusty cloud.

This obscures them in visible and near-infrared spectra, making them prone to mis-identification as distant, fuzzy galaxies, or reddened background stars.

Infrared sky searches like WISE might pick up an infrared signal, but efforts to then confirm the presence of a new planet using standard telescopes fall short of making a correct identification of a local planet. Optical telescopes may show a faint, fuzzy mass, rather than a distinct planetary object. So, the infrared signal (and commensurate visible fuzz) would be attributed to a different phenomenon, like a distant red-shifting galaxy.

Could the powerful JSWT change matters? It is possible that a broad search in infrared will bring forth this elusive object; however camouflaged it might be by accumulated interstellar debris swirling around it. The technique for hunting out planetary mass objects and dark brown dwarfs residing within nebulae might also become an effective method to find the kind of object I am advocating closer to home. To do this, the telescope would need to be specifically trained to hone in on candidate objects. This will depend very much upon whether such a sighting sparks serious interest.

If I am right about the mechanisms of continued planetary accretion in interstellar space, then a sighting by JWST might not be recognized as a Planet X candidate. Without a wider appreciation of the potentially variant properties of externally located planets, the status quo could remain in place: Strong indirect evidence of a massive perturbing body located beyond the heliopause, but with little practical prospects of actually spotting it.

Planets in Interstellar Space

As we have seen, star systems are extensive tracts of real estate. The gravitational influence of the central star is felt deep into surrounding interstellar space and comets as far as a light year from the Sun may be in orbit around it. This represents a full quarter of the distance to the nearest star. These kinds of distances are enormous compared to the limits of what most people would identify as the edge of the solar system. The rest of the solar system beyond the Kuiper belt is vast; peppered by comets within the inner and outer Oort clouds.

Located well beyond the Heliosphere, the comet clouds are technically located in interstellar space, even though the comets remain loosely bound in ancient orbits transiting around the Sun.

Planet Nine would fall into this category. Its existence, if proven, would extend the popular conception of the edge of the solar system from Neptune's distance at 30AU out to perhaps twenty times that distance. The term Planet X is loosely defined, at best. It simultaneously incorporates a sizable Planet Nine body lying halfway towards the inner Oort cloud, as well as a possible Mars-sized body located within the extended scattered disk beyond the Kuiper Belt. It might even describe a sub-brown dwarf, like Tyche or Nemesis, located somewhere out among the comet clouds. The discovery of just one of these planetary possibilities would irrevocably alter the common definition of what constitutes the edge of the solar system.

The zone between the Kuiper belt and the inner Oort cloud is largely empty and lies well beyond what would have been the edge of the Sun's initial protoplanetary disk. Can any planet form at that distance, independently? We have seen that above about 10MJ, brown dwarfs form like stars: They don't need a protoplanetary disk to form out of. Instead they form from the collapse of relatively small rotating nebula.

So, we have two options: Sub-brown dwarfs may form alongside stars in associated nebula pockets within stellar birth clusters. Or they may form alongside other planets, before getting ejected into interstellar space. They may escape the star completely, becoming free-floating planets, or they may get caught in wide, loosely-bound orbits around their parent stars. In these interstellar locations, substantial planets may continue to grow, and accumulate materials around them, captured from their ever-changing environments. These planets mop up materials from interstellar space, which are themselves being re-supplied by inter-galactic interactions.

Imagine this taking place all over the Milky Way. Over time, this increases the overall mass of dark sub-stellar objects in the galaxy, whose presence is essentially invisible to us, and whose lifetimes are essentially infinite in length. If this mechanism is more broadly applied, then it may explain some of the galaxy's missing mass. I think that massive planets are prevalent everywhere, in star systems and interstellar space alike. They act as catalysts for the accumulation of clumps of dark ISM, peppering space with countless pockets of lost matter.

If this is so, then we would expect to see the occasional influx of interstellar materials through the solar system. The appearance of a curious visitor to the solar system has ignited this debate. In the next chapter, we will consider the implications of the flyby of the interstellar object, **Oumuamua**.

CHAPTER

14

'Oumuamua

In October 2017, astronomers using the PanSTARRS 1 telescope on Haleakala, Maui announced that they had observed a fast-moving object moving through the solar system, at a speed of about sixteen miles per second (Beatty 2017).

The object's appearance was that of a point of light in the sky, rather than a traditional comet. So, although the object had obviously arrived for a very distant point, it was not behaving in a comet-like way. There was no coma of volatile gases swept back away from the Sun to form a magnificent cometary tail. This was certainly unusual, and led to a quick reclassification of the visitor as an asteroid.

Hurtling Through the Solar System

The speed with which the mysterious object was moving was much faster than that of any regular, orbiting object in the solar system. At the distances involved, the object was moving faster than the escape velocity of the Sun. It was clearly on course to exit the solar system completely. Rather than moving along a standard elliptical orbit, on a parabolic trajectory as it rounded the Sun, this new zippy little asteroid was traveling along what astronomers call a hyperbolic orbit (NASA 2017).

The object was given the designation A/2017 U1. The "A" stood for asteroid. The object's original designation by the Minor Planet Center had been C/2017 U1 (PANSTARRS), indicating its initial categorization as a comet (Minor Planet Center 2017). A/2017 U1 had entered our solar system from the direction of the solar apex, in the northern celestial hemisphere. It then passed within Mercury's orbit, had crossed the ecliptic and then, on 9th September 2017, achieved perihelion a mere 23 million miles from the Sun. Yet, it emitted no bright, billowing tail of blown-off gases (Meech et al. 2017). As a result, it was deemed to be an asteroid; albeit a very weird one.

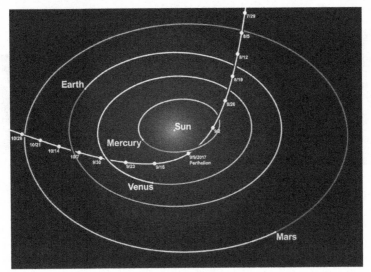

Fig. 14-1 'Oumuamua trajectory diagram

Following its perihelion passage in the southern celestial hemisphere, A/2017 U1 crossed back through the ecliptic along its rapid outward trajectory, traveling at 27 miles per second towards the constellation Pegasus. This strange visitor was determined to be the first confirmed example of an interstellar asteroid or comet observed traversing the solar system. Its designation changed for a second time, and the interstellar visitor became 1I/2017 U1 ('I' for Interstellar). This was just as well, as it turned out, because the question of whether this object was an asteroid or a comet was far from resolved.

Other possibilities about the origin of 1I/2017 U1 had been systematically ruled out. The route it took through the solar system had not brought it close to any of the planets, eliminating the possibility that its rapid speed may have been the result of a sling-shot action (Davis 2017). Far more likely, then was that this object had penetrated the Solar System from outer space. Its velocity fitted nicely with the average speed of cosmic materials within the Sun's galactic neighborhood moving through interstellar space. Its eccentricity was found to be high, too, indicating that 1I/2017 U1 would achieve escape velocity from the Sun and return back to interstellar space. The object's encounter with the Sun had changed its course: It was now rapidly receding into the distance, towards the constellation of Pegasus (The Sky Live 2018).

The Visitor's Point of Origin

This interstellar hypothesis was also supported by the direction from which 1I/2017 U1 had originally came from, a point in the sky known as its "radiant", just 6° from the star Vega. This was close to the same direction the Sun itself is heading as it meanders through interstellar space, known as the "Solar Apex" (Beatty 2017). It is statistically the most likely direction from which an interstellar visitor might emerge:

Imagine a gnat flying directly into the windscreen of an oncoming car. The direction of travel of the car plays an important part in the likelihood of a meeting between insect and vehicle and, therefore, the direction from which the gnat will arrive from the viewpoint of the driver.

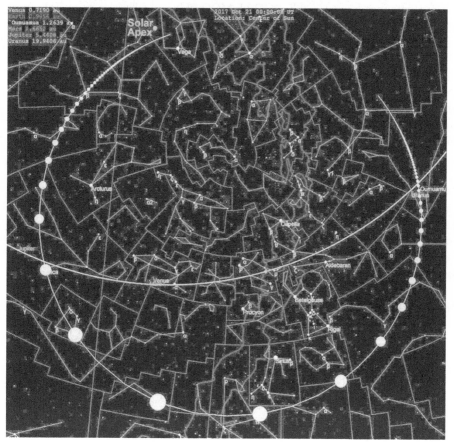

Fig. 14-2 Solar apex of 'Oumuamua diagram

Although this cosmic gnat had originated from the general vicinity of the star Vega, it was not from Vega itself. Like all stars, Vega is also in motion, and so changes its position in the sky over time. If 1I/2017 U1 had been ejected from the Vega system then, given its current rapid velocity, it must have begun its journey 300,000 years ago. At that time, Vega was some distance away from where it is now. This object is not, therefore, an artifact from Vega. More than likely, 1I/2017 U1 was simply an interstellar lump of rock of unknown origin which had, quite by chance, flown through the Sun's windscreen.

Hawai'ian Messenger

After some deliberation about this historic celestial object, 1I/2017 U1 was given the Hawai'ian name "'Oumuamua" by its Pan-STARRs discoverers (Dunford 2017). 'Oumuamua is pronounced "oh MOO-uh MOO-uh", and means "a messenger from afar arriving first". The word has a similar sense to the English word "scout".

On 14th October 2017 1I/'Oumuamua made its closest approach to the Earth, at a distance of some 15 million miles. This was during its outward journey away from the Sun – it was already well on its way towards the solar system exit. At this closest approach, astronomers were able to image the object and obtain a spectrograph of the light from it. This showed a reddish hue, which may indicate the presence of organic materials on its surface. On the face of it – literally – the interloper appeared to have a composition similar to a typical KBO.

Many KBOs are tinted red. This is thought to be due to the fierce action of solar radiation on surface, or sub-surface, ices (Fox 2010). Pluto has a reddish pink tint, too, displaying a diversity of surface features indicative of a complex, possibly violent, history (Choi 2017). Maybe, then, this mysterious visitor was a KBO which had been perturbed from its eternal dance around the Sun to plunge into the center of the solar system at a heady rate of knots?

However, this combination of factors raises a tricky issue. If that was the case, then why had the volatiles, usually associated with such cold, distant objects, not been driven off by 1I/'Oumuamua's encounter with the Sun? The disturbed outer solar system object should have created a cometary tail when it neared the Sun. This argument could be taken further: If the object was derived from the outer zones of any star system, then it should really have behaved more like a comet than an asteroid.

It cannot, on the other hand, be a returning solar system comet which has already had its volatiles driven away by a prior perihelion passage around the Sun. If that had been the case, then how can we explain its rapid velocity?

An Orphaned Object

The object seemed to be pre-stripped of volatile ices – but not by the Sun. Therefore, it must have originated from the inner part of another star system, where the heat of the parent star was sufficient to have robbed it of its volatile ices long before 1I/'Oumuamua exited stage left. In other words, its origin was within the "ice line" of another star system. Any volatile materials were burned off by the action of the young star, leaving it a barren rock; an asteroid.

Another possibility is that this interstellar interloper emerged from a shattering cosmic collision powerful enough to have blown away all its volatile materials. Statistically, this seems most likely to have occurred during the planet-forming period of its original system, when chaos reigned. It is possible that 'Oumuamua was tidally disrupted by a gas giant planet prior to its final ejection. It may once have been part of a larger planetesimal (Raymond 2018).

Another similar suggestion is that 'Oumuamua is a fragment from a disintegrated planet; one that was rendered asunder by the competing forces within a binary star system (Ćuk 2018). It has since been suggested that binary star systems naturally eject a greater proportion of asteroid-like material than single star systems, making the 'chance encounter' with 1I/'Oumuamua more probable than previously thought (Jackson et al. 2018).

What are the Chances?

These various solutions could go some way towards off-setting a tricky statistical problem about the billowing mass of material ejected from young planetary systems (Meech et al. 2017). Generally speaking, cosmic materials in interstellar space are so spread out that the chance of them encountering one other is statistically remote. Astrophysicists can count the rate of objects intruding into the solar system, and extrapolate from that the probable population of interstellar objects in our local region of space. If all interstellar objects began life in star systems, then the interstellar population of objects can provide the basis for calculations about ejection rates. If 1I/'Oumuamua turns out to be an asteroid, then vast amounts of material must be routinely ejected from young star systems, populating interstellar space with immense debris fields.

Ejected interstellar comets should vastly outnumber interstellar asteroids in number (Raymond 2018).1I/'Oumuamua seems to be an ejected rocky exo-asteroid. If so, then it originated from within the ice line of its home star system. However, it is more difficult to eject these inner, tightly bound objects than outer bodies like comets. So, the number of interstellar comets should greatly outnumber interstellar asteroids. The asteroid-nature of 'Oumuamua therefore presents astrophysicists with a significant problem. It should be a comet, but, like a Manx cat, it has no tail. If it really is an ejected asteroid, then this chance encounter with 'Oumuamua implies that the comparative population of interstellar comets is huge.

There are other possible origins to consider. Instead of having been ejected from the birth of a planetary system, it may have been thrown out of a dying one. 1I/'Oumuamua may have been ejected from a dying star system, where a red giant star engulfed planets as it expanded, then cataclysmically collapsed into a white dwarf. The incredible forces involved in these stellar death throes may have driven off the volatiles from this cosmic shard as it was flung into interstellar space (Hansen & Zuckerman 2017).

It should also be pointed out that some comets in the solar system appear like asteroids. Some 'dry' objects have been discovered moving along comet-like orbits. These are known as Manx-type comets; lacking tails. But these are uncommon so, again, this explanation would put 'Oumuamua into a very rare category of objects, defying statistical credibility given its solitary discovery.

The Cosmic Tumbler

The inert nature of 'Oumuamua was unusual, to say the least. But that was not the only bizarre quality to emerge from the observations of this mysterious object as it swept through the solar system. In addition to its reddish complexion (the overall color of 'Oumuamua is neutral with a reddish hue) and lack of out-gassing, the object notably exhibited a highly unusual variation in luminosity. The regularity of this variation led astronomers to deduce that this curious object had a rather odd shape.

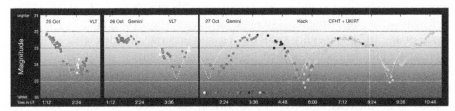

Fig. 14-3 'Oumuamua brightness variation diagram

Calculations have indicated the interstellar asteroid is shaped like a fire extinguisher: Its length is about six times longer than its width. The periodic variation in the light received from this elongated object implies that it is tumbling head over heels every 7-8 hours. The variation in its light indicated that 'Oumuamua was patchy in appearance. The predominant coloration along the length of the object was red, but other aspects of its surface resembled the color of dirty snow (Fraser et al. 2018). Overall, it appeared "slightly red" (Jewitt et al. 2017).

Incinerated Organics

How it is that our alien friend 'Oumuamua had both a reddish tint (indicating surface organics) yet was not seen to drive off its remnant water ices during perihelion?

Solar system minor bodies can vary in color, and this is sometimes dependent upon which population group any particular object belongs to. For instance, there is a precedent for reddish asteroids in our own system – the so-called "P-type asteroids". These dark, reddish asteroids can be found in the outer asteroid belt. They appear to be coated by kerogen related organic compounds. The surface properties of these objects imply that some water-based surface chemistry has taken place at some point in the past. It is likely, then, that these asteroids continue to have some water ice hidden within their interiors.

If we compare 'Oumuamua's coloration to those of various classes of solar system objects, then the interstellar asteroid seems to most closely resemble comets or organic-rich asteroids (Meech et al. 2017), and members of the dynamically excited populations of KBOs. However, 'Oumuamua is less red than the ETNOs whose orbits extend beyond the heliopause (Bannister et al. 2017). Its coloration bears little resemblance to "ultra red" KBOs (Jewitt et al. 2017). One might have expected the opposite if 'Oumuamua is an object that has endured a long journey through interstellar space.

The red appearance of these more distant travelers has been attributed to their exposure to cosmic radiation beyond the protective 'shield' of the heliosphere (Jewitt 2002). As objects move through interstellar space, they are bombarded by galactic radiation. This results in the polymerization of their surface organic materials, creating ultra red compounds. The problem, then, is reconciling that idea with an asteroid from interstellar space. If such an object were to have surface organic materials present, as many outer solar system objects do, then its interstellar flight from its own original solar system should have turned these materials deep red. So, why is 'Oumuamua only "slightly red"? It seems more like an object from our own backyard than a true interstellar traveler.

One possibility is that 'Oumuamua is a shard of material ejected from a star's inner system, where volatile organic materials would already have been eradicated from the object's surface by the action of the star. However there are other possibilities, too, including a complete re-think of the previously assumed correlation between the ultra-redness of some objects with distance from the Sun. Perhaps there was some resurfacing of the asteroid, affecting its color (Bannister et al. 2017). Alternatively, its mottled exterior may simply reflect the interstellar battering it has sustained over time.

There is too little data to go on here to draw any firm conclusions. 'Oumuamua may be an unusually light colored interstellar object, compared to its normally red brethren. This may have made it relatively more visible, providing an element of observational bias aiding its discovery.

The Mysterious Shard

Many have wondered whether 'Oumuamua might be artificial, rather than some oddball asteroid randomly popping in from goodness-knows-where. After all, it more closely resembles a 230 meter-long cigar-shaped mothership than a classic, potato-shaped asteroid. SETI, the multinational team conducting the long-running Search for Extraterrestrial Intelligence, wondered the same.

It wasn't just the unusual shape of the object that drew the attention of SETI researchers. They were perplexed by the lack of cometary gases coming off 'Oumuamua, and wondered whether this might be explained by it being an artificial object.

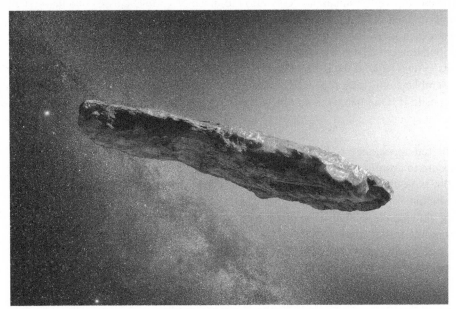

Fig. 14-4 'Oumuamua image

Additionally, this first known interstellar object somewhat resembles the eponymous vessel in Arthur C. Clarke's novel "Rendezvous with Rama" (Clarke 1973) – the fictitious visiting spacecraft initially considered to be an asteroid by astronomers. The scale is very different, though – 'Oumuamua's length is hundreds of meters, compared with Rama's 34 miles.

Objects moving through space do not need to be aerodynamic. A cursory glance at orbiting space station designs, with their massive arrays of solar panels, will confirm this. A thin needle-like shape might lower the number of high-velocity collisions with interstellar materials. However, the interstellar shard 'Oumuamua is tumbling chaotically. This seems to exclude the idea of this object being some kind of well-controlled vehicle conducting a long distance foray through interstellar space. That revolving motion may create centripetal forces within the craft that simulate gravity. The problem with this argument, at least in this example, is that the rate of rotation is extremely slow – with one rotation taking eight hours or so. Not much of a simulated gravity to be felt there…

'Oumuamua is dense, likely consisting of rock and metal. Its interior structural integrity has had to be robust to withstand this chaotic tumbling. Again, this may be suggestive of an artificial structure. In an attempt to find out, SETI's "Breakthrough Listen" team trained a powerful radio telescope on 'Oumuamua to pick up artificial signals (Griffin 2017; Breakthrough Initiatives 2017). However, the elongated asteroid failed to display any obvious indication that it was a product of artificial intelligence (PhysOrg 2017).

One can understand the excitement that swirls around 'Oumuamua, given the list of oddities associated with it: Its hyperbolic trajectory; its radiant near to the solar apex; its cartwheeling motion; its elongated shape; its lack of out-gassing volatiles; its patchy colored appearance. This has led to much scientific speculation about the origins of this weird and wonderful shard of rock and metal. The tumbling motion of 'Oumuamua through space puts a great deal of internal strain on this object. Its internal structure must be very robust for it to have remained intact for what may have been billions of years since its ejection from another star system (Bannister et al. 2017). That implies that it may have originated from the interior of an alien planet, or planetesimal, which was catastrophically torn apart.

'Oumuamua's Many Cousins

As 'Oumuamua is a solitary object in a class of its own, it is impossible to tell whether its properties are typical of interstellar asteroids or not. However, if its solitary arrival on our cosmic shores presents a statistical norm over the time we have been actively seeking out such objects, then calculations suggest there is a much greater population of interstellar asteroids across the galaxy than previously considered possible.

Gregory Laughlin of Yale University and Konstantin Batygin of Caltech (of Planet Nine fame) conducted some numerical analyses based upon the assumption that 'Oumuamua's appearance was not a statistical blip. Their combined number-crunching showed that, on average, each star in the galaxy expels the equivalent of two Earth masses of material into interstellar space (Laughlin & Batygin 2017).

That is an awful lot of ejected material being flung about out there. The currently accepted model of solar system formation – the Nice model – predicts that a degree of expulsion will take place as primordial planets jostle for supremacy. During the early lifetime of a star system, there is a period of dynamical instability resulting from interactions between the star's nascent giant planets and the waning primordial disk from which they emerge. This period of instability result in the loss of chunks of the star system to interstellar space. In the case of the solar system, such interactions resulted in the mass-depletion of the current Kuiper belt, which contains just a tiny fraction of what it may once have had.

That may be fine with small, loosely-bound objects like the KBOs. However, giving larger, inner planetary bodies, like planetessimals, the old heave-ho requires a very special set of circumstances. There is a clear recognition that it is not that easy to eject material from the gravitational field of a star. The 'throw line' of each star system varies with the mass of the star. According to the Nice model, stars would need to have substantial planets placed in the right orbital locations to facilitate these ejections. It is thought that each star system would need the presence of gas giant world/s and, in particular, its own equivalent of a distant Neptune to make this happen (Laughlin & Batygin 2017).

Over the last decade or two, it has become increasingly apparent that our solar system has a different layout of planets compared with known exoplanetary systems. In other systems, small planets seem to flock together with other small planets, like ducks in a row. In other systems, it is the large planets that team up (Howell 2018). The solar system is unusual in that it mixes them up. The chaos-driven mechanisms of the Nice model rest upon the presence of the giant planets Jupiter and Neptune. Yet, the format of the solar system may be unusual compared to others. So, if interstellar objects are very common, but the layout of the solar system is very unusual, then that would require a different method of ejection than the one offered by the Nice model.

The Cosmic Apple Cart

'Oumuamua's apparent lack of volatiles only serves to confuse matters. As noted previously, the question of whether this interstellar object was originally an asteroid or a comet could potentially have a massive impact on calculations aimed at working out the amount of material ejected into interstellar space by young star systems.

In their calculations, Laughlin and Batygin assumed that 'Oumuamua is probably a comet, despite the distinct lack of cometary activity it exhibited during perihelion. This is in keeping with the stance taken by Sean Raymond et al. (2018). Both sets of researchers argued that to conclude that 'Oumuamua is an asteroid would quite literally open up the interstellar floodgates.

When comparing interstellar comets with interstellar asteroids, then the comets, with their perihelion comas, should be much easier to detect. Therefore, there should be a strong selection bias towards comets. Secondly, it should be the case that comets are more easily expelled from star systems, because they are more loosely bound than asteroids. So, if the first and only interstellar object to be observed is, unexpectedly, an asteroid, then there should be a very serious number of corresponding interstellar comets out there.

To justify the appearance of an interstellar asteroid like 'Oumuamua, each gas giant planet would have to be ejecting an absolutely phenomenal amount of material – perhaps more than one hundred Earth masses!

Laughlin and Batygin proposed that interstellar asteroids are probably ejected by long-period 'sub-Jovian planets', like ice giants, or super-Earth-sized worlds (Laughlin & Batygin 2017). Perhaps they are hinting at the role an object like Planet Nine might play in all of this.

A large population of interstellar asteroids would imply two things: There is (1) a much greater population of debris materials in interstellar space than currently believed possible; and (2) there is a correspondingly large population of objects flung out into long-period orbits around stars. If 1I/'Oumuamua is an asteroid, then far-flung Planet X-type worlds should be commonplace. Even though it represents just a single data point, this object could upset the entire cosmic apple cart.

Hint of Comet Activity

Further observations on 1I/'Oumuamua's trajectory, conducted by the European Southern Observatory in 2018, indicate that its motion is being affected by a factor other than the usual gravitational interactions. The object was found to be moving faster away from the Sun than it should be (ESO 2018). Out-gassing seems to be providing some additional locomotion, affecting the object's trajectory. This has led to the conclusion that 1I/'Oumuamua is an interstellar comet, after all. The caveat is that this "comet" must have some "unusual properties" (Micheli et al. 2018).

Fig. 14-5 Comet 'Oumuamua image

Explanations for the lack of observable coma involve the erosion of the comet during its travel through interstellar space. This comet/asteroid has been colliding with materials in the interstellar medium (ISM) for as long as it has been tumbling through interstellar space. The underlying assumption is that these high velocity impacts will chip away at the object's surface. This space weathering will remove the outer layers of dust and ice, leaving behind the dense rock and metal core.

However, the blotchy red surface of 1I/'Oumuamua shows variation in tone and color (Meech et al. 2017), which is probably attributable to thick coverings of organic materials irradiated in interstellar space. How does that relate to the concept of a flayed cosmic bone, whose flesh has been stripped away by the debriding action of cosmic dust storms? Astronomers cannot have this both ways.

Interestingly, the amount of organic 'space grease' dispersed within the ISM seems to be significantly higher than previously thought (Günay et al. 2018). I suggest that a heavy coating of space gloop on 1I/'Oumuamua would readily explain the lack of cometary tail. Through long-term exposure to such interstellar materials, 'Oumuamua became covered by a thick coating of grease. This would have been heavily irradiated, and then toasted nicely during perihelion. Such a covering could seal in the comet's volatiles, largely preventing the expected spectacular comet display at perihelion. This would be rather like tarring a wooden boat to keep it watertight.

As for its origin, the team that described 'Oumuamua's properties, led by Dr Karen Meech, are not absolutely sure that it isn't an object from the outer solar system, perturbed into a hyperbolic orbit by an encounter with an as yet undiscovered planet! Noting that the patch of sky that 'Oumuamua emerged from (near Vega) is different from the favored location of the elusive Planet Nine (near Orion), they recommend that any search for as-yet undiscovered planets in the outer solar system should be directed towards the general direction of the Solar Apex (Meech et al. 2017). I made a similar recommendation some years ago (Lloyd, 2005, p207).

New Simulations Point to Oort Cloud Disturbance in Gemini

Judged to be the first confirmed observation of an interstellar object, 1I/'Oumuamua had an orbital path at perihelion in the shape of a hyperbola, rather than a classic parabola. This hyperbolic orbit is too fast for the Sun to hold onto the object and, unless it is acted upon by another body to slow it down, it will escape from the Sun. Other comets have been found to exhibit hyperbolic orbits, meaning that they are also moving fast enough to potentially escape from the solar system.

It is thought that most observed comets originate from the hypothetical shell of comets which surrounds the solar system, known as the Oort cloud. This spherical cloud of highly dispersed objects might extend out to about a light year distance. The further away an object in the cloud resides, the more loosely bound it is to the Sun. Other forces may become significant at these distances, distorting the spherical arrangement. These forces include the gravitational influence of passing objects, like a star or a sub-stellar traveler (for example, a free-floating planet moving through interstellar space, close to the solar system).

Another significant influence at these distances is the galactic tide. This force, pulling from the direction of the galactic core of the Milky Way, may act to distort the Oort cloud, pulling it into a distended shape in the same way that the Moon affects

Earth's oceans (Matese et al. 1995). These external forces may influence the outer comets, occasionally nudging a given object into a Sun-crossing trajectory. If the nudge, or perturbation, is sufficiently strong, then the comet may be accelerated into a hyperbolic orbit. It may gain sufficient energy to enable it to escape the Sun's gravitational field completely.

The problem for astronomers is that objects may also have an interstellar origin, rather than always coming from the Oort cloud. It is difficult to establish whether a speeding comet's hyperbolic orbit indicates an interstellar origin, or whether it infers that the comet is home-grown, but has been given an extra gravitational kick. Although it may be difficult to establish this on an individual basis ('Oumuamua provided an exact interstellar fit), it may be possible to establish trends from the statistical data as to which source is the more common.

The Scholz's Flyby

An international team of Spanish astrophysicists, who have more than a passing interest in the topic of Planet X, have performed powerful computer simulations to build up a picture of the trajectories and spatial origins of various hyperbolic comets (de la Fuente Marcos, de la Fuente Marcos & Aarseth 2018). Following adjustment for the Sun's own movement through space towards the Solar Apex, interstellar visitors would likely have a more or less random distribution to their radiants (the position in the sky from which they came e.g. meteor showers striking the Earth's atmosphere from a particular direction).

The team carried out statistical analysis on the sky maps of these hyperbolic radiants. They looked for patterns or clusters of these origin points. Statistically significant patterns emerged from the data.

A particularly large source was located in the zodiacal constellation Gemini. There are a number of possible reasons for this apparent cluster. One possibility is a close flyby of a star in the past which could have disrupted the outer edges of the distant Oort cloud, sending comets in-bound towards the Sun. Looking at the tracking of candidate flybys in the (by cosmic standards) relatively recent past, the team concluded that there is a possible correlation between this cluster of hyperbolic orbit radiants in Gemini, and a close flyby of a neighboring binary system known as Scholz's star some 70,000 years ago (de la Fuente Marcos, de la Fuente Marcos & Aarseth 2018).

Scholz's star actually consists of a red dwarf star and a brown dwarf binary. Currently located about 20 light years away, this binary system may be a close neighbor to the Sun relatively speaking, but was discovered only recently. The delay was probably because of a combination of factors: Its proximity to the Galactic plane, its relative dimness, and its slow relative movement across the sky (Mamajek et al. 2015): Its retreating motion is mostly along our line of sight, making it difficult to differentiate from background stars. Seventy thousand years ago, its distance was less than a light year away from us. At that time, Scholz's star may have disrupted the outer Oort cloud, creating the spike in the distribution of hyperbolic comet radiants in Gemini.

Non-Random Patterns

Other patterns have emerged from the hyperbolic data which may have originated from the same flyby event, or as a result of different factors. One of these is the presence of one or more embedded planets somewhere within the Oort cloud, themselves capable of nudging comets out of their ancient pathways (de la Fuente Marcos, de la Fuente Marcos & Aarseth 2018). Comets jilted into a Sun-crossing orbit by embedded perturbers can take a very significant time to get here. Many of them shouldn't reach us for hundreds of thousands, possibly even millions of years (Ananthaswamy 2015).

Seventy thousand years seems like a long time, but it may not be enough to witness the arrival of perturbed long-period comets from the very outer realms of the solar system. However, if the perturbing object gave outer comets enough of a push, then their arrival in the planetary zone of the solar system may be sufficiently rapid. We are considering hyperbolic trajectories here, after all.

Just for comparison's sake, a back-of-an-envelope calculation shows that 1I/'Oumuamua could have covered about six light years in 70,000 years. Six light years is further than the nearest star. But that would assume that 'Oumuamua was traveling straight at us like a cosmic bullet for all that time.

According to Kepler's Third Law, objects moving along an elliptical path speed up as they near their perihelion transit, and then slow down when moving back toward their most distant, aphelion position (Strobel 2010). A perturbed comet from the outer Oort cloud would take longer to get here than my back-of-an-envelope calculation suggests. A stellar flyby would not cause a perturbed long-period comet to come shooting straight at us like a bullet from a gun. Instead, it would take a more leisurely route around the celestial houses before speeding up as it approached the Sun. But that is not to say that 'Oumuamua didn't originate from the outer Oort cloud (Meech et al. 2017): Its 'interstellar' properties might simply be coincidental.

If it was not Scholz's star that caused the clustering of hyperbolic comet radiants in Gemini, then prior stellar flybys might also have set these cosmic wheels in motion. There is something of a back catalogue of stars which likely invaded the Sun's personal space at some point over the last few million years (Bailer-Jones 2015).

Alternatively, perhaps we are returning to similar arguments put forward regarding sub-brown dwarfs in the outer Oort cloud (Murray 1999), like 'Tyche' (Matese, Whitman & Whitmire 1999), and 'Nemesis' (Davis, Hut & Muller 1984). Each of the arguments put forward for these objects relied upon non-random patterns in the ingress of long-period comets from the outer Oort cloud; whether in space, time, or both. Dr Carlos de la Fuente Marcos put to it to me this way:

"As explained in our work, a significant fraction of the objects with inbound velocities above 1 km/s have radiants observed projected towards the area we mention in Gemini. We argue that this may be consistent with the most recent known stellar encounter, but we do concede that other explanations are possible, including the presence of yet-to-be found perturbers (bound to the Sun)." (de la Fuente Marcos 2018)

This new work on comets with hyperbolic orbits has the potential to stoke up the sub-brown dwarf Planet X argument still further, to the dismay of astronomers who might have hoped this whole subject had been put to bed long ago. There is also the potential for more red and brown dwarf flybys emerging from the cosmic woodwork over time, too.

It has been argued that there has been an over-reliance upon the astrometric catalogue, assembled from data gained by the European Space Agency's Hipparcos missions, when conducting systematic searches for such objects in our galactic neighborhood (Mamajek, et al. 2015). This particular database seems to contain a relatively small number of red dwarf stars compared to their wider galactic distribution. Increased use of Gaia should pull up more of these neighboring, dim red dwarfs which Hipparcos may have missed. Perhaps Gaia will bring out more besides, like previous free-floating brown dwarf flybys, and possibly ultra-cool sub-brown dwarfs embedded unseen within our immediate neighborhood. This latter hypothesis remains a distinct possibility.

An Emissary from Planet X

Let us take this argument a step further. Could 'Oumuamua have been flung towards the Sun by a close encounter with Planet X? One of my research colleagues, Al Cornette, thought so very early on into the breaking of the news about A/2017 U1 (Cornette 2017). As we have noted in this chapter, the international group of astronomers who first described this object's properties are holding open the possibility that it originated in the Oort cloud, and may have been given a nudge by an as-yet undiscovered planet lying out there among the comets (Meech et al. 2017). Its radiant, near to the solar apex of the solar system, might give us a very strong clue as to the location of a distant Planet X object, or even a binary Dark Star hidden among the outer comet clouds.

I described this possibility in my previous book, **Dark Star** where I wrote the following:

"We know that the sun is moving in a slightly odd direction compared with its neighbors. It is heading towards the Solar Apex, near the star Vega in the sky, and this may turn out to be coincident with the position of the **Dark Star**." (Lloyd 2005, p23)

A hypothetical axis between the Sun and the Dark Star might align itself over time with the direction of travel of the solar system. In which case, the Solar Apex would be coincident with either the perihelion or aphelion positions along the Dark Star's orbital path around it. Most likely, it would be the aphelion position. If this sub-brown dwarf object was close to or even within the torus-shaped inner Oort cloud during this long aphelion transit, then it might be nudging objects towards us from the cloud. 'Oumuamua may be just such an object. Hence, this is why such an object would be flung down towards us from the direction of the Solar Apex.

As Lord Alfred Tennyson wrote so beautifully:

"The Sun flies forward to his brother Sun;

The dark Earth follows wheel'd in her ellipse." (Tennyson 1846)

If 'Oumuamua was originally an object from the outer solar system which plunged rapidly towards the Sun as a result of an encounter with a distant Planet X body, then the mottle red shard could turn out to be one of many emissaries pointing the way back.

Interstellar Origins

This object's properties hold many ambiguities, but on balance it bears more resemblance to known objects from our solar system than not. These similarities may place it among our own, in which case it is a dislodged, relatively inactive comet traveling at great speed. More likely, though, it is an interstellar asteroid from an alien star.

However, 'Oumuamua did not behave like a comet as it swept around the Sun at breakneck speed. Its appearance was that of a relatively 'dry' asteroid, rather than a water ice-laden comet that spews its volatiles into space as it nears the Sun. Its coloration is similar to certain Trojans and other, inner solar system objects. Assuming that 'Oumuamua is an interstellar asteroid, it has been estimated that there may be as many as 10,000 similar objects moving through the planetary zone of the solar system at any one time, each taking about ten years to make the ~60AU transit (Jewitt et al. 2017). Its identification as an asteroid from another star could have profound implications for the quantities of material ejected by young star systems, and thus populating interstellar space.

However, as noted previously, this exiting object has been deviating from its gravity-driven flight path. If the additional propulsion is due to unseen out-gassing, then 'Oumuamua may be a comet after all. If so, then its trajectory prior to its discovery was also likely affected by out-gassing. Consequently, we can never be entirely sure where it originated from (Micheli et al. 2018). If it is 'just' an interstellar comet after all, then astrophysicists can sleep easier in their beds at night: The projected populations of interstellar objects drop back to 'normal' levels again.

Finding more of these objects will provide us with a clearer picture of these interstellar visitors. As observational techniques and instrumentation improve, more of these interstellar interlopers should be spotted and studied (Najita 2017). Only then can the vexed questions of their population levels and origins hope to be answered.

Jovian Mysteries

In this chapter, we will consider gas giants and sub-brown dwarfs, concentrating upon our best known gas giant neighbor, Jupiter. In particular, we will focus upon the composition of giant worlds, and how this alters their properties. This will build upon our previous discussions about the formation of the solar system, and explore further the concept of planet formation in interstellar space.

My working assumption is that there are one or more massive planets in the outer solar system, with likely a great deal of minors ones, too. I remain optimistic that included in their number is a planet more massive than Jupiter, although skeptics will happily tell you that it should definitely have been discovered by now by sky searches like WISE. However, I think that there is an underlying scientific assumption about the conditions prevalent in interstellar space that needs challenging.

The clear conditions within space inside the heliosphere are attributable to the action of the solar wind, and dynamic forces which operate on an ongoing basis. We take this for granted – it is so obvious that it doesn't come into our thought processes. But that clearing mechanism does not occur outside the heliosphere. Instead, materials outside of that boundary are subject to different, lesser forces, such as the galactic tide.

Giant Planet Formation in Interstellar Space

Scientists can measure the 'fog' of dust and gas across interstellar space, and it is clear that this is variable. For the most part, there is little interstellar medium obscuring our view of the stars. Occasionally, our view is blocked by gigantic molecular clouds and stunning nebulae, which can vary substantially in their size and opacity. Some nebulae are dense enough for stars and brown dwarfs to coalesce out of. In their midst, planets form.

This is all well and good. I am not trying to argue that interstellar space is completely filled with planet-forming fog. That would clearly be nonsense, and could be disputed by any night-time view of the glory of the stars. Instead, I am arguing that once you have substantial objects in interstellar space, that these will act as catalytic seeds upon which to continue to build ever bigger planets. In this way, materials coalesce around these dark interstellar objects, turning fog into patchy rain.

The way planets form remains largely theoretical, although there is supportive observational evidence of extrasolar planets in the midst of protoplanetary disks, creating swirling, tell-tale bands in the dust (Max Planck Institute for Astronomy 2018). In the same way that planets form out of protoplanetary disks, planetessimals in interstellar space are capable of mopping up materials. The more massive they get, the more efficient they should be at grabbing material from nearby space as their gravitational pull intensifies with size. This represents a virtuous circle of accretion: The predatory planet becomes increasingly efficient at drawing in its prey. Massive interstellar planets act like whales in a sea of plankton.

Of course, the available materials are less readily obtainable than in the dense hothouse of a protoplanetary disk. But planets form under those early conditions at breakneck speed. In the case of planetary accretion in interstellar space, the process is ponderously slow. It is also intermittent in nature, dependent upon the prevalent spatial densities at any given time. Like driving a car through patchy mist and fog on a cold winter's night, variable clouds of gas and dust will be encountered by free-floating planetary objects as they move through interstellar space.

We know that when the Sun passes through enormous clouds of gas and dust in space, its heliopause can be pushed back by the external gaseous pressures (Zank & Frisch 1999; Yeghikyan & Fahr 2004). This exposes the planets to the conditions of a dense interstellar medium, causing a period of accretion to occur. However, the balance of forces at work is quickly re-established in the Sun's favor as it re-imposes its heliospherical shield. In so doing, it once again clears the space within. As a result, exposure to these external materials is short-lived.

However, outside the Sun's zone of control (or, indeed any star's magnetic field), the rules are very different. Free-floating planets moving through interstellar space are completely exposed to these materials. The more massive these worlds are, the more likely they are to mop up materials. As material comes into their area of influence, there is no solar wind to drive it away.

A Complex Shroud

I contend that such planets accrue clouds of matter over time. These spin into dust disks, fall down to the planet's surface, or remain in orbit forming a continuous local nebula around the planet. This creates a dark, foggy shroud around the free-floating planet. There are a number of consequences of this line of reasoning:

The first is that planets in interstellar space may look different to those sheltered within heliospheres. Without the action of the solar wind of an accompanying star, they will come with more baggage. Planets in interstellar space may appear more like super-sized comets.

The second is that I think the content of interstellar space may have a granular nature. By this I mean that dust and gas may accrue around planetesimal seeds, creating small zones of relatively dense material: In effect, highly localized, small nebulae. Because these are so small, and so dark, they will not be readily apparent when measuring the "column" of gas and dust between stars. They will get averaged out, effectively, by the vast swathes of nothingness between them. Astronomers assume uniformity. **I am arguing for irregularity**. It is a bit like the weather: The standard concept of interstellar space is equivalent to clear skies with the odd cloud here and there. By comparison, I think that the 'space weather' out there is far more complex, creating localized micro-climates, particularly around free-floating planets.

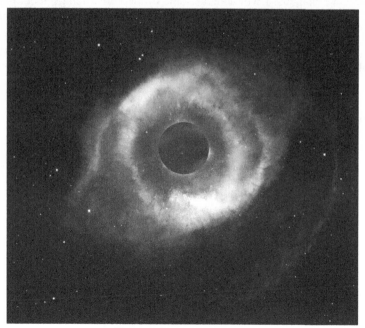

Fig. 15-1 Shroud planet image

The third is that loosely bound planets located outside the heliosphere will be all the more difficult to detect. Their visible signatures will be made fuzzy by the dusty shroud enveloping them. Their heat emission will be absorbed by the materials swirl-ing around them, complicating their emission signatures. They will not be invisible, of course, but any faint light reflected back from them will not behave in the way that astronomers expect. As a result, they run the risk of being misidentified as another phenomenon, or perhaps simply being overlooked.

A fourth is that unless these massive objects break apart or catastrophically collide with other cosmic bodies, they will naturally continue to build mass through their lifetimes. Those lifetimes may be many billions of years long – perhaps far older than the Sun. Because of the appearance of the interstellar asteroid 1I/'Oumuamua, some astrophysicists are starting to wonder whether there is far more interstellar debris swirling around out there than previously thought possible (Laughlin & Batygin 2017). That not only presents a ready source of accretion fuel for hungry interstellar planetary-mass objects, but also extends the argument as to how these voluminous materials are not showing up in the measured densities of columns of interstellar medium. (Answer: they slowly coalesce into massive interstellar planets).

Another issue is how well we really know the objects which populate our own solar system. Can we be sure they originated here? Might some of the bodies in our own system be immigrants from other star systems?

In the following chapters, I will address some of the new findings about planets in our solar system that have come to light in recent years, bearing in mind some of these points. The anomalous nature of some of these findings often provides good reason to think that there is another piece to the cosmic jigsaw needed to complete the overall puzzle. More and more astrophysicists are considering the possibility (even likelihood) that the puzzle piece is Planet X.

Return of the Pioneer Anomaly

One of the many pieces of evidence put forward for the existence of Planet X over the last few decades has been the so-called "Pioneer anomaly". The Pioneer mission sent two spacecraft across the solar system, visiting a number of planets as they traveled. They not only photographed these planets, but used the gravity of the planets to accelerate onwards, deeper into the solar system. This "gravity assist" is often used to allow spacecraft to pick up speed.

As the Pioneer probes traveled across the outer planetary zone and on towards the heliopause, it became apparent that the craft were not moving away from the solar system quite as quickly as the theoretical trajectory projections demanded. Something was essentially slowing them down. Similar effects were duly noted for the Galileo and Ulysses probes. However, there was nothing out of the ordinary for the two Voyager spacecraft or the New Horizons probe following its visit to Pluto.

Many ideas were put forward about the Pioneer anomaly. They included gravitational or physical interactions with clouds of interplanetary dust in the Kuiper belt, and maybe the tug of an undiscovered Planet X body. One of the lead researchers into the

Pioneer anomaly at the Jet Propulsion Laboratory was John Anderson (e.g. Anderson et al. 2001), who, interestingly, also had a longstanding interest in the possible existence of a Planet X body (Wilford 1987). At one point, puzzled physicists began to wonder whether this marginal but well-defined anomaly might require new laws of physics (Matthews 2002). In the end, it was agreed by technical experts that the anomalous deceleration was the result of radiation pressure caused by non-uniform heat loss from the probes (Murphy 1999; Rievers & Lämmerzahl 2011). Previous speculation about missing planets and new physics were dismissed.

Fig. 15-2 Pioneer spacecraft

However, the anomaly seems to persist, turning up in the increasingly accurate navigation and telemetry data returning from various spacecraft performing flybys past the Earth (Acedo 2017). The Juno spacecraft, now orbiting Jupiter, is reported to be slightly misplaced from its expected position (Acedo, Piqueras, & Morano 2018). This anomaly emerged from the Doppler shift calculated from returning ranging data from the probe as it circumnavigated the poles of the great gas giant.

Quixotically, Juno did not exhibit the same anomalous behavior during a previous flyby of Earth. This suggests that this is not, then, the result of an internal machination of the probe itself, as described for the Pioneer probes. Instead, there appears to be an unexplained external effect worthy of further examination. The authors of the Juno study have considered some fairly speculative possibilities, such as the presence of dark matter, tidal effects, and possibly even new laws of physics. However, they remain frustrated in their quest to match these anomalous readings to a single workable explanation (RAS-WEB 2017).

I am intrigued by two possibilities. The first is whether the gravitational tug experienced by these spacecraft is dependent upon a non-uniform cross-section of the planet. In other words, Jupiter might evoke unexpected variation in its gravitational attraction, local to a spacecraft flying over it. This would depend upon whether the interior mass of our planet is slightly offset – or, at least, has areas of high interior density which are not uniformly distributed. While I could see this being a possibility for a rocky planet like Earth, with its diversified, layered internal composition and overall non-spherical 'geoid' shaping, it is more difficult to see how that might be the case for a gas giant, like Jupiter, whose interior would surely be more uniformly distributed.

The team studying the trajectories of Juno's polar orbits of Jupiter considered this too, and accounted for Jupiter's multipolar gravitational fields, which result from the planet's oblate shape. A look through research literature on the subject of how planetary gravitational fields vary with planetary structure and shape quickly descends into a world of mathematical pain. Surely, somewhere in the discussion of zonal gravity coefficients and spherical harmonics lies the solution to these flyby issues?

I wrote to Dr. Luis Acedo, the lead author of the new flyby papers, to inquire further, and received this helpful response to my first question: Could these anomalous accelerations be due to a non-uniform planetary interior creating slight gravitational differences affecting the craft during a flyby?

"Of course, they could arise from the longitudinal non-uniformities in the gravitational field of the planets. These are characterized by the so-called tesseral harmonics and I showed that a rotating planet with longitudinal non-uniformities can transfer energy to the spacecraft in a previous paper [Acedo 2016]. However, they are not enough (in the case of the Earth, at least) to explain away the anomalies. For Jupiter, it is still an open question and we have to await for a full analysis of the Juno's orbital data at JPL because the conclusions in the paper are based on preliminary fits and our knowledge of Jupiter's gravitational field is much more incomplete than that of the Earth." (Acedo 2017a)

Since then, the team conducting this research have posted updates to their paper which indicates that, despite further investigation, anomalies remain (Acedo, Piqueras, & Morano 2018).

My second thought is that probes moving through trajectories off the plane of the known planets might be more susceptible to small gravitational tugs due to unknown planets which are themselves misplaced from the ecliptic. So, for instance, the orbital path of a Planet Nine body is thought to be inclined from the ecliptic by about 30° (Batygin & Brown 2016). Are the anomalous gravitational tugs on the spacecraft more likely to happen during polar flybys than during equatorial ones? Again, I asked Dr. Acebo as to whether this was the case:

"Yes, polar flybys seem more prone to show large anomalies than equatorial orbits but this is a very qualitative classification. For example, the largest anomalies were found for the NEAR flyby on 1998 whose orbit was ... contained in a plane perpendicular to the equator so the orbit orientation seems to play a role in the magnitude and sign of the anomalous velocity change. Altitude over the Earth surface is also an important factor." (Acedo 2017a)

That seems to keep the possibility alive, at least, although it is far from conclusive evidence of a distant gravitational tug. But how massive would a distant planet need to be to provide even this slight deviation? Too large, according to Dr. Acedo, in further correspondence with me:

"...the effect of a hypothetical 'Planet Nine' in the interior Solar System is not easy to measure and it has no significant contribution to the flyby anomaly. In the **Astrophys. Space Sci.** paper I considered the effect of the tidal forces exerted by Jupiter on a flyby of the Earth and dismissed it because its contribution to the anomalous velocity change is not comparable with the observations [Acedo 2017], the same can be said of 'Planet Nine'. I also remember a paper by Lorenzo Iorio in which he studied the perturbations of a new planet in connection with the lingering anomaly of the Moon's orbit varying eccentricity [Iorio 2011]. But to explain this he concluded that it would require a Jupiter-like planet at 200 AU that is much larger than the estimated size of 'Planet Nine'. Summarizing, anomalies in the inner solar system (if they are real and not the result of measurement and modeling errors) cannot be accounted by new trans-Neptunian planets." (Acedo 2017b)

This response discussed an anomaly I wasn't even aware of! Perhaps a sub-brown dwarf at the distance of the inner Oort cloud would be sufficient to create these effects, but we then return to the issue of perceived realism. From the point of view of astronomers, such a body should already be in the can.

The probes which have exited the solar system offer so much beyond the glorious array of images they sent back during their main missions. In the case of Pioneer, questions were raised about fundamental physics due to their slowing motion. Although the Pioneer anomaly was eventually explained by an engineering solution, similar issues keep cropping up with other spacecraft. The Voyager spacecraft have proven incredibly tenacious, as seen through their ability to respond to new thruster demands decades after these systems went into hibernation (Crane 2017). Perhaps the extended missions of these remarkable craft will provide hints of what lies beyond.

The Vision Thing

Jupiter, the solar system's largest planet, is turning out to be as majestic as its ancient name suggests. It is the nearest thing we have to a sub-brown dwarf, and its properties hint at the marvels which its heavier, darker cousins might exhibit.

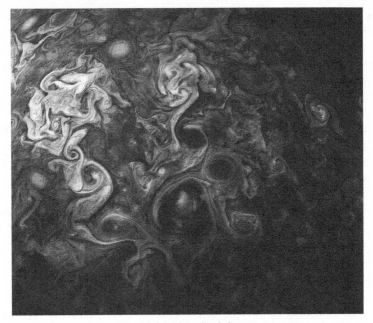

Fig. 15-3 Jupiter clouds image

High definition images taken of Jupiter's poles by the space probe Juno show a vibrant, churning cloudscape so beautiful that it could have been worked in oils (The Space Academy 2017). The gnarly appearance of the storms and tempests woven into this mind-blowingly immense vista seem peaceful enough from space, but the ferocity of their winds can only be imagined. Although the colors have been artificially enhanced (NASA's Juno Mission 2017), Juno's imaging equipment has captured the incredibly beautiful blue colors of the polar zones and the immense set of storms swirling within. (color images are available in ebook versions of *Darker Stars*)

These dramatic regions contrast strongly with the generally dull series of beige bands wrapping around the more familiar equatorial region (although Juno has also allowed us to better appreciate the intricate patterns of these banded zones, too). These blues are more reminiscent of the ice giants, Neptune and Uranus, and perhaps even of Earth – although the constituent gases of the atmospheres of these worlds can differ significantly from Jupiter's. The different colors and properties of Jupiter's clouds can be attributed to their constituent gases – mostly hydrogen and helium, but also water, ammonia, methane and sulfur.

Juno has sent us stunning images of the incredibly complex cloudscapes of our solar system's largest known planet. It seems to me that the solar system is starting to come to life. Not necessarily through the realization of the existence of alien biological life – although that may yet come to be – but instead through our appreciation of the system's rich visual complexity. For their time, the Pioneer and Voyager space probes provided what were incredible images of the outer solar system planets. But limitations in the image-capturing technology also tendered a sense of dull uniformity.

In the decades before the space-probe images had been sent back from the outer solar system, sci-fi writers, film-makers and scientists had built up an amazing array of ideas about what these worlds might be like. This exciting vision became ingrained within the public collective consciousness, and to some extent helped drive NASA's ambitious space program forward. This was enhanced by a sense of mystery, and the hope of the discovery of alien life.

In hindsight, images returning to our television screens in the latter part of the 20th century clearly did not really do these mighty worlds justice. So, although obtaining the planetary images were astonishing achievements in themselves (Hao 2017), the disappointing lack of features within them dashed many hopes, and, I suggest, provided the public with a stagnating view of the outer solar system. The Voyager imagers taught us that, like the disappointments of a colorless Moon, a lifeless Mars and an overheated Venus, the outer solar system seemed to consist of a rather uniform set of giant planets distinctly lacking in the vibrant complexity of our own Earth. How things have changed in recent years!

Jupiter's Northern Lights

It is not just the new visual imagery which is challenging us to look with a renewed sense of astonishment at these immense worlds. Cassini and Juno have sent back valuable scientific data from their on-board instruments, gathered from the highly charged and radioactive regions of space surrounding Saturn and Jupiter. Saturn and its rings have been utterly spectacular in HD. Its main moon Titan has been revealed to be a weathered, complex world beneath an immense, petrochemical-laden sky: More on Titan later in the chapter.

Juno's data about Jupiter's polar aurorae indicates that the acceleration of particles through the Jovian atmosphere – due to its strong magnetic and electrical fields – is being driven by a different, altogether more powerful process than the one we observe here on Earth. The mighty gas giant appears capable of accelerating charged particles within its atmosphere to exceedingly high energies (Agle & Brown 2017).

The Juno spacecraft appears to be observing two processes at play regarding the generation of aurorae:

1. A discrete high-energy, downwardly directed, electron acceleration within Jupiter's polar regions.

2. An upward, magnetic-field-aligned electric potential of up to 400,000 electron volts.

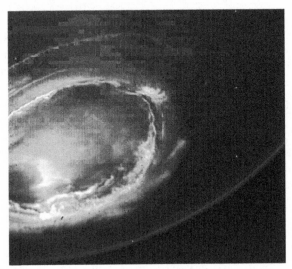

Fig. 15-4 Jupiter southern lights image

This latter electrical potential is an order of magnitude greater than the largest potentials observed on Earth (Mauk et al. 2017). If Jupiter can generate such powerful aurorae, then what visual effects might its more massive, denser cousins, the sub-brown dwarfs, be capable of?

Brown Dwarf Magnetic Fields

Jupiter's ability to accelerate high-energy electrons hints at the titanic internal processes as we scale up to the sub-stellar class beyond. Failed stars should be quite capable of generating even more potent aurora displays from their immense magnetic and electrical fields. Brown dwarfs are thought to generate strong fields by a convection-driven dynamo process within their electrically conducting interiors (International Max Planck Research School 2018). Dynamos occur when the interior of a planetary body rotates faster than its surface layer. This accelerates electrons, creating magnetic forces and subsequently strong magnetic fields within and around the star or planet.

It had been thought that the complex internal structure of stars was very different to that of smaller ultra-cool dwarfs: Where stars have a radiative core surrounded by a convective envelope, ultra-cool dwarfs are fully convective throughout. So, sub-brown dwarfs should be incapable of strong field disturbances, like magnetic field reversals. This category of enormous planetary-mass objects should not be capable of exhibiting the same kind of magnetic field activity cycles observed in stars (Kohler 2016).

However, a recent scientific study which examined radio flares emitted by ultra-cool dwarfs indicated that brown dwarfs and sub-brown dwarfs might be capable of experiencing magnetic field reversals, similar to stars, on decade-long timescales (Route 2016).

Brown dwarfs sometimes seem to be a lot more like stars than planets, even though the processes by which they emit their dim light are very different from the nuclear fusion processes driving stars. For instance, brown dwarfs can exhibit surprisingly strong surface magnetic fields and, more importantly, these fields can show unexpectedly high levels of disturbance (Emspak 2017). In a way, all of this tells us something we already know – brown dwarfs sometimes behave like stars, sometimes more like gas giant planets. Jupiter is the closest cousin to sub-brown dwarfs that we have to study, and its internal complexity is only just starting to emerge from under those immense clouds.

Patchy Weather and the L/T Transition

We have looked at the remarkable cloudscapes of Jupiter, and the processes driving its powerful aurora displays. How these findings extrapolate up to sub-brown dwarfs and brown dwarfs remains less than clear, however, particularly with regard to the composition of their atmospheres and their weather.

Ultra-cool dwarfs are of a similar size to Jupiter, but much more massive. Their density intensifies with rising mass, raising atmospheric pressures, and stimulating ever more powerful displays of aurorae and flares. Young, warm brown dwarfs are capable of giving off white light flares that are even stronger than those emitted by the Sun (Gizis et al. 2017), although this remarkable activity is not apparent with older, cooler brown dwarfs (Bryant 2016).

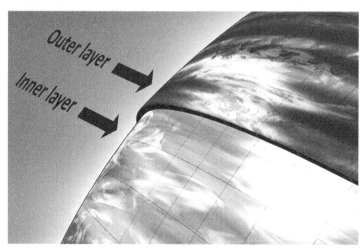

Fig. 15-5 Brown dwarf layers diagram

As brown dwarfs age, they cool. Astronomers have observed that cooling, older brown dwarfs actually start to lose their immense layers of cloud (Carnegie Institution for Science 2018). Their heavy, dark skies begin to clear. The older, cooler brown dwarfs, known by astronomers as "T-dwarfs", have clearer skies than their younger cousins, the "L-dwarfs". This is a little like comparing the crystal-clear skies of Ice Age Earth to the increasingly volatile weather conditions more frequently experienced on our now warmer world. The boundary between the brown dwarf types is known as the "L/T Transition", and it continues to hold many mysteries for astronomers. Perhaps emulating the aging of a fiery, red-headed youth into a cooler, follicly challenged gentleman, there seems to be a transitioning temperature where those fiery locks dissipate.

Scientists have identified an unusually red brown dwarf, functionally named 2MASS J13243553+6358281, lying at the lower end of the brown dwarf mass spectrum, at 11–12 Jupiter masses. Its age, ~150 million years, is deduced from the company it keeps: It travels with the AB Doradus Moving Group (Gagné et al. 2018), and is one of the nearest known planetary-mass objects to us. Its age and mass place it in this tricky transition phase between the young, hot L-type dwarfs and the older, cool T-type dwarfs; an issue that has been subject to theoretical study for a while (Burrows, Sudarsky & Hubeny 2006). There is quite a lot of context here, so let us explore this transition zone in a little bit of detail.

Many sub-brown dwarfs and brown dwarfs rotate quickly, like Jupiter. It was theorized that the hot, dense clouds of the L-dwarfs would eventually thin out and disappear as these objects age and cool. At some point during this slow, cooling process a patchy variation in cloud cover should be evident. At the L/T transition, careful observation of these transitioning brown dwarf objects should show some variability with time. (Note that a tightly bound pair of binary objects first needs to be excluded, as this, too, could create a similar variable appearance in the object's light; a function of their twirling dance. This scenario also had to be ruled out for the tumbling interstellar asteroid 1I/'Oumuamua, which could conceivably have been a binary asteroid).

In our solar system, Jupiter's complex cloud system shows significant patchiness, where breaks in the cloud cover expose warmer layers below. Planet mass objects which are larger than Jupiter would, presumably exhibit more extreme variation still. Nonetheless, the "L/T transition state" is a variable zone where the properties of 'in-betweener' planetary-mass brown dwarfs remain very difficult to predict. Multiple factors are at play, including complicated chemistry within the clouds and weather variations across the brown dwarf's atmosphere.

So, when does this patchiness begin? At what mass? At what temperature? At what age? At what color, even? Caroline Morley writes that the L/T transition between the cooler T-dwarfs and their hotter brown dwarf cousins, the 'L-dwarfs', occurs between about 900 and 1100°C.

As we have seen, 2MASS J13243553+6358281 lies at the lower end of the brown dwarf spectrum, almost being a planetary-mass object (or sub-brown dwarf). So, its properties are helpful for determining the L/T transition point for ultra-cool dwarfs. In this particular case, the transition phase from cloudy to cloud-free seems to occur around 900°C (Gagné et al. 2018), which is about 1000°C hotter than Jupiter.

The transition from hot, dust-laden atmospheres in the L-dwarfs to clearer, cooler skies in the T-dwarfs occurs relatively quickly. The reason for the sudden break-up and disappearance of cloud cover is not well understood (Morley 2012). The colors of failed stars depend significantly upon the properties of the clouds within their atmospheres. Below around 2700°C, "hot" brown dwarfs (L-dwarfs) feature "exotic" dust clouds (Radigan 2015). These clouds are laden with salts, silicate dust (or rock) and scalding-hot liquid iron rain (Clavin 2014).

Not particularly hospitable, for sure, but this dusty property helps explain their distinctly red color. As these L-dwarfs cool and age, they gradually turn into the cooler T-dwarfs. It is thought that during this L-T transition phase, the mind-wrenchingly awesome molten iron/rock-laden condensate clouds gradually dissipate, becoming increasingly patchy until, eventually, they are gone. This seems to be the case with the ultra-cool T-dwarf 2MASS J21392676+ 0220226 (2M2139), for instance. The light from this object varies greatly as it rotates. Astronomers studying this regular variation concluded that its atmosphere is highly fragmented (Radigan et al. 2012). (The way this light variability has been modeled against changes in cloud cover has been neatly illustrated in a short video released by NASA Spitzer (2017)).

From these findings about the L/T Transition, we can safely conclude that any 'old' sub-brown dwarf hanging around in the depths of the outer solar system should have clear, cool skies.

Saturn's Family of Oddballs

In my previous book, **Dark Star**, I focused a lot of attention on the Galilean moons of Jupiter. These mini-planets, with their internal tidal warming and sub-surface oceans, provide a strong precedent for my arguments about how habitable conditions could arise in a binary sub-brown dwarf system (Lloyd 2005). In the intervening years, there have been more detailed explorations of the planets in the outer solar system, as well as their moons. The emerging findings provide us with further insights into how a habitable sub-brown dwarf system might work in practice.

Cassini's long mission to Saturn and its moons has provided the public with captivating images of great beauty. As well as incredible views of the famous rings, polar flybys enabled Cassini to study the persistent hexagonal pattern in the polar cloudscape around Saturn's northern pole, monitoring variation with its patterning and color over time. Larger than Earth, the six-sided vortex of powerful jet-streams that shape Saturn's polar weather system continue to puzzle atmospheric scientists.

Saturn's extensive family of moons has also been the subject of intense scrutiny. Some of these findings have been unexpected, to say the least.

Titan's Origins

Titan is a large moon orbiting Saturn with a super-sized atmosphere that, like the Earth, consists largely of nitrogen. The isotopic ratio of nitrogen in comets has been found to be very similar to the nitrogenous atmosphere of Titan. This implies that the composition of Titan has a similar source as that of the comets from the Oort cloud (Mandt et al. 2014).

The finding is at odds with the more obvious assumption that Titan formed within the young Saturn's own warm protoplanetary disk. Instead, it presents us with two distinct possibilities:

1. Titan was captured by Saturn at some point in its past.

2. Titan was always one of Saturn's moons, but its atmosphere was delivered to it, perhaps as a result of the cometary bombardment during the Late, Heavy Bombardment. In this scenario, all of its nitrogenous ices would have been dumped onto its surface sometime after its formation, by impacting comets originating from the Oort cloud.

In a press conference, Kathleen Mandt, the scientist leading the work on Titan's nitrogen isotopic composition, argued that Titan's nitrogen cannot have naturally evolved within the life-time of the solar system. It simply contains too much nitrogen (NASA/JPL 2014). Therefore, Titan may be older than Saturn. Furthermore, the high eccentricity of Titan's orbit also suggests its origin was not part of the normal process of formation of Saturn's system of moons.

It seems likely, then, that Titan was captured by Saturn, rather than having formed in the gas giant's midst. Was Titan once a planetary comet from the Oort cloud that strayed too close to Saturn? Might Titan even have originated from another star system?

Some years ago, scientists argued that all planets began life in the protoplanetary disk that swirled around the Sun, and argued that Planet X-like bodies were unlikely because there was no theoretical underpinning for how they formed. More recently, that fairly blinkered vision about how planets form has been disrupted by knowledge of the often extreme variation of extra-solar planetary systems. So, we now know that planets can form in the outer zones of planetary systems, allowing for the creation of 'moons' like Titan. If Titan has more in common with the comets than the planets in its midst, then that implies that the solar system has been very severely shaken up in the past. But Titan is not the only curiosity among Saturn's moons.

Enigmatic Enceladus

Cassini imaged incredible jets of water streaming out into space from Saturn's little moon Enceladus. Feeding these plumes of water is a global sub-surface ocean lying beneath about 20 miles of ice (Press Association 2014). The little moon has been found to wobble as it orbits around Saturn. Careful observations of the moon noted that its angular velocity varies over time as it orbits around Saturn. The best explanation for this variation is that Enceladus has a liquid ocean between its ice-covered surface and its core (Thomas et al. 2016). Its rocky core essentially floats within the moon's global sub-surface ocean, and so is capable of moving independently of the surface. This structural fluidity within Enceladus creates the observed wobble in the moon's orbital path (Tamblyn 2015).

The deduced presence of an entire ocean below the icy surface of the moon is surprising. Even with tidal heating by the parent gas giant, thermal models do not provide the kind of extensive sub-surface ocean that Enceladus appears to have. Worse, these effects should work more strongly on another of Saturn's moons, Mimas (Neveu & Rhoden 2017). Yet, Mimas behaves more conventionally than Enceladus. Saturn's moons present an internally inconsistent picture, and continue to puzzle planetary scientists. Their diversity challenges theoretical assumptions.

Fig. 15-6 Fountains of Enceladus image

This finding has some profound implications for the potential for life within a moon orbiting a distant gas giant (European Space Agency 2015). It strengthens the assumptions I based my original **Dark Star** Theory upon (Lloyd 1999): Sub-surface aqueous oceans are turning out to be commonplace among moons orbiting gas giants and ice giants.

Warm water is spewing out of pressurized cracks in Enceladus's icy surface into space, creating spectacular geysers. That water provides scientists with the opportunity to analyze the materials dissolved in that warm water (Hsu et al. 2015), revealing methane within the sub-surface fluid (Bouquet et al. 2015). It is very clear that oceanic conditions below Enceladus's icy crust are ripe for the evolution of life, at a location far from the Sun's traditional "Goldilocks Zone". This increases the potential for a habitable environment on, or within, a sub-brown dwarf moon located in interstellar space. These may be free-floating planetary systems, or binary companion objects loosely bound to stars.

Exotic Ices

We have looked at some of the powerful internal processes at work within the gas giants Jupiter and Saturn; but how about the ice giants, Uranus and Neptune? To date, there has been no orbiter mission to either of the ice giants. This may change in the 2030s, although, due to the paucity of sunlight in the outer solar system, space agencies hoping to send ice giant orbiters will need to make their probes nuclear-powered. They will also need to plan for optimal launch windows around planetary flybys for gravitational assists (Wenz 2017). Until then, we do not have the same levels of data for the ice giants that Cassini and Juno have supplied us for Saturn and Jupiter. But Uranus and Neptune are no less important, despite their distance: Mini-Neptunes are the most common form of exoplanet, after all.

Many scientists speculate about what bizarre chemistry may be going on within these immense worlds. Atoms and molecules are pressed together by the extremely high internal pressure within these massive planets, creating conditions which are incredibly difficult to replicate on Earth.

An experiment undertaken at the Lawrence Livermore National Laboratory may have successfully replicated conditions within the ice giants. Scientists speculate that the interiors of the ice giants are rich in "Ice VII", a crystalline form of ice which is produced under incredibly high pressures. This superionic ice allows hydrogen ions to roam freely within the lattice, allowing this form of ice to become a superconductor. This, in turn, might have a tangible effect upon the ice giant's magnetic fields (Mandelbaum 2018). Super-compressed interior 'ice' within these distant worlds may generate superionic conduction through deep layers within these planets (Millot et al. 2018).

Conditions deep within the Earth's mantle have also been sufficiently extreme to create Ice VII. In the terrestrial case, Ice VII has been found within diamonds ejected from the Earth's depths by volcanic activity (Tschauner, et al. 2018). This exotic form of ice is thought to have been made in a transition layer some 410 to 660 km deep. This finding is suggestive of pockets of fluid water held under extremely high pressure within the moving rock layers deep inside the Earth. Perhaps these fluid-rich areas of exotic water and ice are also superconductive.

Theoretical explorations of phase transitions of ice at high pressures predicted these kinds of superconductive effects thirty years ago (Demontis, LeSar & Klein 1988). This property may help to explain the unusually displaced – and tilted – magnetic field structures of the ice giants. Other theories tend towards the idea that the magnetic fields are mostly created by degenerate metallic fluid hydrogen (Nellis 2015). This would be more in keeping with the assumed interiors of gas giants, where an incredibly deep ocean of metallic hydrogen engulfs a rocky iron core that might be about the size of the Earth.

Bowled Over

It has often been thought likely that an early planetary collision was the reason for Uranus being knocked onto its side (the ice giant has a spin obliquity of 98°). An early impact may also explain the profound differences in the satellite systems, magnetic fields and temperatures of the ice giants. Recent scientific work using computer simulations has strengthened the case for an early impact, and nailed the culprit down to a terrestrial planet of at least 2 Earth masses (Kegerreis et al. 2018). The collision which tipped Uranus onto its side was probably a grazing blow by the impactor, that was capable of reconfiguring/building the rings and moons, but insufficiently catastrophic to rip away more than about 10% of its atmosphere.

Additionally, there is reason to suspect that at least some of the impactor has had a long-term effect upon the planetary properties of the injured ice giant. Impactor materials subsequently embedded into Uranus may have created an insulating layer within the planet which is holding in the internal heat of the planet. Whether this is feasible depends upon the angle with which the impactor struck Uranus. A head-on collision would have plunged the impactor deep into the core of the ice giant. On the other hand, collisions involving a more glancing angle of attack could have immersed the impactor debris into a global layer within the ice giant, creating a thermal boundary layer of impactor ice. Other impactor materials would have crashed out into space, offering the opportunity to build some of Uranus's odd collection of moons (Kegerreis et al. 2018).

If the proposed Planet Nine resembles a mini-Neptune world (Batygin and Brown 2016) evicted from the planetary zone early on, then it may have undergone similar traumas. This impact model may provide a reason to consider the cold temperature of Uranus as a by-product of interplanetary collision. In which case, like Uranus, the projected temperature of Planet Nine drops, along with the ease with which it might be found using infrared telescopes.

Puzzling Pluto

Although there have been no orbiters sent to planets beyond Saturn, NASA's New Horizons probe has followed in the footsteps of the Pioneer and Voyager missions in its voyage of discovery through the outer solar system.

New Horizons flew past distant Pluto in 2015, after a nine year voyage through the planetary zone of the solar system. It sent back images of incredible beauty, capturing the distant dwarf planet's coat of many subtle colors (Talbert 2015). Pluto has proven to be a far more interesting world than was anticipated by pretty much anybody. It is not the grey, dull world so reminiscent of, say, most of the dwarf planets in the asteroid belt. In parts it is reddish or, more precisely, salmon pink in color.

Fig. 16-1 Pluto 1315109 image

Its color may be reminiscent of Mars, but the reason for the color is entirely different. The pinky red color of the dwarf planet is thought to result from the chemical action of faint sunlight generating red compounds in Pluto's rather tenuous atmosphere. These then fall down onto the tiny planet's surface.

Before the New Horizons flyby, the Hubble Space Telescope had picked up hints of Pluto's pink complexion. This reddish hue has become more dramatic over the last ten years compared to previous decades. This change in hue is thought to be due to the solar wind stripping away hydrogen from the methane surface ices, leaving dark, red-colored organic compounds exposed (Borenstein 2010).

Red Planetscapes

It is not the first time that reddish dwarf planets have been discovered in the outer solar system. Neptune's largest moon Triton is also pink. Triton exhibits a retrograde orbit, suggesting that it is a captured object, possibly originating from the Kuiper belt (MIT Technology Review 2011).

The reddest of the known KBOs is the enigmatic Sedna (Lloyd 2005a), which is second only to Mars in redness. Sedna is remarkable for a number of things, the most bizarre being its highly elongated orbit (Howell 2014). It is a member of the cluster suggestive of the presence of a Planet X object (Trujillo & Sheppard 2014; Batygin & Brown 2016). However, it should be noted that this redness is certainly not common to all TNOs. In contrast to Sedna's wayward orbit, most of the red TNOs make up a grouping within the Kuiper belt known as the "Cold Classicals" pitched beyond ~40AU. Behaving themselves, they move around the Sun in relatively circular orbits, and predominantly lie in the plane of the planets.

So what makes this red color appear on the surface of these distance worlds? There is likely to be a complex mix of processes underway, both in terms of organic chemistry, and planetary science (Doressoundiram et al. 2008). The red color itself likely comes from "tholins", which are complex macromolecular compounds synthesized from irradiated mixtures of gaseous and solid hydrocarbons, nitrogen and water.

Galactic radiation and ionic plasmas from the Sun can easily toast exposed organic materials, and bombardment by micrometeorites can also reshape surfaces. In contrast to the blackening of surface organics among bodies nearer to the Sun, the TNOs may be located at a more optimal distance to undergo 'safe' organic chemistry on, or just below, their surfaces. Exposure of these compounds then provides the minor body's reddish hue (Fox 2010). One could speculate that this helpful distancing from the Sun may actually facilitate the synthesis of the building blocks for life on, or just below, the surfaces of TNOs.

Another solution for a dwarf planet in the outer solar system is the deposition of organic materials onto the surface, initiated by the Sun's dim UV light acting upon volatile compounds mixed up within its atmosphere. For instance, it is now known that Pluto has an atmosphere of nitrogen, methane and carbon dioxide. This atmosphere probably strengthens during Pluto's long perihelion passage: The dwarf planet's eccentric orbit brings it closer to the Sun than Neptune for about 20 years (Pluto takes 248 years to orbit the Sun).

Fig. 16-2 Back-lit Pluto image

Truly remarkable images sent back from New Horizons in July 2015 show a back-lit Pluto. A hazy hydrocarbon atmospheric could be seen extending to about 80 miles above the planet's surface – significantly greater than scientists had anticipated (Talber 2015a). This haze layer bears similarities to Titan's atmosphere. The new atmospheric data suggests that reddish tholins are deposited onto Pluto's surface from this hazy atmosphere, leading to the reddish coloring across some of the dwarf planet's surface.

Pluto's Sub-Surface Ocean

Pluto's rich and often bizarre topography has been revealed in far greater detail by images sent back by the New Horizons probe. The dwarf planet is pock-marked with any number of remarkable surface features: chasms, craters, canyons. The surface features of Pluto are incredibly varied: Mountains, glaciers, dunes, gullies, ridges and scallops; all vying for the attention of geophysicists. Many of the formations and variations are not only highly intriguing, but also incredibly beautiful. Clearly, Pluto is a vividly complex world.

One of the most prominent features is a bright, extensive heart-shaped area. This is the clearest example of a number of areas across the surface of the dwarf planet, each of which has been substantially re-worked over time. These re-surfaced zones are suggestive of interior geophysical processes that are on-going. There are even hints of a sub-surface ocean.

Fig. 16-3 Pluto size comparison image

Pluto has a binary moon, known as Charon. As with the relationship between the Earth and its oversized Moon, the presence of Charon hints at a major collision between Pluto and another celestial body in the distant past. This collision may even have placed Pluto into its current eccentric orbit. This catastrophic event heated Pluto's interior, enabling a liquid ocean to exist below the dwarf planet's surface (Stephens 2016).

Evidence for this sub-surface ocean comes in the form of the orientation of the heart-shaped feature on the Pluto's surface. It faces towards Charon, coincidentally aligned with the tidal axis between these two minor worlds. However, this re-surfaced, depressed basin, known as Sputnik Planitia, provides insufficient extra bulk on its own to account for a positional reorientation between Pluto and Charon. A sub-surface ocean beneath this feature would provide the additional mass required, however, to cause the worlds to slowly align themselves along this axis (Nimmo et al. 2016). The concept of a sub-surface ocean beneath this zone is further supported by the presence of cracked fault lines in the heart-shaped basin's 1000km-wide crust (Keane et al. 2016).

Shifting Surface Ices

Sputnik Planitia is also rich in methane and other chemicals. Although scientists had been fully expecting to find frozen nitrogen on the surface of Pluto, the variable extent of frozen methane across the planet's surface has been a real surprise. The Sputnik Planum zone (as it was previously referred to) is rich in methane, and also shows a strong presence of carbon monoxide.

These volatile snows and ices create a frozen surface 'sea' which gently shifts position across the surface of Pluto. These dynamic processes are partly due to the dwarf planet's eccentric orbit, which causes intermittent periods of cold and warmth as it moves further away from, and then closer to the Sun. The ices sublimate into the atmosphere and then freeze out again during cyclical seasonal patterns. Atmospheric winds also shift these snowy ices about, resulting in drifts.

Given the remarkable complexity evident upon Pluto's surface, what, I wonder, is going on beneath the surface? Does a richer concoction of hydrocarbons lie beneath these surface features, warmed by Pluto's traumatized – and radioactive – inner core? In other words, does the presence of these simple compounds on the surface, methane and carbon dioxide, lead us to expect that there is plenty of organic chemistry going on within the planet?

The likely presence of a sub-surface aqueous ocean would surely enhance that potential. If such an ocean does exist, then it probably lies many miles below the surface, in a similar way that other moons in the solar system have hidden aqueous depths below their icy surfaces. The presence of ammonia and salt impurities in Pluto's hidden waters may help to drive down the sub-surface ocean's freezing point.

For Pluto and Charon, their mutual tidal heating likely continues to drive internal heating mechanisms, helping to explain the extent of re-surfacing over ancient craters across the landscapes of both. It seems that Charon's orbit around Pluto had previously been more eccentric, exacerbating the effect of tidal friction in the past. This finding correlated well with predictions about a sub-surface ocean made in 2014 (Barr & Collins 2014).

One might be forgiven for thinking that sub-surface oceans are the norm among moons and dwarf planets in the outer solar system. It has been recognized for a long time that icy moons (like the Galilean moons Europa, Ganymede and Callisto orbiting Jupiter), have massive oceans underneath their billiard ball surfaces. When I first wrote about the potential for life in the outer solar system, some 20 years ago, it was in the context of a similar scenario playing out among the comets: I envisioned a large, icy moon in orbit around a massive, hidden planet. I went further, imagining a hidden sub-brown dwarf whose power was such that the moon could even maintain a strong atmosphere, and liquid water on its surface: In other words, the conditions for life (Lloyd 1999).

Since then, discoveries in the solar system have indicated that this scenario not only may extend upwards to sub-brown dwarfs and failed stars, but also trends downwards. Far from requiring a planet larger than Jupiter to drive the internal heating required within its moon to support these kinds of environments, the prevalence of sub-surface oceans in moons and binary couplets (like Pluto/Charon) suggests that we should now expect this kind of effect, rather than be surprised by it.

Distance from the Sun appears to be a less important factor than had been considered. As such, we might now expect to find sub-surface oceans among paired KBOs. These oceans may be warmed further by radioactive isotopes in the rocky cores of KBOs which provide "radiogenic heating". This internal heating may be a sufficient driver to create liquid ocean layers even within single large KBOs, although these layers would be located under very significant layers of ice (Hussmann & Sohl 2007).

The scenario I sketched out twenty years ago is becoming a normal feature of many objects in the solar system. This lessens the need to depend upon an object quite as large as a sub-brown dwarf to create the conditions for life on one or more of its moons (although a breathable atmosphere on one of its moons would clearly be nice). A gas giant Planet X would certainly be sufficient to create the sufficient forces to provide a lunar ocean. Perhaps even a super-Earth Planet X, like Planet Nine, would suffice.

I suspect that a Planet X super-Earth would not, on its own, provide the right conditions for a habitable lunar atmosphere, though. Unless, that is, the properties of interstellar space itself create sufficiently weird local conditions around these massive planets to facilitate atmospheres. A combination of primordial heat, planetary tidal forces, radioactive lunar cores and local nebulae might be sufficient to harness life-supporting conditions around planetary mass objects in interstellar space. This is highly speculative, but the more we learn, the greater the chances of such a scenario seem to be. We assume that the classical "Goldilocks Zone" of a star system provides the template for habitability. But it is not the only possibility. It is just what we know, and can readily relate to.

Ice Mountains

Let us return to Pluto. KBOs are predicted to contain a great deal of water. So it is with this dwarf planet. Its varied surface includes dramatic mountains which are likely to have formed from water ice relatively recently. These youthful ice mountains probably formed in the last 100 million years – an estimate based upon a surprising lack of craters in their vicinity. Given Pluto's extreme cold temperature, this water ice constitutes a much stronger building material than methane/nitrogen ice.

One would expect that the most distant objects in the solar system are the most pristine examples of primordial planetary objects. Yet, in Pluto's case, these icy geological features are some of the youngest observed amongst the planets. The proximity of Charon is likely to be a significant driving force for the internal dynamics at work within Pluto.

Fig. 16-4 Block ice mountains

It is thought that the icy mountains (known informally as the al-Idrisi mountains) arose from the water-ice bedrock below Pluto's surface. As we have seen, this icy layer may, in places, cover a deep aqueous ocean. This suggests, very weirdly, that the mountains are essentially icebergs floating on seas consisting of water, nitrogen and carbon dioxide ices. For instance, the broken terrain around the western edge of the 4-km-deep Sputnik Planitia features jumbled blocks of ice of immense proportions which, together, create mountainous chains of ice. Some of these mountains are ~2,500m high, with one aspect covered in a dark material.

These mountain chains contrast strongly with the adjacent flat plain, which exhibits traced outlines reminiscent of cells. The plain itself seems to be incredibly young – perhaps just 10 million years old (Drake 2015). The glacial nitrogenous ices spread across the plane's heart-shaped surface seem to move around like buoyant clumps in a lava lamp (Perkins 2016). This is hardly the pristine model of primordial quiescence expected of this 'dead' world.

Other parts of Pluto are covered in wide fields of glaciers flowing across the planet's surface. The intense cold at this distance from the Sun (-235°C) means that these are highly unlikely to be made of water ice: As previously noted, the properties of extremely frigid water ice, which helped to form the rugged ice mountains, do not lend themselves to malleable ice flowing across a surface. Instead, these glaciers are likely to consist of nitrogen, carbon monoxide and methane, and are probably still slowly flowing over Pluto's ancient craters today (Moore et al. 2016).

Pluto's active features are thus providing insight into a multi-layered, complex world. Perhaps life, in the form of extremophiles, might also exist deep within the planet, submerged in dark, warm oceans. What is clear from all this is that Pluto is far from being a dead lump of frozen rock. Pluto is not in the vicinity of a gas or ice giant planet. It has only diminutive Charon for company. All the same, it is displaying features which imply that some radical internal processes are going on. These include cryovolcanic activity, which is thought to be driven by radiogenic heating within Pluto.

Pluto's Catastrophic Past

Pluto's moon Charon has played (and may continue to play) a part in all this. Charon's existence points to a prior catastrophic collision. The cataclysm that created this moon boiled Pluto's interior and set the consequent binary planet system onto its irregular orbital path. Pluto's almost perfect spherical shape indicates that at some point in its past it must have been spinning very quickly, probably as a result of the giant impact which led to the creation of Charon. Charon's orbit around Pluto is now fairly circular, pointing to a settling of its orbit over a long period of time. Similarly, Charon's tug on Pluto gradually slowed the larger world's own spin, and tugged the worlds into a harmonious alignment.

The planetary collision which affected Pluto must have been a very long time ago. What caused it? Was it anything to do with Sedna's odd orbit, or that of the other so-called scattered disk objects beyond the regular Kuiper belt? Do these distant cosmic footprints of catastrophism provide more evidence for the existence of a much larger body further out in the solar system? I think so. The outer solar system continues to surprise everyone. It demonstrates greater diversity and complexity than previously dreamed of.

Yet, on the face of it, the environment of the outer solar system should promote quiet stability – a retreat for minor bodies that were dispelled from the heart of the action early on in the life of the solar system. So, why aren't they behaving like that? It is a curious thought that the formation of the Pluto/Charon planetary couple essentially mirrors that of the Earth/Moon coupling. What are the odds that two planets in the solar system both get whacked early on, forming comparatively over-sized moons?

N-body simulations to determine the probability of an early cataclysm between Earth and its collisional partner Theia calculated the most advantageous orbital arrangement to facilitate their fateful meeting (Dvorak, Loibnegger, & Maindl 2015). Setting a series of orbital scenarios up for Theia, that centered upon a semi-major axis of 1.1AU, provided a healthy 25% chance of a collision between the two neighboring worlds. However, it did not take much of a shift further away from the Earth for the chances of a collision to drop off significantly.

The potential for Pluto to have been hit by another KBO depends upon the early population of the belt: If it was substantially greater than it is now, then a moon-forming collision between Pluto and another ~1000km KBO would be more plausible. Similarly, the early migration of Neptune, which is thought to have taken place (within the framework of the Nice model) should create a flux of resonance relationships within the belt that brought once-stable bodies into the path of one another (Canup 2005).

Even so, the Kuiper belt is highly dispersed, and so the statistical possibility of Pluto colliding with another body, even in an earlier, more densely populated environment, cannot be very large. However, the very fact that Pluto is where it is may allude to this object being a statistical blip in the overall collisional stakes. The dynamical scars of prior collisions so far out in the outer solar system are likely to be rare because debris from collisions will readily disperse away from the scene of the celestial battle. Nonetheless, it is not unheard of. One example of the aftermath of a previous collision in the Kuiper belt is the Haumea collisional family, but, again, this is the exception rather than the rule (Marcus et al. 2011).

Impact craters extant on lunar and planetary surfaces across the solar system attest to the violent nature of the solar system, particularly during tumultuous periods of heightened catastrophism, like the Late, Heavy Bombardment. Earlier on, towards the end of the planet-forming period, the Earth and Uranus both seem to have been substantially whacked by planetary impactors. The creation of the Pluto and Charon couplet shows that the Kuiper belt region is not immune to such activity. Add a Planet X object into the mix, and the potential for disruption and collisional activity increases accordingly.

Where to Next for New Horizons?

Following on from its flyby of Pluto and its moons, New Horizons continues its foray into the Kuiper belt. Astronomers used ground-based telescopes to try to locate KBOs along the probe's projected trajectory. However, these efforts were not met with success. As a result, the Hubble Space Telescope (HST) was brought into the search, and managed to locate several candidate objects early in its search.

The best candidate, 2014 MU69, named "PT1" by the New Horizons team, belongs to the afore-mentioned "Cold Classical population" of objects in the Kuiper belt. The New Horizons probe is set to approach 2014 MU69 late in 2018, making its fly-by on New Year's Day 2019. 2014 MU69 was originally thought to be about 20 miles across. However, attempts to determine its size, as it passed in front of a background star, proved difficult. Although detected by several groups of astronomers during this transit, 2014 MU69 appears to be much smaller than originally thought: It could turn out to be a binary object, or possibly even a swarm of smaller objects (Bennett 2017).

New Horizons has enough fuel and electrical power from its plutonium generator to keep it powered up until perhaps the early 2030s. It is potentially capable of adjusting its course to visit at least one other KBO after 2014 MU69, as well as hopefully carry out long-range observations of several more. For instance, in December 2017, New Horizons sent back long-range images of 2012 HZ84 and 2012 HE85, taken by its Long Range Reconnaissance Imager (LORRI). At the time of writing, these two KBOs are the furthest solar system objects ever to have been imaged by a spacecraft (Keeter 2018).

The New Horizons spacecraft's continuing mission also provides scientists with an extended opportunity to see if the resurrected Pioneer/flyby anomaly becomes apparent in its trajectory over time. Perhaps it will also retain sufficient power to send back scientific data when it eventually negotiates the heliopause around 2040.

New Horizons' flyby of Pluto revealed a multiplicity of surface features which few had dared to dream of. There is the promise of further discoveries out in the distant reaches of the solar system. It feels as though the solar system is beginning to reveal itself in a new way – showing a complex and unpredictable side to itself which is both tantalizing, and also rather wonderful.

Meandering Mars

Mars is an odd world in many ways. It has held a fascination for space buffs for decades, if not centuries. Two of its more curious features are its eccentric, meandering orbit, and its chaotic axial tilt. These two factors work together to significantly alter the Martian climate and environment over time. Given the chaotic nature of the early solar system, which we have been examining at some length, it seems likely that the extreme nature of Mars' orbital cycles also hints at the influence of a perturbing body in the planet's distant past. Mars, too, has been subject to injurious catastrophism.

I think it likely that the intervention of a Planet X body within the planetary zone, triggering the Late, Heavy Bombardment 3.9 billion years ago, caused widespread catastrophism throughout the inner solar system. The maverick theorist Zecharia Sitchin proposed that this rogue body swept through the area now occupied by the asteroid belt. He believed that the Earth formed at this distance, but was destabilized from this orbit by the passage of the rogue planet, and subsequently migrated inwards (Sitchin 1976).

The erratic nature of Mars' orbit and tilt, as well as the destruction of its surface habitat, suggests that the red planet was similarly affected by a significant, cataclysmic event in the past. It may even have been the same event. One thing is for sure, the Mars of today bears little resemblance to its prior glory. With these ideas in mind, I would like to explore some of the more chaotic elements of the Martian surface topography, before examining the relationship between Mars and other bodies in the solar system.

Gullies

The Martian surface is a dry, barren wilderness whose total surface area is similar in size to the landmass of Earth. Scientists have long disputed any notion that water, and life, might exist upon the Red Planet. However, over the last 20 years, the "life on Mars" skepticism has gradually thawed in academic circles, as anomalous evidence has emerged of water in Martian rocks (found by the Opportunity, Spirit and Curiosity rovers), dry river beds, gullies and even ancient shorelines of long-vanished seas (Malin & Edgett 2000).

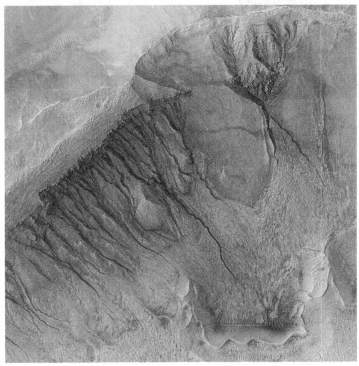

Fig. 17-1 Martian gullies image

Martian gullies are sinuous channels found on steep slopes in mid and high latitudes, located across both the red planet's southern and northern hemispheres. They are characterized by an alcove at the top, followed by a channel, and end with an apron of deposited material at the bottom. They are mostly located at higher altitudes, where the Martian atmosphere is thinner, and tend to face out towards the poles. Unfortunately, their locations on precarious slopes make them unlikely targets for future exploration by rovers. The issue of getting the rover on site is made worse by the thin atmosphere at these high altitudes, as landers require a modicum of atmosphere to assist aerobraking during their descent.

The prevalence of Martian gullies is extensive enough to provide a pattern of evidence that weaves its own story about water on Mars. An examination of orbital reconnaissance images suggests that Mars probably has its own large-scale climate cycles. There is a cyclical swing between Martian Ice Ages and interglacial periods. This cycle is the result of the red planet's variation in tilt, and other orbital fluctuations (Dickson et al. 2015). The variations in the tilt of Mars can change the amount of sunlight striking various parts of the planet's surface over time. Where the current conditions are unfavorable, water may have flowed in the past.

However, some scientists argue that the right conditions may have existed just a half a million years ago (Howell 2015). Other work has brought the dates closer still – a mere 200,000 years. A young Martian crater in the southern hemisphere exhibits well-preserved gullies and debris flow deposits, according to a team led by Andreas Johnsson of the University of Gothenburg. This is strongly suggestive of the flow of liquid water in relatively recent geological time, perhaps as recently as 200,000 years ago (Johnsson et al. 2014). This contrasts sharply with the usual ballpark figure that is cited, of several billion years ago. If these gullies are caused by flowing water, then this crater indicates the relatively recent movement of water on Mars.

The Curiosity Rover has discovered rounded pebbles in a crater that also imply the presence of flowing water at some point in the Martian past. The lander had touched down within an alluvial fan, a geographical feature which is associated with the flow of water. Water appears to be the top contender for the liquid flow, largely because of the presence of water found within local pebbles. However, in this case the estimated age of this geological feature extends back billions of years (Grossman 2012).

So, although the more dramatic Martian gullies are in high altitude regions, and therefore inaccessible to landers and rovers, Curiosity has had the good fortune to provide corroborating evidence for surface water flow in a part of the Martian landscape. Why haven't these dry river beds been buried by the accumulation of debris and dust, frequently swept around Mars within its famous dust storms? Erosion and burial do indeed play a part in the life-cycle of these anomalous Martian features. Many gullies located at low latitudes have been eroded and removed, while gullies at higher latitudes have been buried at some point in the past through the movement of the Martian regolith (Dickson et al. 2015). Indeed, the movement of ground materials in the absence of water might actually be forming the gullies.

A Shifting Landscape

How can rocks and pebbles get moved around without the presence of flowing liquid? Ice, particularly carbon dioxide permafrost, plays an important role in the composition of surface materials in the areas where these gullies are more common. Some of the movement of dry surface material that is filling in the gullies may result from periodic sublimation of frozen carbon dioxide. This sudden change of state from ground ice to atmospheric gas destabilizes the rocks, dirt and pebbles that make up the ground surface around these features, causing movement and in-filling.

There is an ongoing scientific debate about whether this mechanism is capable of actually creating the gullies themselves. On the one hand, compositional evidence of ground features seems to suggest water played no part in recent times (Núñez et al.

2016). On the other hand, the Martian gullies exhibit "sinuosity values" in excess of what is measured for dry channels on Earth (Mangold et al. 2010).

It seems possible that the erosion and burial of the gullies may have been periodically off-set by a re-emergence of liquid water once again cutting its way through the ancient, half-buried valleys. The word "periodically" is important here, because this geophysical process is driven by changes in the tilt of Mars. During times when the planet's tilt is most extreme, substantial quantities of ice can form at the gully locations (Dickson et al. 2015a).

The main driver for this is the changing position of the Martian poles over long time periods, as a result of the change in the tilt planet's spin axis, or obliquity. The Earth's tilt also varies over time, between 22° and 24.5°. This can have profound effects upon the planet's climate over time. But it could be a lot worse: We have the Moon to thank for providing a stabilizing influence.

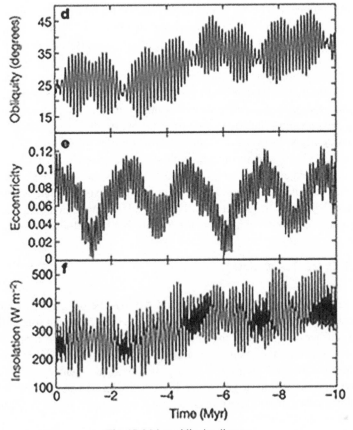

Fig. 17-2 Mars obliquity diagram

In the case of Mars, the variation in obliquity is far more extreme. Its axial tilt varies between about 15° and 35° over a 124,000 year cycle. The obliquity cycle is itself subject to chaotic variation (Touma & Wisdom 1993). At points in the past when the cycle was even more extreme, the obliquity of Mars may have swung chaotically between as much as 0° and 60° (Laskar & Robutel 1993)!

Martian Ice Ages

This extreme shift in the position of the Martian poles could have a massive effect upon the red planet's climate as the cycle progresses, and evolves. Ice caps of water and carbon dioxide warm as they shift towards the equator. The ices then sublimate into vapor. The carbon dioxide which is released into the atmosphere as the planet tilts onto its side not only adds to the thickness of the Martian atmosphere, but also provides a global warming greenhouse effect, providing a virtuous circle of atmospheric thickening and warming. The water released into the atmosphere increases humidity, and can accumulate as frost and ice across different zones across the planet. Through these changes in obliquity, Martian ice caps shift position with respect to the light from the Sun, and new layers of ice can form in completely different locations across the planetary surface.

Fig. 17-3 Mars ocean image

There is strong evidence for an ancient ocean on Mars, capable of covering about a fifth of the surface of the planet (Villanueva et al. 2015). Measurements of the composition of various leached metals across the surface of Mars are suggestive of a vast ancient ocean, marked by shorelines. Massive bodies of flood water once occupied the northern lowlands of the planet, lasting for a significant proportion of the lifetime of the planet (Stiles 2008).

Was this due to a phase in the Martian climate that came to an abrupt halt a few billion years ago? Or, could the oceanic conditions be a transient phenomenon, re-appearing at certain times when the fluctuating environmental conditions on Mars are just right? The very cyclical nature of this process insinuates that Mars may have undergone these changes many, many times over its lifetime. Like Earth, Mars has cycled through Ice Ages and interglacial periods.

Like so many bodies in the solar system, oceanic water may lie below the Martian surface. Could fluctuations in the planet's tilt have released liquid water across the planet's surface? This seems possible, but surface conditions would need to be very different from how they are now.

It is thought that Mars behaves differently from Earth during its Ice Ages. Ice Ages on Earth are marked by accumulations of ice at the poles. On Mars, it is the loss of polar ice which causes the spread of ice across the landscape. Mars may have experienced a very significant Ice Age period between 0.4 and 2.1 million years ago. When Mars' tilt changes, there is a subsequent sublimation of polar ice (where the ice evaporates directly into gas). This releases water vapor into the Martian atmosphere, and subsequently freezes out onto the surface of the 'new' polar zones, which have shifted to what are now the mid latitudes.

Scientists argue that this renewing ice is blended in with plenty of Martian dust. The result of this process is a several meters thick regolith-laden ice sheet. It extends from the thinning poles across the Martian landscape, down to ~30° latitude (Head et al. 2003). Like a glacier, this new icy mantle scours and shapes the land. As the planet's tilt alters again, the top layer of the ice sheet evaporates away.

This sublimation of surface ice leaves behind a blanketing layer of dust over the re-maining ice below. This dusty regolith serves to cover and protect a mantle of ice just below the Martian surface in many mid-latitude regions (Schon, Head & Milliken 2009). Over time, repeated climate cycles will multiply these accumulating effects.

If there is a weak point in this argument, it is how the dust works its way into the forming ice. Mars is famous for its dust storms, which fluctuate seasonally and regionally but are regular enough, and often spectacular. These storms self-evidently shift and subsequently dump dust onto the surface. The dumping of Martian dust is evidently non-negligible, as it can pose an annoyance to the solar panels on NASA's rovers (Webster 2012).

Accumulating dust might not be particularly significant over a few months or years, even if scientists NASA worry at night about the effect dust might have on Opportunity's solar panels. However, over extended periods of time, when Mars is undergoing a periodic shift in its axial tilt, the dust downfall could steadily mix in with the gradual formation of new surface ice.

Although this mechanism might explain the depositing of sub-surface ices onto mid-latitude regions, it does not explain the Martian gullies. These seem to indicate the movement of liquid water across the surface since the last Ice Age (Johnsson et al. 2014; Howell 2015). This opens the door to re-thinking what might be going on below the Martian surface, and the atmospheric conditions that might be prevalent during these extreme moments in the Martian climate. Might liquid water play a part after all? In which case, is there the potential for life 'below decks'?

Methanogenesis on Mars?

Since the Viking lander days, it has been widely assumed by NASA, and most planetary scientists, that Mars is sterile. The gullies in the southern highlands, the ancient oceans in the northern lowlands, and the discovery of water in rocks made by Curiosity, have all fueled speculation that Mars may once have hosted life. The present conditions on the surface of the red planet are not remotely hospitable, offering a major challenge to any astronauts hoping to colonize another world. Surely life is quite impossible in this cold, poisonous and heavily irradiated environment?

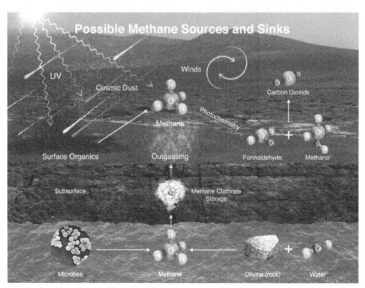

Fig. 17-4 Mars methane diagram

Surprisingly, perhaps, pockets of methane gas have been found on Mars. This has fueled speculation that there may be subterranean life lurking under the Red Planet's surface (Sample 2014). It has been known since 2009 that methane, a by-product of life, exists in the Martian atmosphere (Sample 2014a). Atmospheric methane is known to degrade over time due to the action of the Sun. The implication is that organic gas could not be a remnant left over from a wetter, livelier age. Instead, it must be part of an ongoing process. The origin of that process has been keenly debated ever since.

In late 2013/early 2014, the Curiosity rover noticed a significant spike in the level of methane in the air around the Gale Crater. This unexpected spike was about ten times greater than the usual background level of methane detected in the Martian atmosphere. Curiosity has also provided data about the seasonal variation of these methane pockets, which may reflect changes in ultraviolet light depleting atmospheric methane. The big spike of atmospheric methane picked up during that first autumnal season has not been seen again in subsequent seasons (Webster, Brown & Cantillo 2016). There is, as yet, no explanation for it.

There remains a lack of clarity about what the fluctuating presence of atmospheric methane actually represents. It may be derived from biological processes – like microbes underneath the Martian surface – or it could be due to more mundane geological processes. So, while methane may be suggestive of the presence of biological processes taking place below the Martian surface, it does not constitute conclusive proof.

Briny Water

Water in pure form would find it very difficult to exist as a liquid in the current Martian environment. However, if you contaminate water with impurities that lower its freezing point, then it could exist in liquid form at or near the surface – at least, under certain circumstances (the principle is the same as putting salt on frozen, icy roads).

Speculation about liquid water and life on Mars is supported by controversial evidence involving the movement of briny water just below the Martian surface. The movement of dark, salty water is thought to be the reason for the appearance of seasonal streaks across parts of the Martian landscape, studied by NASA's Mars Reconnaissance Orbiter (Brown et al. 2015). Following topographical analysis of the regions involved, other scientists have argued that the inferred sub-surface movement was instead caused by granular flows of sand and dust slipping downhill. However, even these more skeptical voices concede that it is still possible that briny liquid water played a part in getting these seasonal flows going (Dundas et al. 2017). Until landers and/or rovers can take ground samples for analysis, no one can be entirely sure whether briny sub-surface water is indeed behind the appearance of these seasonal streaks.

An interesting off-shoot of the issue of liquid water on Mars is the potential for humans to contaminate a habitable Martian environment during a manned mission. Ironically, scientific findings suggest that Mars has no hope of maintaining indigenous life provide a green light to a manned mission. Signs that there could be extant habitable environments on Mars create ethical dilemmas for future manned missions. After all, our presence could inadvertently destroy any indigenous Martian ecosystem. Conceivably, this consideration could create a conflict of interest for NASA scientists currently assessing the capacity for Mars to support life. How objective can they be if positive results risk jeopardizing a future manned program, with all the kudos and funding that comes with it?

The Curious Waters of Mars

Adding even more fire to the debate about water on Mars, the Curiosity rover seemed to discover briny liquid water below the red planet's surface (Devlin 2015). Up until very recently, the suggestion that liquid water could exist on the surface of Mars was dismissed as ridiculous by scientists: The climate was too cold and arid, and the atmospheric pressure too low to provide the right environmental conditions to allow liquid water to exist. Permafrost below the regolith was as good as it could get.

Yet, as we have seen, satellite surveys of the planet have provided evidence for the movement of water across the Martian terrain, and rounded pebbles have been found in dry river beds. This implies erosion by the flow of river water over long time periods. The Martian climate may fluctuate wildly over time due to the high variation in the obliquity Mars' axis, and orbital parameters. Although Mars might seem settled and even rather dull at the present time, this might just be a phase it has been going through, similar to Ice Epochs on Earth.

Liquid water can exist on the surface of Mars in the form of salty brines. These brines contain calcium perchlorate that can draw down atmospheric moisture and then percolate down through the regolith – the dusty covering of gritty topsoil on a planet's surface. The presence of caustic perchlorate salts in the Martian soil attracts frosty water from the atmosphere during the cold winter nights.

The presence of these hydrated salts also lowers the freezing point of water substantially. As a result, water can remain in a liquid state at temperatures well below its normal freezing point on Earth. Furthermore, perchlorate salts liberally cover the Martian surface, in regions much 'wetter' than the desiccated Gale Crater region. If this briny liquid water exists below this crater's surface, then it is likely to be found more abundantly elsewhere across the planet, especially where the atmospheric pressure is greater (Martin-Torres et al. 2015).

So much for the arid desert we were all promised. However, the high perchlorate salt content would preclude the presence of sustainable life under these circumstances. So, these findings, in themselves, to do not jeopardize any planned manned missions.

That is not to say that more habitable conditions aren't available deeper down below the soil, though.

Deuterium-to-Hydrogen Isotope Ratios

The NASA rovers are equipped with mass spectrometer and other bits of neat kit to allow them to do chemistry in situ. By taking samples and roasting them at various temperatures, then analyzing the resultant gases, scientists can work out the composition of the soil. Analysis of some Martian soil has also provided data about the red planet's Deuterium-to-Hydrogen ratio, which varies from planet to planet. This ratio will play an important role in the chapters to follow, so it is worth a short explanation.

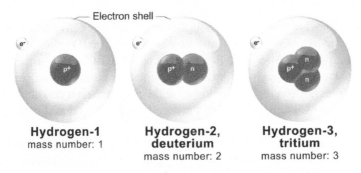

Fig. 17-5 Hydrogen deuterium diagram

A water molecule is composed of one oxygen and two hydrogen atoms. Each of these atoms can have a variable number of neutrons in its atomic nucleus. In the case of hydrogen, a normal hydrogen atom would have just one proton in the nucleus. Add another neutron to the atomic nucleus and you create a deuterium atom. Deuterium (D) behaves just like hydrogen (H), but is heavier: Deuterium is an "isotope" of hydrogen. Add a second neutron to the atomic nucleus, and you have the tritium isotope.

On Earth, the proportion of hydrogen to deuterium in water is over 6,400 to one. In a natural terrestrial water sample, there is one HDO molecule ("semi-heavy water") to ~ 3,200 regular molecules of H20. Finding a natural water molecule with two deuterium atoms is much rarer (one D2O molecule to ~41 million H2O molecules (~6400 squared). If D2O is separated from regular H2O, the heavier distillate is known as "heavy water". This has uses in the nuclear industry as well as the laboratory. If a chemist distills water, then the distillate is 'heavier' once the water with the lighter hydrogen atoms has been removed. The deuterium-to-hydrogen ratio increases accordingly.

It is a very curious fact that the water on Earth shares a similar ratio of deuterium-to-hydrogen isotopes as that of water on bodies in the outer asteroid belt. The early solar system had a lot of water in it, mixed in with all the gas, dust and rock in the protoplanetary disk. As the Sun sparked into life, the solar wind blew some of these lighter gases (the volatiles) away from the Sun. In a way, the early solar system distilled water. Regular, lighter H2O would be removed more easily than heavier water molecules containing deuterium.

As a result, the solar system acted like a natural "fractionating column", with deuterium progressively more concentrated closer to the Sun. Therefore, it is rather odd that the Earth's water has a similar deuterium-to-hydrogen (D/H) ratio to that of asteroids in the outer zones of the asteroid belt.

A popular theory about the origin of Earth's water is known as "the late veneer theory". The inner solar system was robbed of its primordial water by the action of the Sun. Early planetessimals lacked the protection afforded by well-developed magnetic fields around planets, so bodies in the inner solar system should all be dry husks. Yet, the Earth clearly has a lot of water. So, where did it come from?

The late veneer theory proposes that the inner solar system received water later on from impacting icy comets. The Earth has been particularly good at hanging on to these aqueous deposits. All well and good, except that it has become clear that the Earth does not share the same deuterium-to-hydrogen ratio as that of most comets. Instead, many comets have deuterium levels which are considerably higher than in terrestrial water. This brings into question the accepted comet scenario for the acquisition of terrestrial water.

Unfortunately, **nothing** about the deuterium-to-hydrogen ratio relationships between bodies in the solar system is straightforward. One complication is that the solar system has been enriched in deuterium, partly through processes in the protoplanetary disk, but also due to the arrival of deuterium-rich water carried by interstellar comets (Cleeves et al. 2014).

This raises serious questions about the format of the early solar system. Earth's D/H ratio could either provide supporting evidence that planet Earth has migrated across the inner solar system, or indicate that just about everything got mixed up at one point or another. We will explore this complex topic some more later on.

Martian D/H Ratio

Back on Mars, NASA's Curiosity rover made the discovery that soil samples are highly enriched in deuterium compared to natural samples on Earth. These levels are consistent with the deuterium-to-hydrogen isotopic ratio attributed to the Martian atmosphere, which was found to be about five times that of Earth (Leshin et al. 2013). It is clear from this that the deuterium levels on Mars are consistently greater than those on Earth, by a factor of five.

Why would the Earth have more in common with the asteroids in the outer asteroid belt than Mars? You could argue that if the asteroids, Mars and Earth all had relatively similar isotopic ratios of water, then that might be explainable as a generalized property of planets forming in the inner solar system. They are part of a fairly tightly-knit family after all, sharing a similar zone within the solar system. But, instead, there is significant variation… and not in the orderly manner one might expect.

There are many variables at work, listed here in no particular order:

- The initial distribution of waters in the early solar system, as the planets formed.

- The varied effect of the solar wind on those waters across the solar system, driving away the lighter volatiles.

- The protective effects of planetary gravitation and magnetic fields.

- The migratory movement of objects (and their accompanying waters) across the solar system during its lifetime.

- The influx of water from interstellar space.

Nonetheless, the D/H ratio in the inner solar system clearly seems out of sequence. You would expect Earth, being much closer to the Sun than Mars, to have water with a more concentrated mix of deuterium in it. But this assumes both planets still have remnants of their primordial waters. If the waters on Earth, Mars, and the asteroids are the same that they formed with, and they hang onto their water supplies equally well, then you would expect an orderly sequencing of deuterium-to-hydrogen ratios from the Sun outwards. Earth would have had the highest value, then Mars, then the asteroids.

However, a great deal depends upon how well planets and other bodies in the solar system can hold onto their water. The bigger the planet, and the more influential its protective magnetic field, the less atmospheric degradation it suffers over time. So, because of their immense size, the water in gas giants contains a lot less deuterium than the water in all of the other solar system bodies. Jupiter's low D/H level, at 1:38,000, is probably close to the primordial level in the Sun's early protoplanetary disk. Jupiter's gravity and powerful magnetic field has allowed it to hold onto its volatile gases, preventing the fractionating effect suffered by smaller, less-well-protected bodies in the solar system.

Much can happen over 4.5 billion years. Even if the planets started life with an orderly sequence of isotopic ratios with distance from the Sun, the conditions on each world varies, and, if local conditions allow, the stormy solar wind continues to rob these worlds of their volatile elements and molecules.

Without a substantial magnetic field, Mars is heavily exposed to the action of the solar wind in a way that the Earth isn't. This will drive off lighter forms of volatile molecules in the Martian atmosphere: Hydrogen in water vapor will be lost more than its heavier deuterium cousin. This will leave the red planet with a higher proportion of deuterium compared to our well-protected Earth… Assuming, that is, that a vast amount of the water on Mars has been driven off into space over the lifetime of the solar system.

Water on Mars

Despite its appearances, Mars is hardly a desiccated husk of a world. On top of the voluminous water ice bound up in the two ice caps, new studies of the literally thousands of glaciers known to exist across the Martian surface have shown that these icy aqueous reservoirs alone hold enough to cover the entire surface a meter deep in water (Karlsson et al. 2015). The fact that these non-polar rich deposits of water have not been lost over time is remarkable. As previously discussed, this may be due to the general covering of Martian dust, protecting the sub-surface ices from sublimating out into the thin atmosphere and becoming exposed to the aggressive action of the solar wind.

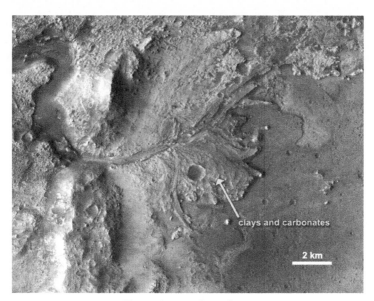

Fig. 17-6 Jezero Crater image

Moreover, there is increasing evidence of ancient lakes across the landscape – an example being the 4 billion year old lake system near Jezero Crater. There are features of this paleolake system which imply that there has been episodic movement of Martian water in the past (Goudge et al. 2015). The lake basin features fans and deltas where sediments have been transported across the landscape at some point in the past. These sediments contain clays and minerals that infer chemical alteration by water.

The Mars Reconnaissance Orbiter has conducted shallow radar (SHARAD) experiments to gauge variability across the non-polar regions of Mars, and has uncovered substantial deposits of water ice below the dusty surface (Stuurman et al. 2016; Campbell & Morgan 2018).These may have accumulated during periods of high axial tilt, and become permanent, buried icy reservoirs below the regolith during a Martian Ice Age (Schon, Head & Milliken 2009).

The scientific consensus on Martian water has changed dramatically in recent years as the evidence for hidden reservoirs has emerged. This is to be welcomed, but the picture remains puzzling.

On the one hand we have the need for the water reservoirs (in the form of glaciers and generalized sub-surface salty brines) to be protected from exposure and thus loss from the planet; and on the other hand we have clear evidence of episodic water flows across the landscape, and a chaotic climate model linked to extreme variations in axial obliquity and orbit. These concepts do not sit together well.

This fluctuating pattern of change is at odds with the general view that Mars is now a fossilized world whose features can only give clues to a glorious past long since gone. It may imply that weather conditions on Mars can vary significantly over time, and that there is plenty of water still on the planet to carve new features, and mix things up. Given all of these factors, it seems reasonable to argue that optimal conditions of obliquity and orbital eccentricity can kick-start Mars back into life – if only for a while.

This then presents ethical problems for the planners of manned missions to the red planet, worried about our capacity to contaminate and destroy. Mars may not be a dead world after all. Instead, it may simply be in a state of hibernation.

A Close Encounter

The presence of atmospheric methane is suggestive of some kind of interesting mechanism going on, which is either geological or biological, or conceivably both. Compared to the Earth, Mars suffers from a highly degraded atmosphere. The atmospheric pressure at the surface of Mars is less than one percent that of Earth's. If the gullies and other river channels evident on the Martian surface indicate that water once moved freely across the landscape, then it has to have been the case that in the past Mars had a much denser atmosphere.

One of the goals of the NASA's Maven orbiter has been to investigate why the Martian atmosphere has eroded so badly over time. It is thought likely that Mars lost its atmosphere to the solar wind because it lacks a protective magnetic field. If so, that process should still be going on today.

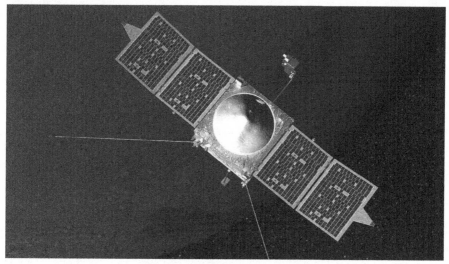
Fig. 17-7 Maven spacecraft image

During Maven's orbit, it dipped towards the Martian surface to collect air samples from different altitudes (Amos 2014).The results seemed to confirm the idea that much of the Martian atmosphere had been lost to space around 3.7 billion years ago, and was still subject to erosion even now due to a combination of solar wind and ultraviolet irradiation (Kramer 2017). Scientists working on the mission noted that the rate of erosion accelerated when strong solar storms hit the red planet (Brown et al. 2015a). This amounts to a drip-drip effect of atmospheric erosion.

Large quantities of the Martian atmosphere may have been lost in a more catastrophic manner. For instance, it may have been ripped away early on in the planet's history, along with the planet's magnetic field. A planetary magnetic field (essential to protect a planet from the harmful effects of the solar wind) is generated by a dynamo of circulating liquid metal within the core of the planet. It is thought that Mars once had such a dynamic core, but lost it sometime before the Late, Heavy Bombardment which began 3.9 billion years ago.

If the dynamo had still been churning deep within the planet during the bombardment, then the sizable impact craters from the time would have become re-magnetized (Phillips 2001). They haven't, so the magnetic field must have been lost early on, leaving the red planet open to atmospheric erosion for well over four billion years. This in turn led to a drying out of the Martian landscape.

What could have knocked out the Mars' dynamo in the first place? A hint comes from the red planet's close encounter with Comet Siding Spring, which was monitored by Maven early on in its mission. Mars was engulfed within the tail of the comet for several hours, and the planet's atmosphere did not fare well during this time. Its weak magnetosphere was distorted by the comet's proximity, and it lost atmospheric materials to space as a result of its encounter with cometary plasma. The overall effect was similar to that invoked by a significant solar storm (Espley et al. 2017).

If such an effect can be registered by the flyby of a simple comet, what might a passing planet do? Mars appears to have lost both its magnetic field and much of its atmosphere sometime after the formation of the planets. The process of Martian atmospheric depletion began about 4.2 billion years ago, and continued for several hundreds of millions of years. This is coincident with the beginning of the Late, Heavy Bombardment, which is thought to have begun about 3.9 billion years ago, or earlier. Indeed, a study of major impact craters across the Martian surface supports the notion that Mars was directly affected by this bombardment, and that it had a severe effect upon the early climate conditions of the planet (Bottke & Andrews-Hanna 2017).

If a massive planet had crossed Mars' path at that time, then it would simultaneously solve a number of problems. Its magnetic strength would be sufficiently strong at close distances to knock out Mars' dynamic core. The unprotected red planet would then be catastrophically stripped of its atmosphere by the action of the intruder's massive magnetosphere. Mars could also have seen its orbital path affected, throwing it into a more erratic path, and sending its tilt into a spin. In a single stroke, the red planet could have been fatally broadsided.

Although the present condition of Mars speaks of such an encounter, the issue is the lack of evidence of the assailant. This scenario is in keeping with Zecharia Sitchin's model for the early solar system, however. In his version of events, the intruder became a Planet X body in the depths of the outer solar system (Sitchin 1976). But, to have caused such an effect, this planet would have to be massive, with a very powerful magnetic field: Like a sub-brown dwarf.

As is the case with many of the other anomalies in the solar system discussed in this book, scientists are struggling to piece together the puzzle about what really happened during the early solar system. It is clear to me that major catastrophism played a role common to the Earth, asteroid belt and, as suggested here, Mars. As a result, the cards in the solar system's deck were reshuffled. The isotopic ratios of water across the solar system may provide us with an important clue to piecing the whole thing back together.

More significantly, though, is the realization that Mars lost a great deal of its atmosphere, and its water, around the time of the Late, Heavy Bombardment. Where one would expect the impactors prevalent during this time to have delivered water to the red planet, instead this event seems associated with Mars losing its magnetic field, its atmosphere, much of its water, and its natural orbital stability. A close encounter with a massive intruder planet during that period would tick a lot of these boxes.

Water World

In the last chapter, I looked at the water on Mars. I discussed the relative abundance of elemental isotopes, particularly hydrogen and deuterium in water, and why this can vary across the solar system. Potentially, this can tell us a great deal about how the solar system formed. With respect to Planet X, the changing isotopes of water across the solar system may provide evidence for the catastrophic appearance of an outside planetary body. However, these variations are muddied by a great many factors, and it is important to try to unpick them first.

As the Earth formed in the early solar system, the heating of the Sun on the primordial Earth should have driven off most of its volatile gases and liquids, including water. The theory is that terrestrial planets in the inner solar system should have become dry husks fairly early on. That seems self-evident when we look at the Moon, Mercury, Venus and Mars. Given that we on Earth are fortunate to have an abundance of oceanic water creates a dilemma. How is that possible? There are two competing theories to explain this:

THEORY 1: The Late Veneer Theory explains that water was deposited onto the Earth by impacting icy comets, as well as asteroids (Albarede et al. 2013). The cataclysmic period known as the "Late, Heavy Bombardment", beginning some 3.9 billion years ago, was characterized by a tremendous pummeling of the Earth and other planets. It remains unclear as to what triggered this event. It is thought that the silver lining of this rampantly destructive phase was the acquisition of our oceans. The bombardment is evident on the pock-marked, cratered surfaces of other rocky worlds. The Earth's geophysical changes in the intervening four billion years, driven by the movement of its tectonic plates, has refashioned the landscape and eradicated the craters. We are left with the deposited water. This is the theory favored by many astronomers.

THEORY 2: The Earth carries within it an immense internal repository of water obtained from the pre-solar nebula. This birthing water became a part of the Earth during the planet's formation, and somehow survived the stripping action of the Sun during the early phases of the solar system. Over time, the Earth's original waters have seeped up to the surface and formed the oceans. This is the theory favored by many geophysicists.

Fig. 18-1 Blue marble image

These theories need not be mutually exclusive. The Earth could have held on to an abundance of water from its formation, as well as getting top-ups from impacting comets and asteroids. How could anyone tell the difference? This is where the deuterium/hydrogen ratios come in. If our water was brought here exclusively by comets, then its D/H ratio should closely resemble that of the water on comets that has been analyzed by space probes, or that has been observed out-gassing from comets as they swing around the Sun.

Astronomers had expected to find that the water of the comets would closely resemble those on Earth. However, this has not proven to be the case for most comets. Of equal concern for advocates of the late veneer theory, comets do not appear to carry as much water as was once thought. Originally imagined to look like "dirty snow-balls", they instead appear to look a lot more like dust-covered asteroids.

As it turns out, the Earth's water more closely resembles that found on some main belt asteroids. On the face of it, these bodies should be much drier than their cometary cousins, and so do not appear to be good candidates from transporting water further into the inner solar system. An additional complication is that the orbits of these outer asteroids do not cross that of the Earth. How did this all get so mixed up?

The Epic of Creation

Let us turn our attention to Zecharia Sitchin's highly controversial ideas about the origin of the planet Earth. He essentially provided a third hypothesis about the origin of water on Earth. His rather unusual suggestion, based upon his reading and interpretation of ancient Mesopotamian texts, is that the Earth began its life where the asteroid belt currently resides. According to Sitchin, Earth was initially a larger, water-covered planet, equated in the myths with an aqueous monster named Tiamat. Tiamat was attacked by the Babylonian usurper god, Marduk (Sitchin 1976).

Fig. 18-2 Tiamat, Marduk image

Sitchin proposed that Marduk was actually a huge, fiery planet of unknown origin. In the Babylonian creation myth known as Enuma Elish, Tiamat was rendered asunder by Marduk's wrath (King 1902). Sitchin argued that the ensuing debris formed the asteroid belt and many of the comets. The remaining hulk of Tiamat was shunted further towards the Sun, taking residence between Venus and Mars. In other words, the remnants of this watery world, which had originally formed between Mars and Jupiter, now became the Earth/Moon system.

Marduk's orbit was also substantially altered by this encounter, and the planet took up a highly elliptical orbit around the Sun. Marduk (which was also referred to by the earlier Sumerian epithet "Nibiru" by Sitchin) became a Planet X body (Sitchin 1976).

The basis of this theory is highly contentious. The Babylonians were fairly decent astronomers, for their time, but they could not have worked this out scientifically. Sitchin advocates that the ancient Mesopotamians were provided with their advanced knowledge in astronomy (and other sciences) by visitors from another world. They then codified this knowledge into a form which they understood: Myth.

The vast majority of scientists reject this notion outright, and Sitchin's theories have been picked apart by skeptics. Nonetheless, this mythical "celestial battle" between Tiamat and Marduk provides us with a third hypothesis to consider, one where the Earth and the asteroids share a common origin. Could Sitchin have alighted upon a possible solution, either wittingly or unwittingly?

If his theory has any basis in truth, then one would expect that the Earth's water would be:

1. In place from the beginning.
2. Resemble water contained within the rest of the debris in the asteroid belt.
3. Have very little in common with comets from the outer Oort cloud.

Some shorter period comets may, like the asteroids, turn out to be remnants of this ancient collision. However, long period comets which reside in the outer comet clouds should have taken up residence before this catastrophic event took place, and so should be have a different isotopic composition.

Sitchin's proposal about the intervention of a rogue planet from interstellar space creates an alternative to the Nice model. Instead of planets and minor bodies in the solar system migrating around due to gravitational interactions between fledgling planets and the dispersing remnants of the pre-solar nebula, the chaos was triggered by the catastrophic arrival of an alien planet some time later.

Testing the Third Hypothesis

If Sitchin was right, water was abundant on the surface of the early Earth. Sitchin argued that the Earth was further away from the Sun than it is now, potentially beyond the young Sun's snow line. It was also larger, carrying a great deal more water. "Tiamat" was an enormous water-world, with a deep, global covering of water. Sitchin had invented the concept of a terrestrial super-Earth back in 1976!

Located in the current zone of the asteroid belt, it may have more closely resembled a massive snowball whose icy exterior formed a billiard ball-like crust over deep sub-surface oceans. This would be rather like the Galilean moons Europa, Callisto and Ganymede – only on a much grander scale. Its size, substantial magnetic field and distance from the Sun reduced its vulnerability to the action of the solar wind, and it was capable of retaining a decent atmosphere, in contrast to demagnetized Mars. Some of that abundant early water was lost during the "celestial battle" between Tiamat and Marduk, spilled out into the solar system to become asteroids, comets, Centaurs and KBOs.

Long period comets do not play a significant part in this contentious version of events. The mass collisions of objects which bombarded the inner solar system about 3.9 billion years ago would not have been comets, but instead the retinue of objects associated with the 'fiery' interstellar planet Marduk, as well as the subsequent scattering of debris across the solar system. The chaotic aftermath of this cosmic encounter also helps explain why water is turning up in all kinds of unexpected nooks and crannies in the inner solar system – like deep within lunar craters, for instance (more on this later).

You can appreciate how the deuterium-to-hydrogen ratios in solar system water could apply to this scenario. This line of reasoning has been of great interest to myself and fellow researcher Lee Covino for many years (Lloyd 2002), and was discussed in a chapter in **Dark Star** (Lloyd 2005). The relative D/H isotope ratios do not present us with absolute proof in support of any of these hypotheses, and caution is needed before jumping to conclusions too readily.

For example, isotopic ratio values for water can vary on Earth, too, depending upon the source. So, what is the true value? Oceanic water, atmospheric water vapor, subterranean water or water locked up in ancient minerals? Our planet's deep water has been found to be chemically different from oceanic water, which means that scientists may be using the wrong D/H isotope ratio for Earth (Meech 2015). This finding raises concerns that the loss of hydrogen into space over billions of years has unhelpfully raised the relative deuterium level in the Earth's surface waters, rendering oceanic water a poor marker for investigations into the origin of terrestrial water (Crockett 2015).

If this wasn't already complicated enough, a "whole-Earth water cycle" appears to be at work, moving waters around between the oceans and internal terrestrial reservoirs deep down in the bowels of the Earth (Schmandt et al. 2014).

The waters found deep within the Earth's mantle are different from the oceans in terms of their deuterium-to-hydrogen ratio (they are lower in levels of deuterium), so it is likely that these deep primordial waters were formed directly from the proto-planetary disk, along with the rest of the Earth, and remain untarnished by evaporation into space of lighter, atmospheric H2O. This means that the Earth's primordial waters, found deep below its surface, bear even less resemblance to that of comets than oceanic water.

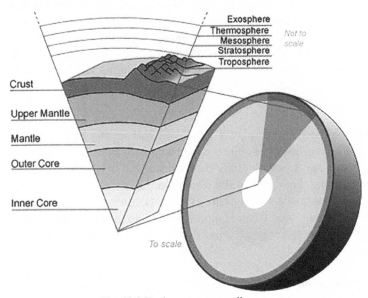

Fig. 18-3 Earth crust cutaway diagram

Given the issues faced by the Late Veneer Theory, astronomers are considering the possibility that water was probably abundant on the Earth early on. Yet, the prevalence of water in the interstellar medium and in protoplanetary disks is in stark contrast to the dry inner worlds of the solar system, which formed within the water-ice snow line. Scientists have started to argue that water present in the primordial soup of the protoplanetary solar nebula may have chemically adhered to dusty grains, preventing expulsion of these volatiles when the Sun sparked into life (Meech 2015).

The Life of Crystals

If Sitchin was right, and the Earth initially formed where the asteroid belt is currently located, then, being at the ice/water boundary of the protoplanetary disk, it could readily have formed with its own internal waters in situ. It did not need the deep waters to be delivered by asteroids, or to have been created by internal chemical reactions. It was simply here all along.

As previously noted, geophysicists also have more than a casual interest in the Earth's water. Their own perspective is generally closer-to-home, and the theories they generate about the origins of the Earth's water tend to be more Earth-centric, and less cosmic. An example of this was isotopic analysis carried out on carbon traces found in 4.1 billion year old zircon crystals. This work revealed that photosynthesis was probably taking place on Earth at the time when the crystals formed (Bell et al. 2015). How these crystals managed to form at this early stage in Earth's history is itself the subject of some debate. Plate tectonics may have been at work on Earth over 4 billion years ago (Hopkins, Harrison & Manning 2008). Scientific analysis of some even older zircon crystals – which were obtained from a sheep ranch in Australia – dates the Earth's crust to a time at least 4.4 billion years ago (Valley et al. 2014).

Geophysicists are becoming increasingly confident in asserting that the early Earth appeared a lot like it does now – rather than the hot, dry primordial lump of rock that astronomers have claimed for decades.

Isotopic measurements of water found in ancient lava magmas in Iceland suggest that the water has been here all along (Cesare 2015).The deuterium level in this pristine primordial water, untainted by later processes, is much lower than Earth's oceanic water. It suggests an origin within the inner solar system. The authors argue that the water came from water-laden dusts within the protoplanetary disk, which locally accreted into the Earth (Hallis et al. 2015). This solution seems to be the current remedy used to explain how water managed to remain within the early inner solar system hothouse.

Early Earth: A Watery World

The fact that primordial terrestrial water is so low in deuterium implies that the early Earth held water accreted directly from the inner protoplanetary disk. But how much water are we talking about? Was the early Earth a water-covered world saturated with the stuff? If so, this would be in keeping with Sitchin's description of the watery planet, Tiamat.

Fig. 18-4 Galilean moons image

Consider Jupiter's Galilean moons, Europa, Callisto and Ganymede. These perfect billiard balls appear to comprise a spherical ice exterior under which lurks oceans internally warmed by the gravitational influence of the nearby gas giant. If Jupiter was brought closer to the Sun, then these moons would develop their own atmospheres, and their global ice caps would melt way, leaving global seas. Might Earth have once have appeared the same?

The rock that was the early Earth was born from an aqueous, muddy birthing pool, and remained immersed within it as the planet coalesced and cooled. This is a hypothesis that was historically rejected because scientists thought that these waters should have been driven off by the early Sun's heat and the action of the solar wind (as well as the violent collisions commonplace in the early solar system), leaving the early Earth a dry, barren place. The presence of water on the Earth has effectively been set back to a point 135 million years earlier than previously thought.

Vesta and Earth

The Earth's capacity to hold onto water has been strengthened by other research. Important evidence for a common primordial source of water within the inner solar system comes from isotopic examination of carbonaceous chondrites in "eucrite" meteorites, which originated from the asteroid 4-Vesta. At over 500km in diameter, Vesta is the second largest asteroid in the solar system, residing in the asteroid belt between Mars and Jupiter. The aqueous isotopic ratio from the asteroid is very similar to that of water on Earth. This has been taken to imply that water was accreted to Earth directly from the protoplanetary disk (Sarafian et al. 2014). Alternatively, water was delivered to the Earth/Moon system by asteroids later on (Albarede et al. 2013). Or, perhaps, both mechanisms were at work.

Within the context of the third hypothesis, I would suggest this finding is also in keeping with the notion that the Earth once inhabited the region of space currently delineated by the asteroid belt; at the very edge of the protoplanetary disk's initial ice-water snow line.

Could Vesta and the Earth have once been part of a larger super-Earth? If this was the case, they should share compositional similarities. Unlike most other minor bodies in the solar system, Vesta is a complex, differentiated body with a basalt rock mantle of cooled lava covering an iron and nickel-rich core. Its layered structure implies that it was once a planetesimal, and so has more in common with terrestrial planets than its fellow asteroids. Unlike other asteroids, it has a dark and a light patches on its surface, like on the Moon. Its crust is strewn with a mixture of additional materials. This rocky mélange contains the remains of a diverse array of impactors that have smashed into this sizable asteroid over billions of years, including one that formed an impact basin, named Rhea Silvia, which covers much of Vesta's southern polar region. Some of these impacts have exposed lower layers of crust and mantle. Vesta provides a good example of "impact gardening", where the object's indigenous surface features have been gradually turned over, and are mixed in with a top-soil of deposited materials (McCord et al. 2012).

Fig. 18-5 Vesta image

So, were the origins of Earth and Vesta actually directly related? Were they both surviving fragments of a cosmic collision which took place in the asteroid belt location very early on in the history of the solar system?

Earth seems to share much in common with Vesta. Despite the distance between them, Vesta and Earth share a common source of water. However, Vesta is very ancient: It formed 4.6 billion years ago, pre-dating the LHB that took place some 500-700 million years later. So, rather like the Moon, it would already have been around when the "celestial battle" took place. Vesta seems to be a planetesimal whose growth was somehow interrupted early on in the planet-forming process (perhaps due to the influence of Jupiter, or due to the collision which created the extensive Rhea Silvia crater).

If Sitchin was on to something, then there was once a watery super-Earth between Mars and Jupiter whose presence changes the entire dynamic of this part of the early solar system. Its fragmentation, he suggested, led to the migration of the largest remaining chunk inwards, along with the Moon (originally "Kingu"). The asteroid belt constitutes the remnant debris field.

One of the most surprising discoveries made during the Dawn spacecraft's exploration of Vesta in 2011 was the presence of gullies and fan-shaped deposits within some of its craters. As with Mars, this suggested the movement of liquid water, or at least the shift of ground materials initiated by the presence of sub-surface water (Landau 2015). Ice is clearly present below the surface of this strange asteroid, and is capable of causing changes to the asteroid's landscape. Arguably, the unexpected presence of this water hints at a different origin for Vesta than previously considered. This also fits with the realization that Vesta's mantle contains the water-hugging mineral "olivine" (Lunning et al. 2015), much like the Earth and Moon. I would suggest that, like the Moon, Vesta may have been a chip off an ancient watery block.

Water-World "Super Earths"

The Sun's 'normal' family of planets is turning out to be anything but normal. The abundance of "hot Jupiters" and "super-Earths" in other planetary systems do not reflect the patterning of planets in our own solar system. Many exoplanets exhibit wildly eccentric orbital paths, not at all resembling the regularity and order seemingly prevalent in the solar system. There are plenty of hot-Jupiters whizzing around their parent star's ankles, but none here. Weird is the new normal as far as planetary science is concerned.

Most other planetary systems favor either collections of large planets, or collections of small planets. Our system has a mix of both. Super-Earths are the most common type of exoplanet elsewhere, but none is to be seen orbiting the Sun. Why?

"Super-Earths" are planets whose mass lies somewhere between that of Earth and Uranus. Their common detection may be a function of their greater mass: Selection bias may skew the statistics because of the greater likelihood of finding these larger planets compared with Mars or Earth-sized worlds. Super-Earths might be terrestrial, like Earth, or they might be predominantly gaseous, like Uranus.

Although gravitational conditions on these worlds are likely to be greater than those on Earth, oceanic life would still be entirely possible on suitably endowed aqueous worlds, as the buoyancy of the waters would counteract the higher gravity. The exoplanet Kepler-62e is a good example of a watery super-Earth.

However uncomfortable this might be for his many skeptics, Zecharia Sitchin may have been onto something with his description of the watery monster planet Tiamat (Sitchin 1976). Sitchin proposed the presence of a watery super-Earth located between Mars and Jupiter during the early stages of the solar system. This was ahead of its time, as the concept of a super-Earth did not feature at all in the standard models of planet formation. Where modern astronomers wonder why the solar system doesn't have its own super-Earth, the late Zecharia Sitchin's writings echo down to us: "It did!"

Water Within

The Earth may have vast swathes of deep water located far below the surface of the planet, some 400 miles beneath the crust. This area is known as the "transition zone" of the Earth's mantle. Estimates of how vast these reserves of deep water might be are truly staggering. It is impossible to say with any certainty how much of the rocky mantle might actually be composed of water, but there is the potential for there to be significantly more water below decks than in the entirety of the world's oceans (Davey 2014). Examination of seismic data from the United States has provided evidence of a transition zone water reserve of around three times the volume of the world's oceans (Schmandt et al. 2014). Further studies conducted on other continents may reveal more subterranean reserves elsewhere.

The transition zone is now known to include a mineral named "ringwoodite", a form of olivine. This mineral acts like a sponge, soaking up and trapping water. As noted earlier, water trapped in the Earth's mantle in the transition zone may be part of a "whole-Earth water cycle", allowing movement of water to the Earth's surface via a process that is driven by geological activity (Schmandt et al. 2014). This has been deduced following the examination of the interiors of diamonds expelled by volcanoes. Hydrated ringwoodite is included within these diamonds, suggestive of a highly pressurized aqueous environment deep beneath the Earth (Pearson et al. 2014).

It is a sobering thought that if all of this subterranean water were to find its way to the surface of our world, all but the tops of mountains would be covered in water (Davey 2014). his brings us back to Sitchin's notion of Tiamat. Could this deep reservoir of water be just a fraction of what once covered this planet? After all, the Earth will have lost a great deal of water to the solar wind over time. It is highly likely that the Moon formed as a consequence of a collision with a Mars-sized object, and such an impact would have displaced a very significant amount of primordial water. Perhaps Vesta, too, was the off-shoot of a similar, very early encounter between planetessimals, one of which went on to become the early Earth.

Migrating Earth

The Earth is currently thought to be ~4.54 billion years old. The Earth appears to have cooled dramatically early on in its life, forming an early crust (Devitt 2014). This early cooling may have allowed water to form on the Earth's surface, leading to the formation of simple life over four billion years ago.

The Earth was struck very early on by a Mars-sized planet, inflicting a massive wound and resulting in the formation of the Moon. That occurred ~4.53 billion years ago, before the date when the terrestrial zircon crystals formed. The lunar crust had formed by 4.51 billion years ago, as analysis of zircon crystals in lunar rock has shown (Barboni et al. 2017). So, that earliest period of 140 million years in the

Earth's life included the age of the molten magma oceans, a colossal planetary impact which formed the Moon, and a dramatic downward lurch in the Earth's temperature, allowing its own crust to form. This rapid cooling is despite the energy gained during the Moon-forming collision.

I suggest that this early cooling of the Earth-Moon system challenges our assumptions about Earth's place in the early solar system. The Earth's current location in the inner solar system is akin to being sat near the living room fire. As we have seen, the solar wind drives off volatiles, including water vapor, meaning that the early Earth should have struggled to hold onto its water – especially during a time when the planet was still raging hot, and subject to planetary collisions. Up until recently, the conceptualization about the early Earth was that of a hot, dry hulk. However, emerging geophysical evidence indicates that it retained its water and cooled rapidly. This third hypothesis suggests that the early Earth formed three times as far from the Sun as it is now, at the edge of the protoplanetary snow line. Based upon ideas that have emerged from Zecharia Sitchin's contentious theory (1976), I first put this forward in scientific terms in my earlier book (Lloyd 2005). A similar scenario has also been suggested by Alberto Saal, of Brown University (Grossman 2013).

This positioning within the solar system enabled the Earth to hold onto its volatiles, and to cool rapidly. The planet then migrated inwards following a catastrophic encounter with another planet at the beginning of the Late, Heavy Bombardment. As a result, the Earth arrived in its current location as an already cooled planet covered with water.

This combination of initial formation at or beyond the solar system's snow line, followed by inward migration, allows for a life-bearing planet with full oceans and protective atmosphere relatively close to the Sun. This is a simple solution which potentially explains how Earth managed to tick so many habitability boxes. It is also in keeping with the similarity of the deuterium-to-hydrogen isotope ratios of the Earth and the asteroids.

However, the concept of a large planet being able to form in the vicinity of Jupiter has been challenged. Offering a skeptical voice, Nick Moskovitz, a planetary scientist at M.I.T., has argued that there is no evidence that a planet larger than Mars ever existed in the zone currently occupied by the asteroid belt. He contends that the presence of Jupiter in this region makes the probability of the formation of a terrestrial planet not just low, but virtually impossible: The gravitational field of Jupiter is so vast that it should disrupt the accretion of a planet in this zone, leaving only dwarf planets and asteroids (Dorminey 2013).

I am generally cautious when scientists offer such an unambiguous line. The history of science is awash with examples of such bold statements that are quietly dropped years later when new evidence emerges. One only needs to review the remarkable variety of extra-solar planets discovered in the last decade or so to see how strong

theoretical positions previously held by astronomers led to a degree of uncomfortable post facto revision. There is already some evidence for planet-forming processes in the vicinity of the asteroid belt.

Asteroids and Catastrophism

Many of the meteorites that fall to Earth represent leftover building blocks of the
solar system, including many of the comets and asteroids. Unlike Vesta, most rocky
asteroids (and therefore meteorites) are generally composed of materials named
chondrites. Asteroids represent the primordial building blocks of the solar system,
and formed some 4.55 billion years ago. Current theory suggests that these building
blocks did not have the opportunity to form into larger, planetary masses, which
could then have undergone internal processes common to accreting planets – like
melting and planetary differentiation. This thinking is in line with arguments about
Jupiter's over-riding gravitational influence on the asteroid belt. Theoretically, Jupiter's presence at 5AU rules out the formation of sizable planets at 3AU.

However, there is evidence to suggest that some asteroids have taken part in internal
planet-forming processes. Some meteorites that have come to us from the asteroid
belt appear to have been the by-products of planetary formation. They often contain
within them tiny spherules known as "chondrules". The relative abundance of chondrules implies that catastrophic processes were common in the early solar system.

These minerals had previously experienced a molten phase, implying that they probably formed within impact plumes. Rather than simply being chips knocked off
leftover building materials in the solar system, the presence of chondrules implies
planet-forming processes followed by cataclysmic events (Johnson et al. 2015). The
age of formation of chondrules goes right back to the beginning of the solar system,
when planets were accreting out of the protoplanetary disk (Connelly et al. 2012).

Chondrules seem to have required relatively high temperatures to form, and many
mechanisms have been offered to account for their abundance. The favored view is
that they formed during violent impacts of planetessimals that were in the process of
being formed. There were likely large numbers of these protoplanets milling around
in the chaos of the early solar system.

So, rather than meteorites emerging from a dysfunctional building site where nothing
really got going, it seems likely that planet-forming in the vicinity of the asteroid belt
was well advanced, despite the presence of Jupiter.

Water in Asteroids

Water ice has been found on and in main belt asteroids, potentially in significant
quantities. Internal water ice may make up a quarter of the mass of the asteroid belt's

largest resident, the dwarf planet Ceres (Thomas et al. 2005). Although it is difficult to be sure about how much water is contained within asteroids, astronomers have found a stunning precedent for a watery asteroid elsewhere – one that has literally "spilled the beans".

This ex-asteroid is in the white dwarf system GD 61, located some 170 light years away. Its discovery results from its catastrophic tidal destruction at the hands of the nearby star. Its remains now enfold the white dwarf in a sea of rubble. Clearly, such a tiny object could not have been directly imaged. Instead, by examining the chemical constituents which provide the distinct spectrum of the light emerging from the star, astronomers realized that a large asteroid, whose rubble is strewn about the GD61 system, had deposited a massive amount of water onto the surface of the white dwarf. Working backwards, the astronomers concluded that the composition of the disintegrated asteroid, which had originally been the size of a dwarf planet, had been about 26% water (Farihi, Gänsicke & Koester 2013).

Fig. 18-6 Ceres layers diagram

This composition is similar to what is projected for the interior of the dwarf planet Ceres in our own system. Ceres may have once have had a global ocean according to scientists working with data about the minor planet, which had been sent back by the Dawn spacecraft (Fu et al. 2017). There may still be some sub-surface ocean remnants deep down within this asteroid.

A collision with a water-laden asteroid about the size of Ceres might, on its own, have provided sufficient water to account for the Earth's oceans. One could speculate, then, that the Moon may have formed following a collision between a Mars-sized, volatile-rich planet, and the Earth. In other words, much of the water on the

Earth and Moon originated from a colliding planet which had, itself, formed beyond the solar system's water-ice snow line.

This is the new mantra in astronomy: The Earth's water was delivered not by comets, but by water-rich asteroids.

The Moon's Terrestrial Water

Given the relationship between the waters of Earth and the asteroids, does lunar water also exhibit similar isotope ratios? Although the Moon appears to be a desiccated husk of a world, it does contain reservoirs of ice. Measurements of the deuterium-to-hydrogen isotope ratios of water found in lunar rock samples returned by the Apollo missions confirmed a similarity with terrestrial samples (Saal et al. 2013). This implies that the water in lunar rocks shares a similar origin to that of the interior of the Earth, and the asteroids.

If the Moon did indeed form from debris thrown out into space around the Earth after the impact of another planet, then the water found in lunar rock samples is likely terrestrial water. The Moon's water would have been derived from the Earth's mantle, if the cosmic impact had been sufficiently deep. If the Earth was located at 3AU at that time, then conceivably the asteroid belt represents the debris field of this early Moon-forming collision. At a later stage, the Earth/Moon system migrated inwards to 1AU, possibly as a result of an encounter with another, more massive object during the Late, Heavy Bombardment.

In the next chapter, we will examine new findings about water on the Moon in greater detail in an attempt to unpick this mystery.

Moon Mysteries

The generally accepted theory of the Moon's origin has been around for a while. The "giant impact theory" involves the early Earth colliding with a Mars-sized body within 100 million years of the birth of the solar system. The ejected debris from this colossal impact was not sufficiently energetic to escape the Earth's gravitational field, and instead became a ring or field of debris orbiting the Earth. This eventually accreted into the Moon. The theory is facing some difficult questions, and significant issues remain unresolved. We will be exploring some of these in this chapter, and considering a radical new solution.

The Mysterious Origin of Lunar Water

Until recently, the surface of the Moon was thought to be entirely devoid of water, very much in keeping with its long-standing image as a featureless, grey desert. However, a shock finding in 2009 rocked that consensus. India's Chandrayaan-1 lunar probe, orbiting 100km above the surface of the Moon, conducted chemical, mineralogical and photo-geologic mapping of the Moon's surface. It was looking particularly for water ice at or near the lunar poles. The Chandrayaan-1 mission, operated by the Indian Space Research Organization, dropped an impactor onto the lunar surface near to the Moon's southern pole. This impact ejected dusty debris, which was then analyzed for water ice.

The mission was a huge success for India's space program. It revealed that many of the minerals on the lunar surface are infused with water (Pieters et al. 2009). At the time, this begged the question as to why the old superpowers had not included instrumentation capable of detecting water on the lunar surface during their extensive explorations.

It turned out they had; inadvertently, anyway. Some NASA probes which had passed the Moon en route to other destinations had tested out their instruments by surveying the lunar surface. It seems that no one checked the data for lunar water at the time, assuming none would be there. In the light of India's shock findings in 2009, NASA re-examined that stored data and quickly confirmed the presence of water ice on the lunar surface.

Fig. 19-1 LCROSS separation image

NASA also launched the Lunar CRater Observation and Sensing Satellite, or LCROSS, which found water ice within the permanently shadowed region of Cabeus crater, near the Moon's southern pole. Again, they achieved this by allowing the mission's upper stage rocket to fall onto the lunar surface, ejecting surface materials for analysis. Following these confirmations, the remarkable discovery of lunar water was quickly accepted by the scientific community.

Hydrated Lunar Soil

Where had the water come from? Perhaps water ice had been deposited onto the lunar surface by comets or asteroids over the lifetime of the Moon itself. Or, perhaps, these were leftover aqueous remnants from earlier times. Either way, against the odds, the water had survived the Sun's bombardment of high energy radiation. That made sense when considering water ice located in permanent shadow in deep, polar craters. It was more puzzling for water ice mixed into the general lunar soil.

In contrast to the longstanding deposits of water ice found in deep, polar craters, the water discovered within the lunar soil in 2009 seems to have a more dynamic characteristic. It appeared to be coming and going gently with the lunar day. This implied that some kind of chemical process may be at work on the lunar surface.

As odd as it may seem, water found on or just below the lunar surface might have originated from the solar wind (Stephant & Robert 2014). It was thought that the solar wind carries charged protons which interact with oxygen rich molecules on the lunar surface, synthesizing water (shades of the Eley–Rideal reaction mechanism described in Chapter 11).

The initial thinking was that water then 'moves' about through sublimation, and perhaps even water flow, as the lunar day/night cycle heats and cools the surface. As a result, synthesized water could move around the surface of the Moon, and end up pooling in deep, dark craters near the poles. But further investigation since then has shown that water is present widely across the Moon's surface, irrespective of location, and even during the heat of the lunar day. Because this surface water stays put, it cannot be the origin of deep, polar crater ice (Bandfield et al. 2018).

Its origin may lie in chemical processes driven by the delivery of charged protons to ancient materials lying on the lunar surface, or perhaps this surface water has emerged from the Moon's interior. Or, maybe it is a bit of both. Perhaps these considerations play a part in the seasonal appearance of water on the surface of Mars, too – a possible consequence of its atmosphere being routinely "sputtered" by the action of the solar wind.

Orbital Issues

Before delving further into these thorny questions of water on the Moon (highlighted by plucky geochemists), it is a good idea to look at other issues it flings up for astronomers to consider. There are a number of outstanding anomalies, including the lunar orbit's inclination from the Earth's equatorial plane, the Moon's axial tilt (obliquity), and changes in its magnetic field history.

The first of these issues has come out of more recent ideas about how planets form: The Moon's orbit around the Earth should be more or less aligned with the Earth's equatorial plane. Instead, the lunar orbital tilt is set at about five degrees from that plane, which is an order of magnitude greater than theory allows. This is known as the "lunar inclination problem".

This anomaly may bring into question the long-held theory of how the Moon formed. There are other ways to explain this problem, though. Some astrophysicists have suggested that the Earth once might have had a second smaller moon – now lost –

whose influence was enough to disrupt the Moon's orbital inclination (Ćuk & Stewart 2011).

Alternatively, perhaps there were other gravitational influences at work that caused the Moon's orbit to become more inclined. Computer simulations have been conducted by astrophysicists, based in the south of France, to determine whether collision-free encounters might be capable of causing the Moon to shift its ground. It turned out that you don't need an actual collision with a planetesimal. Instead, the close approach of a series of planetessimals would be capable of nudging the Moon into its inclined orbit (Pahlevan & Morbidelli 2015). Their combined mass could be as low as one percent of that of the Earth.

So what is thought to have happened to these fly-by-night perpetrators? They may have eventually collided with the Earth/Moon system, suggest the astrophysicists from Nice, and so disappeared into the still-forming planetary bodies during a late point in the accreting process. It seems to me that this boils down to some tidy house-keeping. The easiest and neatest arrangement to explain their absence is for these bodies to have been absorbed by the Earth-Moon system once their perturbing work was done.

Fig. 19-2 Moon craters image

There are other possibilities, of course. As we have discussed, ~3.9 billion years ago, the inner solar system was subject to a massive bombardment of planetessimals. The consequences may have extended beyond lunar impact craters, to have actually adjusted the Moon's orbital inclination around the Earth. However, one might imagine that interactions with bombarding objects would have been largely statistically random, such that any influence on the orbital path of the Moon would have been smoothed out over time. In other words, the passing showers of comets, if that is what they were – were hit-and-run affairs passing the Earth-Moon system on all sides and from all directions. Or were they?

A Colossal Debris Field

Another possibility is that the perturbing influence was not the repeated passing of a mass of small bodies, but instead a catastrophic passage of something much greater in mass – an event of great significance ~3.9 billion years ago which, in one fell swoop, knocked the Moon's orbit out of line with the Earth's. In other words, the self-same event that caused the Earth to migrate inwards, closer towards the Sun, was responsible for nudging the Moon's orbital path out of kilter with the Earth's equatorial plane.

It is increasingly accepted that the apocalyptic Late, Heavy Bombardment (LHB) began about 4.1 – 3.8 billion years ago, following a period of quiescence. It took place long after the solar system had stabilized into an orderly state, and occurred over a period lasting at least 100 million years. Some evidence suggests that the bombardment lasted as long as a billion years (Lowe, Byerly & Kyte 2014). The impacts were each cataclysmic in nature, involving multiple objects of the order of 50km diameter each (Sleep & Lowe 2014). The layers of impact spherules left across the globe at that time dwarfed that left by the impact that killed off the dinosaurs (Marshall 2014).

Two words spring to my mind to explain such a long and sustained bombardment: "Debris Field". It seems likely to me that the Earth routinely passed through a colossal debris field over a protracted period of time. Consider: When we get meteor showers, the meteors are a collection of minor debris from a destroyed comet arranged in a massive cloud that the Earth moves back through annually. Extrapolate that notion upwards significantly. If the Earth was periodically passing through a massive debris field of asteroids, some of which were 70km in diameter, then I would suggest that there are two possibilities to consider:

1. The disrupted Earth/Moon system was moving through the main asteroid belt field, causing immense disruption, for nigh on 1 billion years. This implies that the Earth's orbit was very different at that time. It may have been highly eccentric during its inward migration, before settling down into its current orbit following these self-same encounters.

2. The Earth may have been moving through a cloud of debris left over from one almighty collision that occurred around 4 billion years ago. It involved, presumably, a destroyed rocky planet. If so, then the field of debris has long since completely dissipated – cleared out of the planetary zone. (Recall the definition of planets compared to dwarf planets). For Earth/Moon system to be moving through that field of debris over a prolonged period of time implies that the Earth was somehow involved in that initial collision: It was essentially re-visiting the scene of the crime every year, enduring further pummeling until the field of debris had dispersed.

If the Earth had started its life in the area currently occupied by the asteroid belt and had been nudged into a new, eccentric orbit following a catastrophic encounter with a massive planet ~4 billion years ago, then both of the above notions may be correct. The Earth's new, erratic orbit might have caused it to move periodically through an extended asteroid belt for up to a billion years. That debris field would have been strewn across the entirety of the inner solar system, affecting all of the inner planets.

The resulting cosmic encounters could have pummeled the Earth and Moon repeatedly during that time, nudged the Moon's orbit into its eccentric inclination, and eventually jolted the Earth's dynamically unstable orbit into something more satisfactory. In other words, following a period of turmoil in the inner solar system resulting from this encounter (from which Mars would not have been exempt), things eventually settled down into stable, segregated patterning. Mars and Earth would clear their 'zones' over time, leaving just a residue of the collisional debris field between Mars and Jupiter.

In this way, then, the catastrophic arrival of a rogue planet ~4 billion years ago set in train a series of events over the course of hundreds of millions of years, displacing the Earth/Moon system, misaligning the Moon with the Earth, thrashing Mars, and cratering the surfaces of the inner planets.

There are many other hypotheses about what caused the LHB. Planetary migration of the gas giants is a current favorite. This is part of the hypothesis known as the Nice model. Their collective migration caused Neptune to graze the Kuiper belt, sending swarms of asteroids and comets into the inner solar system.

Instead, I suspect that an outside entity was to blame, crashing through the solar system, and leaving behind a debris field of world-shattering proportions, of which the asteroid belt is the surviving remnant. My bet is that this object was a massive planetary body which encountered a young, watery Earth. The cataclysms that resulted pockmarked the inner planets (normally protected by Jupiter) and sent the disruptive usurper deep into the outer solar system. It remains there as a mysterious force, pulling seemingly invisible strings.

Unraveling the Mystery of the Lunar Craters

The Moon's surface provides the most startling visual record of the LHB, as well as the lunar rock samples which have enabled planetary scientists to date the catastrophism to this period. On Earth, tectonic movements and geological activity during the intervening three to four billion years have disrupted the cratering which was also doubtlessly visited upon the Earth at the same time.

It turns out that the cratering patterns on the Moon have changed somewhat over time. The bombardment cracked the lunar surface and made the lunar crust much more porous. This has made the history of the Moon more difficult to determine than one might have expected (Soderblom et al. 2015).

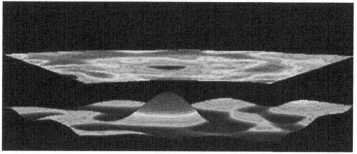
Fig. 19-3 Mascon image

NASA's GRAIL (Gravity Recovery and Interior Laboratory) mission has revealed much about the materials buried beneath the surface. This mission used twin spacecraft to map gravitational anomalies within the Moon. It picked up concentrations of mass below the planetary surface which scientists call "mascons". These were caused by ancient impacts. A pattern has emerged regarding the nature of these lunar mascons, suggesting that impact craters below the lunar surface were filled in by lava or iron-rich mantle seeping upwards from the Moon's deep interior (NASA/JPL 2013).

So, even though the lunar surface that we can see appears considerably trashed by impacts, still more craters lie hidden beneath the lunar surface. Some of them are truly colossal. Studies of these sub-surface craters may provide a clearer dataset of collisions during this period, and perhaps illuminate the nature of the LHB itself.

Tipping the Moon

The Moon always points the same face towards us. This is due to its tidally-locked orbit around the Earth. It is therefore counter-intuitive to think that it may have rolled over onto its side at some point in the past. Yet, some evidence suggests that the

Moon was indeed subject to a catastrophic event of such proportions that it was tipped over from its original axis.

The Moon is thought to be a 'chip off the old block' following an almighty collision. It formed from the orbiting debris field around the wounded parent Earth. From about 4.2 billion years ago, the Moon had its own magnetic field – one that it subsequently largely lost ~3.5 billion years ago.

Analysis of magnetometer data sent back from Lunar Prospector and Kaguya lunar orbiters has shown clusters of magnetic fields in two distinct areas of the Moon. The first was at the poles, where one might imagine it to be. However, the second cluster is significantly off-set from the polar axis. This implies that at some point the Moon's magnetic field suddenly jumped positions. There appears to have been a reorientation of the Moon with respect to its axis of rotation, by between 45° and 60° (Takahashi et al., 2014). This may have been due to a particularly hefty collision, or perhaps as a result of some of the external forces already explored in this chapter.

The reader will note that the timing of the loss of the Moon's internal core dynamo, at ~3.5 billion years ago, is well within the long period LHB starting ~4 billion years ago, and running on to ~3 billion years ago (Marshall 2014). So, the polar wander event may have occurred towards the beginning of the LHB ~4 billion years ago. This catastrophe involved the Earth/Moon system directly, causing it to migrate inwards from its prior position in the asteroid belt into an eccentric, dynamically unstable orbit.

A later encounter with one or more substantial asteroid bodies during its periodic sweep through the debris fields in the inner solar system (as per Pahlevan and Morbidelli's simulations (2015)), caused the Earth/Moon system to be nudged into a more stable location just one astronomical unit from the Sun. This fly-by somehow also demagnetized the Moon, an event dated to ~3.5 billion years ago.

Off-Axis Moon Causes Scientific Headache

There is further evidence that the Moon has experienced tilt issues in the past, and this time it is related to the ice found in lunar craters near to the lunar poles. A careful examination of the distribution of water ice at the lunar poles has shown that the Moon experienced axial adjustment at some point in the past.

It came as something of a surprise when it was realized that lunar ice showed a precise, off-axis relationship between the lunar poles. Collections of water ice near to the north and south poles are off-set from the Moon's polar axis to precisely the same degree: 5.5°. They are "antipodal" to one another. This implies that the position of these collections of water ice were once the positions of the Moon's poles. The implication suggests that the Moon's poles have wandered. Once again, the event which caused this shift has been dated to ~3.5 billion years ago (Hand 2015).

The reason the scientists initially missed this, of course, is that they didn't expect to see it. Just like NASA missed water on the Moon for decades, despite the evidence under their noses. India's ISRO Chandrayaan-1 lunar orbiter first detected water on the Moon in 2009.

The Moon is tidally locked in its orbit around the Earth, such that it rotates on an axis perpendicular to its orbital plane around us. A shift in its axis would presumably create a very different dynamic relationship between the Moon and Earth. However, in terms of survival of ice on the lunar surface, it is the Moon's axial relationship with the plane of the ecliptic that really counts. This is because it is the action of the Sun's radiation which will sublimate off exposed volatile water ice. As a result, most water ice will be removed by the Sun's bombardment of radiation, leaving only those small deposits left in the permanent shadows of deep craters located near to the lunar poles.

This leads us to the main problem: If the off-center ice formed early on, and the axis shift occurred, say, billions of years ago, then how have the off-center deposits of ice survived, even if they are holed up in deep lunar craters? As the Moon spins on its axis, these craters should be more readily exposed to the action of the Sun's glaring heat and the relentless solar wind than ice in purely polar craters. Five degrees may be insufficient tilt to make a fundamental difference but, as we have already noted, there was a more dramatic polar wander event at some point in the past, shifting the lunar ground between 45° and 60° (Takahashi et al. 2014). There is no way ice could have survived in equatorial regions for any length of time. Yet, despite these tilt variations, and the continuous action of the Sun, the Moon's crust appears to still contain substantial quantities of water.

It now seems likely that the early Moon had significant supplies of water which became part of its basalt rock (Hui et al. 2013). The Earth and Moon are intimately related. They either formed as a double planet, or, as most scientists believe, the Moon formed as a result of a massive collision between the early primordial Earth and a Mars-sized body. Over time, additional water ice could have been added to the Moon by comet and 'wet' asteroid impacts, or by chemical action on the lunar surface.

Nonetheless, there should still have been an erosion of that water ice if exposed to the action of the Sun over time. There could be a cyclical process of deposition and then removal over time through sublimation into space. The presumption implies that extant non-polar ice was either relatively recently deposited, or is hidden from the Sun's eroding glare.

The Moon's current obliquity is about 6.5 degrees. As things currently stand, the Sun cannot rise more than 1.5 degrees above the horizon at the lunar poles. This keeps the ice in the deep polar craters safely in the shadows. When the Moon was much younger, conditions may have allowed the Moon to collect and retain more water ice at its polar regions. However, during the period of high lunar obliquity some 3-4 billion years ago, when the Moon was about half the distance from Earth than it is now, the Moon lost much of this sub-surface ice (Siegler et al. 2011).

I would argue that the Earth/Moon binary was located at the edge of the solar system's water-ice snow line before the beginning of the LHB. There was plenty of opportunity to obtain, and hold onto, water ice at that distance, perhaps three times further out than now. Once the Earth/Moon system migrated inwards, the vast majority of this ice would have been blasted off the surface of the Moon by the action of the solar wind (the Moon lacks a protective magnetic field). However, during this inward migration, the Moon also steadied itself, allowing deposits of lunar ice to be retained in deep craters near to the poles for the next 3.5 billion years.

Earth Collision Fragments Found in the Asteroid Belt

The early collision between Earth and a Mars-sized body that created the Moon apparently scattered debris materials into the asteroid belt. Many of the meteorites that have undergone alteration through exposure to extreme heat have been dated to around 105 million years after the solar system's birth, some 4.47 billion years ago. This is coincident with the timing of the Moon-forming impact (Bottke et al. 2015).

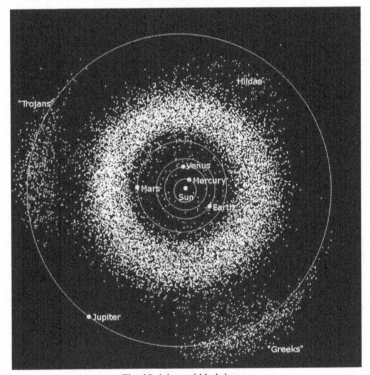

Fig. 19-4 Asteroid belt image

You might wonder how the asteroid belt managed to find itself in receipt of so much debris from such a distant collision. As we have discussed, this puzzle is readily solved through the realization that there was once a more intimate Earth/Moon/Asteroid Belt relationship. I would argue that all of these materials occupied the asteroid belt zone up until a catastrophic event ~4 billion years ago. If we work on this premise, it is natural to infer that the Moon-forming event, which took place 4.47 billion years ago, would have scattered materials into that neighborhood, contributing towards the current asteroid belt.

This boils down to a very simple bit of logic. For the asteroids to be intimately connected with a bombarded Earth so early on, a conclusion can be drawn that many of the asteroids, too, were created from that same Moon-forming event. That could only be true if the Earth began life where the asteroid belt is today.

The later catastrophism, occurring ~4 billion years ago and involving a rogue planet, propelled the Earth/Moon system inwards and created an extended asteroid belt well into the inner solar system. Continuous interaction between this huge debris field and the inner planets created the late, heavy bombardment cratering effects. The gravitational action of the planets eventually cleared this debris field, leaving just the main asteroid belt located in the gap between Mars and Jupiter.

Water in the Moon's Interior

The Nice model sets out, in expansive detail, how the giant planets are thought to have migrated across the solar system – their orbits varying significantly over time. This seems to be accepted as quite normal. Yet, few in the field of astronomy seem willing to suggest that the Earth might also have started its life in a different location. If the Earth formed between the orbits of Mars and Jupiter, then we might readily explain why the waters of the asteroid belt, the Earth, and the Moon, all seem to share a common origin.

Water found on the Moon surprised NASA scientists, who unwittingly sat on the data for decades whilst erroneously assuming that the Moon was bone dry. Not so, as it turned out (Hui et al. 2013). Furthermore, lunar water has been found to share a common origin with Earth, and also with carbonaceous chondrites – minerals found in meteorites that originate from the asteroid belt. The fact that the Moon, having formed following a cataclysmic impact between the Earth and a Mars-sized object, is not bone dry after all indicates that the early Earth was more aqueous in content than previously thought. Because comets are being ruled out as a source of terrestrial water, we are left with one of two possibilities:

1. The Earth formed further away from the Sun than it currently is. It needed to be at the edge of, or beyond, the water-ice snow line of the early protoplanetary disk to retain its water. Then, due to an encounter with another sizable planet, it migrated inwards.

2. The Earth was pummeled not by comets, but by 'wet' asteroids that had themselves migrated into Earth-crossing orbits. Again, that collision is reliant upon the intervention of a sizable planet to nudge Earth-crossing asteroids inwards. The Nice model would advocate an inward migration of Jupiter to set this series of events off, as well as an outwards migration of Neptune to perturb the Kuiper belt.

Either way, planets need to move around to account for the Earth's extant water. So, why shouldn't it be the Earth? If the Moon began life three times further from the Sun than it is now, shouldn't it hold more water internally?

Fig. 19-5 Lunar rock image

As we have discussed, some of the lunar rocks have been found to hold deposits of water within their constituent minerals, often trapped within primitive basalts contained within volcanic glass beads (Saal et al. 2008). It had always been assumed that the interior of the Moon was dry, but evidence is steadily emerging to challenge that cast-iron premise.

Recently, satellite imagery and thermal data has been analyzed to determine the distribution of these water-laden deposits across the lunar landscape. It turns out that many volcanic deposits spread over the surface of the Moon appear to contain large quantities of water (Milliken & Li 2017). This may suggest that the magma within the Moon has a fair amount of water within it, on a global scale.

This flies in the face of the standard concept of the Moon forming from a very hot event where the volatiles were blown away into space. Instead, it suggests that substantial quantities of water may have been added into the Earth/Moon mix immediately after the Moon-forming impact (Saal et al. 2008). The Earth/Moon system does seem to have been water-rich at that time (Lin et al. 2017).

The discovery of water within the interior of the Moon is in keeping with the increasing realization that the Earth also contains vast amounts of water deep within its interior. In fact, many dwarf planets and moons across the solar system seem to have water trapped deep within their interiors. This seems to be the new norm.

In the case of the Moon, given the tidal heating generated by the presence of the Earth, one could be forgiven for speculating whether the Moon itself contains a deep subterranean ocean, deep below the surface. It would need to reside below the Moon's now-cooled global magma ocean, which is a hardened crust some 34-43km thick.

How did the water get to the Earth and Moon, and why is there so much of it, seemingly against all the odds? It is not just water that is giving planetary scientists headaches. Other anomalous volatiles are to be found in lunar rocks as "melt inclusions". These are volatiles which should also have been driven off early in the history of the solar system. They include elements like lead and sulfur. Although these volatiles should have been lost during the Moon-forming event, they could have been delivered into the mix by the later impact of meteorites.

There may have been a "hot start" to the Moon's development (the catastrophic impact upon the Earth which flung material into a debris belt around the planet), followed by a cooler and longer period of accretion, during which the molten Moon was bombarded with water-bearing meteorites over a relatively long period of time (Hauri et al. 2015). This type of thinking raises the possibility that the Moon's formation was not the result of a single cosmic collision, but that multiple impacts over a long time period led to this complex picture. Or, the entire system shifted its position.

An Anti-Anthropocentric Viewpoint

I would argue that it is not necessary to resort to a complex series of events to explain the puzzling biochemical signature of internal lunar rocks. The problem here is the overarching assumption by nearly all scientists working on this, that the position of the Earth is cast in stone. I believe that this assumption is borne of an intrinsic human need to assume an anthropocentric view of the universe. Our species was once dragged kicking and screaming to the realization that the universe does not revolve around the Sun. This is the last foundation stone of our anthropocentric self-importance: The Earth is immutable, the one great constant in an ever evolving tableau around it. But why should this be? Surely there is sufficient evidence to consider the possibility that the Earth started life further away from the Sun, beyond the solar system's snow line?

Bear in mind "Occam's Razor" and consider this: If the Earth formed at the edge of the water-ice snow line of the solar system, and was then struck by Mars-sized planet during the Moon-forming event, the then rich quantities of volatiles would not have been so readily lost to space. It seems obvious. Fragments of the collision would become parts of the asteroid belt, providing similarities between these rocks and those on Earth and the Moon. This happened 4.55 billion years ago.

Later, once things had settled down, and with the Earth/Moon system still located ~3 AU away, a second cataclysm happened: A sizable planet grazed this zone (Planet X) about 4 billion years ago, perturbing it and dislodging the Earth/Moon system inwards, closer to the Sun. This late event spread debris right across the inner solar system, creating the LHB. The encounter significantly affected both the Earth and Moon, and other bodies in the inner solar system, like Mars. The Earth and Moon retained their earlier internally-located waters, and our binary planet system settled down into its present location of 1 AU by about 3.5 billion years ago, having cleared out its zone of asteroid debris. Similarly, the Moon's axis straightened out as a result of these multiple encounters with objects in the debris field, allowing some of its original surface ices to be retained in deep lunar craters over the long term.

Comets and Asteroids

Until recently, comet bodies were hidden from view within their spectacular tails. Earthbound telescopes could not probe their mysteries. To better understand what were universally assumed to be "dirty snowballs", probes have been sent to comets to investigate.

However, this preconception of an ice-laden traveler through the solar system took a bit of a battering when the first images of the main bodies of comets were returned by spacecraft. Far from being pristine balls of glimmering ice, these bodies looked more like dusty rocks. Although water is clearly present in the comae emitted by comets as they approach the Sun, it was not as apparent upon the actual surfaces of these objects.

As we have discussed in previous chapters, for a long time scientists held the view that the Earth's water originated from comets. The argument went that the Earth's relatively close proximity to the Sun means that its early water should have been driven off with other volatiles by the Sun's heat. Yet, the Earth is covered in oceans, providing something of a mystery.

The Late, Heavy Bombardment, as well as the arrival of interstellar comets over billions of years, may have provided the Earth with its veneer of water. If that is the case, then it follows that the distinguishable properties of the water (i.e. the now-familiar deuterium-to-hydrogen isotopic ratio of water) on Earth should be similar to that of the comets.

However, unexpectedly, most cometary water exhibits deuterium levels around twice that of the Earth's oceans. All things being equal, that implies that comets could only have contributed a fraction of the water in our world's oceans. The rest of the Earth's water has to have come from elsewhere.

Comets can be subdivided into different classes. The vast majority of comets are thought to reside in the Oort clouds, and are known as long period comets. They usually have orbital periods lasting up to millions of years, and visit the inner solar system extremely rarely, if at all. Those visits are thought to be caused by a gravitational nudge from a passing star, or the action of the galactic tide. Upon passing the Sun during their spectacular perihelion passage, long-period comets may then take up residence closer to the Sun.

In contrast, there is a different grouping of short period comets which orbit the Sun within the planetary zone of the solar system. These may have originally been long period comets whose trajectories have changed over time. The migration of comets across the length and depth of the solar system is a complicating factor here, not least because comets are considered to have first originated within the protoplanetary disk of the early solar system, and migrated outwards. Let us turn our attention to our neighborly short period comets.

Jupiter Family Comets

In 2011, the Herschel Space Observatory observed the peanut-shaped comet, 103P/Hartley (sometimes known as Hartley 2) during one of its flybys. Comet 103P/Hartley belongs to a group known as Jupiter-family comets, and has an orbital period of 6.46 years. On board Herschel was a bit of kit named HIFI - the most sensitive spacecraft instrument yet devised for detecting water in space.

Fig. 20-1 Hartley 2 image

HIFI revealed that Comet 103P/Hartley's deuterium-to-hydrogen isotope ratio is almost exactly the same as the water in Earth's oceans (ESA 2011). Further work indicated that the comet formed in warmer conditions than long-period comets. In other words, some short period comets, like Hartley 2, probably formed relatively near to the Sun. This makes sense of their Earth-like deuterium-to-hydrogen isotope ratio, which is also the same as asteroids from the outer belt.

Around this time, the European Space Agency was in the process of sending the Rosetta spacecraft on a ten year journey to rendezvous with Comet 67P (more accurately known as 67P/Churyumov-Gerasimenko). It arrived in November 2014. Rosetta dropped its Philae lander onto the surface of the comet. Like Comet 103P/Hartley, Comet 67P is also a short period Jupiter-class comet; this time with an orbital period of 6.45 years. You will note that this is almost exactly the same orbital period as Hartley 2. Comet 67P is now very much at home among the planets, having had its orbit affected by a close pass with Jupiter in 1959.

Fig. 20-2 Comet 67P image

However, unlike the Earth-like water found around Hartley 2, the deuterium-to-hydrogen isotope ratio for Comet 67P was more than three times greater than that of the Earth's oceans (ESA, 2014). In fact, it was the highest deuterium-to-hydrogen isotope ratio yet recorded in the solar system – greater even than that of long-period comets derived from the Oort cloud.

You will have noted, then, that even though these two comets currently have almost identical orbital periods, their constituent water appears to be very different in composition. This glaring difference, combined with variation seen in other comets, seems to argue against the Jupiter family of comets being a convenient reservoir of water for the Earth's oceans (Altwegg et al. 2015).

Given the difference in their water compositions, Comet 67P most probably had a different origin than Hartley 2. Through carefully examining the comet's orbit, astrophysicists now think that Comet 67P originated from the Kuiper belt (Galiazzo et al. 2016). Because this is a much colder zone than Comet 67P's current location, it likely has more in common with longer period comets than Hartley 2. But, given its extremely high deuterium-to-hydrogen isotope ratio, I wonder whether Comet 67P actually originated from further away still. (If 1I/'Oumuamua had out-gassed some water, we would have an interstellar D/H ratio for comparison by now).

The Asteroid/Comet Identity Crisis

The solar system seems to be full of shifting sands. It is unsurprising, in many ways, given the different planetary forces at work, and the sheer timescales involved that the potential complexity is staggering. Interventions in the form of interstellar planets, brown dwarfs and passing stars, or the intrusion of vast quantities of nebula material, can all have caused disruptions in the past.

Then there is the chaos generated by the purely in-house dance-mix that is the Nice model: Planetary migration on a grand scale, capable of flinging around pretty much anything beyond the Earth's orbit. It is also true to say that most of the moons and irregular satellites in the outer solar system seem to have been captured by the ice and gas giants somewhere along the way, perhaps during the upheavals predicted by the Nice model. So, there is plenty of scope for things to get shaken about in the solar system, no matter which model you favor.

Within the Kuiper belt, some KBOs resemble the irregular satellites captured by the outer planets during solar system upheavals. Other KBOs are redder. Do the different patterns of coloration of objects in the Kuiper belt provide clues to their different origins? Should it surprise us if comets, KBOs, Centaurs, Trojans and asteroids have all gotten mixed up along the way, too? Who is to say that these categories are so precisely defined anyway?

As we have seen, Comet 67P may have been a KBO at some point. We have considered whether objects in the extended scattered disk beyond the Kuiper belt may originally have been KBOs pulled up by Planet Nine, or were inner Oort cloud objects dragged down. We have met the first confirmed interstellar visitor, which seems to show properties of both asteroid and comet. The old certainties are fast receding. Perhaps we should take nothing for granted at all.

Asteroidal Comets

Let us take a look at the so-called "asteroidal comets", which seem to straddle the ever- thinning demarcation line between these classic groupings. The possibility that a fraction of the objects populating the Oort cloud might have originated from the inner solar system has been around since the discovery of a dark asteroid orbiting the Sun in a cometary orbit back in 1996.

The object was given the name 1996 PW. It may be an extinct comet, or it may have originated in the inner solar system before being flung out beyond Neptune by a planetary encounter. After its discovery, theoretical work indicated that perhaps one or two percent of comets in the Oort cloud could have originally been similar, rocky bodies from the planetary zone (Weissman & Levison 1997). That amounts to about 8 billion cometary asteroids!

This proposed variability of source for the countless Oort cloud objects has recently been tested theoretically, using computer simulations, and these have come up with a similar estimate for the population of these stray asteroids. That implies a lot of crashing about in the early solar system. Indeed, you could argue that the entire solar system is a complex debris field of scattered rocky objects, including the comets within the Oort clouds.

It is also recognized that there are asteroid-comet transition objects. The Damocloids are such a grouping, but there are others, too, including some main belt asteroids (e.g. 7968 Elst-Pizarro) and Near-Earth Objects (e.g. 4015 Wilson-Harrington) which have been seen to occasionally erupt with cometary activity. This may imply that the volatiles have been sealed into the object's interior, only to be released following an impact with a small body which chips away at the sealant layer (Toth 2005). Such subsurface icy glints have been seen on Comet 67P, as we have noted. Exposed sub-surface layers of volatiles can then sublimate off, providing a brief cometary coma. But the natural processes that seem to cover the asteroid-comet in its sealant surfacing eventually fill in the cracks again. The body then returns to its dormant state.

That process does not seem to hold true for long period comets emerging from the Oort cloud, which seem only too ready to release their volatile waters into space. However, might it be true for a comet from interstellar space? I have considered whether 1I/'Oumuamua might be covered by a layer of sealant, defying any de-stabilizing eruption of internal volatiles. 1I/'Oumuamua may have been so thickly covered by dark, irradiated gloop that it was incapable of spilling the content of its interior out into space.

In general, then could interstellar comets be darkened like this, with their ices caged in beneath a solid shell of silicate and greasy organic matter? Until very recently, comets were self-selecting – the more spectacular their erupting gases, the more likely they would be observed and recorded. Dark, dormant comets that defy the heat of the Sun, and remain completely intact, may have passed through the outer solar system without astronomers even noticing.

The comet/asteroid debate is becoming increasingly like the classic "chicken and egg" argument. Until recently, it was thought that the asteroids are dry rocks because they were relatively close to the Sun and had already lost their volatiles during the early solar system. By contrast, the comets were 'dirty snowballs' that had formed in the cold depths of the outer solar system from a mixture of materials still rich in water ices. They then spread out into the Oort clouds. Now we find water-bearing asteroids, and inactive comets. The demarcation lines are blurring.

Dirty Snowballs

The most cursory examination of Comet 67P by the Rosetta spacecraft shows that it looks a lot like a standard asteroid. The Philae lander, which had been dropped down onto the surface of the comet from Rosetta, sent back images that looked like they had been taken in a moonlit quarry. The comet looks nothing like a dirty snowball: There was no surface ice in sight!

Fig. 20-3 Champollion image

Given that evidence suggests that Comet 67P began life as a Kuiper Belt Object, which had somehow been displaced towards the Sun, perhaps that is less of a shock than one might have expected. Additionally, you might argue that this short period comet has been through its perihelion passage enough times for its ices to have been driven off by the Sun's heat.

However, this imagery was not the expectation when Rosetta was first sent towards the comet. In a presentation paper to the 37th Lunar and Planetary Science Conference in Texas in 2006 (Andrews et al. 2006), the official artist's impression of Philae and Rosetta shows a comet surface which is covered in ice and snow. This picture succinctly reflected the "dirty snowball" view of the composition of comets widely held at the time. A similar, but ultimately slightly more realistic illustration was prepared in 1997 regarding the proposed Champollion spacecraft (pictured). Again, the "dirty snowball" concept predominated.

The Philae lander's system of ice screws and harpoon clearly demonstrates what was assumed about the surface of the comet when the mission was in the design stages: There would be plenty of surface ice for the harpoon to attach itself to when stabilizing the lander's position (Cornette 2014).

Instead of a dirty snowball, Comet 67P looks more like a barren rock. Its coma is sourced from ices within, with internal volatiles sublimating during perihelion and violently out gassing from the comet. This is in keeping with Hartley 2, where out-gassing of volatile carbon dioxide spews grains of water ice out into space to form the comet's coma. To understand these objects, scientists need to find out more about what is going on inside comets, and what they are made of. Given the unpredictable way comets behave, that is easier said than done! However, as far as the superficial surfaces of the comets go, they do seem to look a lot like classic asteroids.

A Shifting Landscape

Objects in the solar system are subject to many forces that can alter their trajectories over time. What we might see, and then make straightforward assumptions about, is not necessarily how things have always been. Two comets that appear to belong to one family have very different properties, and probably quite different origins. Comet 103P/Hartley seemed to start life in a relatively warm part of the solar system, perhaps as an inner Centaur (an object located among the outer planets, between Jupiter and Neptune), before becoming displaced into its current short-period cometary orbit. By contrast, it now looks like Comet 67P is a displaced KBO, and so originated in a colder zone of the solar system.

But their proximities to the Sun over many cycles, during which both have behaved as short-period comets, will also have changed their properties. Volatile materials located on and within Comet 67P will have been driven off by the Sun over time. In so doing, chemistry may have taken place. This object may no longer present as it once did in the Kuiper belt. Nonetheless, finding out about it may tell us a great deal about the conditions in the Kuiper belt.

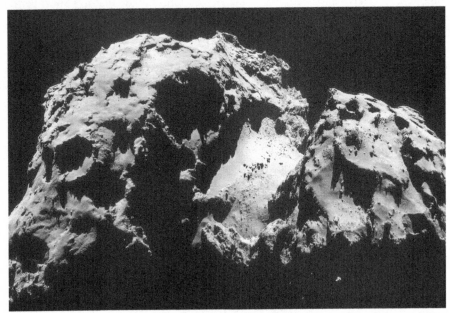

Fig. 20-4 Comet surface image

Over time, ESA has noticed changes across the comet's surface. Careful imaging by the Rosetta probe has revealed breaks in the dark, rugged silicate and organic laden surface of Comet 67P, showing interior water ice gleaming within. The dark materials covering the object are what remain after surface ices have been driven off by solar radiation and wind. These "refractory materials" are silicate rich (dust, sand, rocks) and are full of organic, carbonaceous materials.

The bright marks in the cometary landscape appear to be sub-surface ices exposed after debris falls on the comet (Filacchione et al. 2016). The implication is that there may be plenty of water ice lying under a fairly superficial surface layering of refractory materials. It seems reasonable to assume that repeated perihelion passages around the Sun have roasted the surface, leaving a dehydrated residual covering of rocky and organic materials which has darkened the comet's surface significantly (Capaccioni et al. 2015).

After some difficulties associated with its initial positioning on Comet 67P, ESA's Philae lander returned data from the comet confirming that its surface is rich with organic compounds. Its instruments were also able to provide information about the chemical composition of these compounds. What a rich soup it is! Organic polymers were found that contain similar spreads of carbon, hydrogen and oxygen as found in biological molecules. An example is ribose sugar, which is a constituent of biological molecules like DNA and RNA (Connor 2015).

Panspermia

Missing from the concoction was nitrogen, however. Even so, the complex chemistry taking place on the irradiated surface of the comet shows the potential for the creation of a "primordial soup", which might have the potential to seed planets with the essential building blocks of life. Remember that any comet whose orbital path takes it beyond the heliopause finds itself immersed in interstellar space, and therefore prone to encounter the greasy aliphatic hydrocarbons contained within dusty interstellar medium.

The potential for life to be transmitted to planets via comets has been a longstanding idea that is finding greater acceptance. It is one possible mechanism among several falling under the umbrella of the "panspermia theory", which involves the transportation of microbial life between cosmic locations (Wallis & Wickramasinghe 2004). The discovery of organic compounds in interstellar space put the panspermia theory onto a secure footing (Wickramasinghe & Allen 1980). The frequent movement of star systems past, and even through, star-forming nebulae provides ample opportunity for cross-fertilization (Napier, 2007).

If simple chemical organic compounds are in place on comets derived from the outer solar system – essentially frozen in time for 4.5 billion years – and even more complex chemistry is apparent on the reddened surface of the KBO/dwarf planet Pluto, then it stands to reason that a much larger Planet X body would provide a viable environment for even more complex chemistry, and/or biochemistry, to take place further out.

Oxygen on a Comet

To some extent, the discoveries of rich soups of organic materials on objects from the Kuiper belt, like Comet 67P and Pluto/Charon, were considered theoretically possible. What was less predictable was the presence of oxygen on Comet 67P. After all, oxygen is a reactive gas which should have jumped at the chance to make water, especially when in contact with the immense amount of primordial hydrogen swilling around in a star-forming nebula. In other words, any molecular oxygen (O_2) at large when the comets were formed at the very beginnings of the solar system should have been used up long ago. Nevertheless, there it was, in abundance: Molecular oxygen made up a sizable 4% of the coma enveloping Comet 67P (Bieler et al. 2015).

Oxygen is obviously a key component of the Earth's atmosphere, and has been found in the atmospheres of Jupiter and Saturn. But this was the first time it had been discovered out-gassing from a comet. The gases coming out of comets generally include water vapor, carbon monoxide and carbon dioxide, as well as a mixture of other molecules, including sulfur-based compounds and hydrocarbons. For example,

Timeless Voyager Press

out-gassing carbon dioxide was the main driver of expulsion of grains of water ice from the interior of Comet 103P/Hartley out into its coma.

But oxygen should not be there. It is too reactive a chemical. Planetary scientists have been challenged to explain this anomalous finding. Molecular oxygen may have survived against the odds on a comet, having possibly been trapped within grains of ice crystals and rock. However, this seems improbable within the constraints of the currently accepted model of solar system formation (Bieler et al., 2015). Teams of scientists tweaked their calculations of the Sun's pre-solar nebula to show how this oxygen might have emerged unscathed within a comet. But this all seemed a little far-fetched, a point which they seem willing to concede.

This impasse lasted for a couple of years, until a fascinating laboratory experiment was conducted by a chemical engineer. The experiment simulated how oxygen might be produced on the surface of a comet. This made use of a technique for extracting molecules from a suitable surface by battering it with high-speed, charged molecules – in this case, of water (Yao & Giapis 2017). Ionized water from the comet's coma is thrust back onto its surface by the solar wind. This high energy bombardment picks up oxygen which is bound to materials on the silicate-rich surface, releasing oxygen molecules into the coma (Clavin, 2017).

Generating Planetary Atmospheres

One might speculate that such a mechanism could be occurring on other bodies in the solar system, too. For instance, the geysers of water pouring out of Saturn's little moon Enceladus might be capable of producing similar effects, as the ionized water makes renewed contact with the small moon's surface.

Other environments subject to high speed molecular materials could also be producing atmospheric gases in this way. For instance, 1I/'Oumuamua had a cruising speed through interstellar space of about 26 km/s, and this was considered typical for objects moving through the Sun's interstellar neighborhood.

The interstellar medium (ISM), which includes gas and dust, is moving at similar speeds, as it, too, revolves around the galactic core. When ISM encounters a planet beyond the heliopause, like a loosely bound Planet X body, then we can imagine that, without the protective sheath offered by the heliosphere, this planet would be taking a routine micro-battering. This, through the mechanism discussed above, could produce atmospheric gases. If grains of interstellar water ice are causing the pummeling, then those atmospheric gases will include oxygen.

Of course, in the freezing depths of interstellar space, these gases will eventually freeze out onto the planet's surface, or else be lost to space. In the case of larger free-floating worlds, an extensive enshrouding nebula of dust and gas may form around the planet, held in place by its commanding gravity. This cloud would obscure the planet with dust and gas held within a dynamic equilibrium between surface generation through pummeling, loss to space of fast moving ejected materials, and material slowly descending back to the planet's surface that failed to escape its sizable gravitational well (Lloyd 2016b). The result would be a rather nebulous atmosphere, somewhere between that of a traditional planetary atmosphere, and a comet.

In the case of tiny 1I/'Oumuamua, it may have been exposed to the radiation present in interstellar space for billions of years. Beyond the safe confines of a star's extended magnetic field, the cosmic radiation could combine with the kinds of processes described above to thoroughly cake the object's surface in organic and silicate materials.

The field of "astrochemistry" recognizes the many examples of chemical reactions taking place within gigantic molecular clouds and other interstellar environments, including the synthesis of surprisingly complex organic molecules (Ehrenfreund & Cami, 2010). Like a car windscreen, billions of years of flight through diffuse interstellar material could readily cake a great deal of muck onto the surface of a comet. Cosmic radiation bombarding the comet could work to seal this muck in through further chemical reactions on its surface.

The Case of Titan

As an interesting side note to this, measurements of the nitrogen isotope ratio in cometary ammonia have proved similar to those in the atmosphere of Saturn's moon Titan. This provides evidence that the nitrogen in Titan's atmosphere originated under conditions similar to that of the comets (Mandt et al 2014). Titan has a high orbital eccentricity, too. Its size and its atmosphere both seem out of place.

Did Titan originate from elsewhere completely? Is it actually a captured exoplanet? This is a contentious suggestion, but if it is correct, then Titan potentially tells us a great deal about the kinds of advanced chemistry going on in interstellar space around free-floating planets. Titan's hydrocarbon soup is well advanced, enveloped within a very thick nitrogen and methane atmosphere. It has inexplicably accumulated a lot of nitrogenous material. It therefore stands in sharp contrast to the rest of the solar system's "home-grown" moons.

An Ice-Laden Asteroid

In 2015, the Dawn spacecraft imaged mysterious "lights" in the craters of the dwarf planet Ceres, the largest object in the asteroid belt. On first impression, these seem to be impact marks from meteoritic bombardment. Brighter materials lying below the darkened surface had been exposed. Even so, these "glints" were uncommonly bright for an asteroid.

Speculation about the nature of the nature of the bright materials was rife within the planetary science community. The bright spots, now widely thought to be salt deposits, were named after ancient Roman festivals: Cerealia Facula and Vinalia Faculae. Wine is not thought to have been involved.

However, that is not the only mystery on Ceres. There is a growing consensus that the asteroid may have been experiencing geophysical processes in the relatively recent past. Ice sits in permanently shadowed craters on Ceres and is widespread below the surface. In fact, Ceres is something of an "ice palace". This has raised the question of whether Ceres might have an abundance of active "cryovolcanoes" lurking in the darkness of some of its craters.

There appears to be spectrum of color among these Cerean ices, ranging from a bluish tinge for the recently exposed/erupted sub-surface ices, to a more reddish tinge for the older crater ices (Schmedemann et al. 2016). It is not known why this color differential is evident, but it seems to imply some kind of process at work on the surface of Ceres. Cryovolcanic activity is the mostly likely theory to explain this effect. There would need to be an internal source of heat and pressure driving these volcanic processes. On the face of it, this seems unlikely for an isolated planet the size of Ceres, yet water vapor was seen emerging from Ceres back in 2014. It was thought that this effect might be due either to:

a. Cryovolcanoes;

b. The sublimation of ice into water vapor. Like comets, Ceres is out-gassing. This would blur the distinction between asteroids and comets still further.

Ceres seems to be more geophysically active than previously thought possible for an isolated dwarf planet located in the asteroid belt. There is a 4,000m high ice mountain on the surface of Ceres, named Ahuna Mons, which is thought to be an ice volcano (ASU 2016). Ceres has an abundance of water ice (Prettyman et al. 2016), much of it lying close to its surface, and some of it wrapped up in this huge natural structure.

Incredibly, liquid water may once have flowed across the Cerean surface. Remember, this dwarf planet is still an asteroid! The bright, salty patches, seen glinting on the surface of Ceres, appear to be the leftover residues from evaporating water emerging from within the planet. Icy patches noted within high-altitude craters hint at ice layers just below the superficial surface. Or, they may be deposits from water driven off the surface elsewhere across the dwarf planet's surface. There appear to be on-going geophysical and mineralogical processes across the Cerean surface. Some of the minerals hint that Ceres may have originated from the outer solar system (Carrozzo et al. 2018).

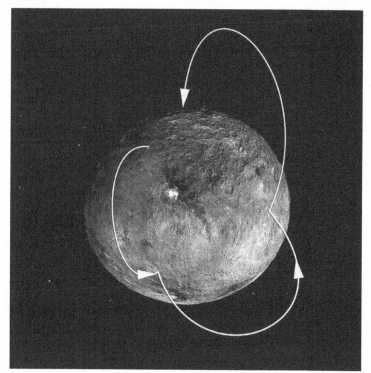

Fig. 20-5 Ceres water hopping image

Ceres seems eerily reminiscent of the sub-surface water on Mars, whose own atmosphere is too weak to sustain liquid water on the planet's surface. Looking at the bigger picture, the solar system is providing us with evidence of more active geophysical processes than previously thought possible. Below the seemingly dull surfaces of various worlds, there is plenty going on. This flies in the face of theory, which previously considered these worlds to be essentially dead lumps of rock.

Migration of planets seems a key factor, helping to explain why planets located within the snow line have so much water ice buried beneath the surface (which should have been driven off early in the lifetime of the solar system). Are other asteroids similarly laden with sub-surface reservoirs of water ice? In that case, what structurally differentiates asteroids from comets? Is the only difference analogous to that between an active and a passive volcano?

Our System's "Impossible" Configuration

The old demarcation lines separating the solar system's minor bodies are becoming less and less clear. The old certainties about how they attained their current configuration in the solar system are in doubt. The standard model of planet formation has failed to reconcile the distribution of rocky planets and the asteroids. Indeed, it may never achieve such a resolution (Tsiganis 2015). Although computer simulations are able to model the fairly speedy accretion of planets from trillions of pebbles forming out of the protoplanetary disk, and falling in towards the Sun, the main sticking point for the Nice model appears to be the existence of the asteroid belt.

Attempts to get all of the solar system objects to end up in their respective current boxes have been fraught with difficulties. Proffered solutions can involve some pretty convoluted scenarios. According to the Nice model, the outer planets undergo extensive migrations during a period of intense instability caused by gravitational interactions with the dispersing protoplanetary disk (Levison et al. 2011). This pattern of migration may then bring about the observed structure of the Kuiper belt (Levison et al. 2007). It may even be consistent with the presence of Planet Nine (Batygin & Morbidelli 2017).

However, for all its seeming success, the Nice model seems to struggle to replicate the structure of the inner solar system – in particular why the inner asteroids are so different in their composition to the outer belt asteroids (which themselves resemble Earth).

It is now realized that there is a rich diversity of objects across the asteroid belt, and beyond (Chu 2014). Many asteroids which seem to have formed in warmer parts of the solar system have been found further out than other asteroids which originated in colder zones. This swapping of positions of many asteroids indicates that the asteroid belt is not the static entity it was once believed to be. Instead, it is increasingly being considered to be an important part of a dynamic structure of minor bodies spread across the entire solar system.

Substantial tweaking of the Nice model is necessary to simulate the distribution seen in the main belt (Deienno et al. 2016). This tweaking might require Jupiter to actually migrate right through the belt, driven by changing dynamics in the dissipating protoplanetary disk. Its combined inwards-then-outwards migrations initially clear the belt, before pulling in debris from both the outer and inner parts of the solar system (Walsh et al. 2011). Ceres may then be part of this in-wash.

A different variation on this theme is that the divided asteroid belt came together from opposite parts of the solar system. According to this hypothesis, no materials formed in the current location of the asteroid belt. Instead, the C-type asteroids migrated in from the outer solar system following the formation of the gas giants, and S-type asteroids migrated outwards following the formation of the terrestrial planets in the inner solar system. Both selections of object were leftover building blocks, according to this idea (Raymond & Izidoro 2017).

We are indeed confronted by an "impossible" configuration, along with an increasingly elaborate series of dynamic processes to explain it. Astronomers once thought the solar system was a stable, orderly place. Now, it is viewed as a jumbled mix that has slowly settled down following a great deal of internal disruption.

What a contrast to the thinking a couple of decades ago! At that time, when specu-lative ideas about rogue planets sweeping through the solar system were proposed, mainstream astronomers were quick to point out how stable the current solar system is, and how that stability would not be possible if catastrophic passages of such plan-ets had ever taken place. Yet, now we have scientists at the forefront of solar system studies demonstrating that the mass migration of the giant outer planets across the solar system are actually necessary for the solar system to have taken on the compo-sitional appearance it has today!

There is no effective difference between a double sweep of Jupiter through the early inner solar system, and the passage of a rogue sub-brown dwarf or planetary-mass object through the same zone at a later stage. They essentially produce the same effect. The subsequent movement of the rogue planet into the far reaches of the solar system can then explain the issues with the Kuiper belt and beyond.

It is that simple!

CHAPTER 2

Nibiru

The phenomenon that is Nibiru is equated with the ancient symbol of the Winged Disk. For good reason, according to the writer Zecharia Sitchin (1976): The rogue planet that passes through our solar system's planetary zone every 3600 years or so moves through an orbital path similar to a comet. Readers of my previous non-fiction title *Dark Star* (Lloyd 2005) will be familiar with some of the difficulties that emerge from this scenario.

Fig. 21-1 Zecharia Sitchin

I believe that Sitchin was on to something when he wrote about the 12th Planet. However, his primary interest was in interpreting ancient texts, and not in astronomy or astrophysics. As a result, the scientific credibility of his particular Planet X scenario was low. Issues include the habitability of a lone, terrestrial planet in the outer solar system; the regular return of this planet to a pristine asteroid belt, which is seemingly unaffected by this perpetual intervention; the provenance of such an object; the long-term stability of its orbit; and the painful lack of an observed Planet X object located within a maximum distance of 465AU.

Yet, many aspects of his theory seem to be borne out by scientific discoveries over the last few decades. The whole concept of Planet X was denigrated for decades, and is now fashionable once again, thanks to prominent academics at Caltech. Because the topic is taken more seriously, astrophysicists are less concerned about their reputations being tarnished by association. They have churned out any number of academic papers to try to explain, or refute, the emergent evidence in the outer solar system regarding the presence of another planet. Low probability scenarios that could explain how Planet X got there are now seriously considered, rather than being immediately dismissed. There is a palpable sense of urgency to finish the jigsaw and find this thing.

However, the goal posts have also shifted. Credible scenarios for Planet X place it well beyond the zone described by Sitchin. Instead of a few thousand years, its orbit looks like it encompasses tens of thousands of years, perhaps more. It does not visit the planetary zone during that vast sweep. Instead, it keeps its distance somewhere in the gap between the Kuiper belt and the inner Oort cloud. It is probably a sizable planet of the order of a mini-Neptune (I would argue larger, and more distant, than even this). This does not reconcile well with 'Nibiru'. A world like Planet Nine could not have made an appearance in our skies in ancient times (and certainly not during out current times!)

As Planet X has once again become the object of serious study, it has pulled away from the realm of sensational speculation. Those intrigued by the notion of a Doomsday World find this change disappointing, perhaps even worthy of suspicion. I suspect that this closes their eyes to the real potential of Planet Nine. As we have seen, beyond the numbers, there are some striking parallels between its orbital arrangement and that of Sitchin's Nibiru: The location of the axis of aphelion/perihelion is the same; the tilt is the same. These are two remarkable coincidences which neither side seems comfortable to acknowledge.

Let us set aside any knee-jerk dismissal, and consider whether Sitchin somehow got some of this right. His theory was based upon his interpretation of ancient Mesopotamian texts. His description of the orbit of Nibiru was derived from qualitative accounts involving ancient constellations. If he was going to get anything right, then, it should be the location of the perihelion passage through the sky. This would give him the perihelion and, by default, aphelion constellations, and the inclination of the planet's orbit with respect to the ecliptic.

What is Nibiru?

"Nibiru" is an Akkadian word meaning a crossing point, generally of a river. It might be a ford or a ferry. In this context, then, it appears to be a means of transiting from one cosmic location to another. That might be a comet or minor body moving between Planet X/9 and the Sun. To make the complete round trip in 3,600 years, however, is clearly problematic, at least for a natural body moving under its own volition around the Sun.

The true nature of Nibiru remains a point of conjecture. It is undeniably referred to in the ancient Babylonian creation myth, Enuma Elish, in a cosmic context (King 1902). Sumerologists have provided many explanations for the name, usually arguing that Nibiru refers simply to the planet Jupiter. In so doing, they perennially ignore Nibiru's association with the color red. This is self-evidently not the color of Jupiter.

The great cosmic battle between Marduk and Tiamat lies at the core of this mystery, for which Marduk earned the name 'Nibiru':

"... the star, which shineth in the heavens.
May he [Marduk] hold the Beginning and the Future, may they pay homage unto him,
Saying, "He who forced his way through the midst of Tiamat without resting,
Let his name be Nibiru, 'the Seizer of the Midst'!
For the stars of heaven he upheld the paths". (King 1902)

Marduk's presence in the heavens was truly awe-inspiring. No pottering white planet this:

"He set the lightning in front of him,
With burning flame he filled his body". (King 1902)

Personally, based upon this description, I think that "Marduk" is a substantial planet. As a result, I proposed that this mysterious celestial object was actually a failed star, or sub-brown dwarf (Lloyd 1999 & 2005). This idea became an important revision to Sitchin's controversial theory over the last decade or two. It has also spawned its own Doomsday Planet scares in various derivative books, articles and YouTube videos.

Even if the dramatic account in Enuma Elish describes the collision between a massive rogue planet (Marduk) and a watery super-Earth (Tiamat) some 3.9 billion years ago or so, then we are still left with the problem of the exact nature of Nibiru. Realistically, any significant rogue planet must be located well beyond the planetary zone. As a result, a considerable underlying problem for Sitchin's theory is the orbital period.

3,600

The figure of 3,600 is a core component of the late Zecharia Sitchin's vision and argument for the properties of his Planet X body, Nibiru. In an attempt to understand where this number came from, I would like to explore in some detail some of the source material he may have drawn upon.

Fig. 21-2 Cylinder seal Oannes image

Sitchin drew upon many comparative sources from ancient Near and Middle Eastern traditions. He considered the Levant to be melting pot of religious and mythological thought. For instance, when the Babylonians sacked Jerusalem and dragged much of the population of Israel off into slavery, they likely imparted much of their knowledge (which they themselves derived from the Akkadians, and Sumerians before them) to the teachers and leaders of the enslaved clans of Israel. The ancient symbolism thus found new life in Jewish texts, many of which were compiled around that time.

Sitchin seemed absolutely convinced that the ancient Mesopotamian numerical concept of the Sar described the length of Nibiru's orbit. In comparison to other planets in the outer solar system, it represents a sizable number of years. Neptune orbits the Sun in 165 years, while Uranus keeps track with a standard modern human lifetime, at 84 years. Even the demoted dwarf planet Pluto – which usually keeps its distance beyond Neptune – whizzes around in a comparatively speedy 248 years. Even so, it seems very likely to me that 3,600 years is a considerable underestimate.

The number 3,600 stems from the unusual sexagesimal (order of 60) numbering system used by the Sumerians (O' Connor & Robertson 2000), and was known to them as the sar, or shar in Akkadian. I have often wondered why Sitchin decided on this number for the orbital period of Nibiru. Sitchin carefully thought things through before committing his ideas to paper, and such a central tenet of his work had to have something more behind it than an arbitrary choice from a series of important sexagesimal numbers. Aspects of his writings that mesh with this number pertain to:

1. The Nippurian Tablets of Destinies (Sitchin 1998, p176)

2. The Sumerian King Lists (Sitchin 1993, p10), where the length of the reigns of the antediluvian kings was measured in a combination of sars and ners (1 ner = 600 years)

3. Stages in human biological and social development over long periods of time, which Sitchin often pointed to as important markers for a return of the planet Nibiru (Sitchin 1976, p415)

These, you might think, provide a fairly strong pointer towards cycles that might integrate with the return of Nibiru. However, note that the antediluvian King Lists, which seem to offer the most concrete correspondence with higher numbers in the Sumerian sexagesimal numbering system, require the inclusion of the ner, or 600 years.

If the reigns of these early rulers were connected entirely to orbital periods of 3,600 years, then those of the last two antediluvian kings, En-men-dur-ana and Ubara-Tu-tu, were measured in part-orbits. So how would that work? Furthermore, after the purported Flood, Kingship came to the ancient Sumerian city of Kish, and the length of the reigns began to be measured in hundreds, rather than thousands of years.

There appears to be more flexibility built into these ancient chronologies than is suggested by simply alluding to the periodic return of a planet. Perhaps the central importance of the sar as a "princely" or "royal" number (Sitchin 1990a, p212) was the crux of the matter? In that case, why choose 3,600 and not, say, the rather grander 36,000 (the Sar'u in Old Babylonian) or even the Sar'ges (60x60x60=216,600)? These numbers could just as adequately describe the long orbital period of a planet in a comet-like orbit.

I should be absolutely clear about this: There is nowhere in Mesopotamian writings where it is written "The orbit of the planet Nibiru is 3,600 years". To arrive at this conclusion, Sitchin had to do some detective work, and make some assumptions. In doing so, were there any other factors guiding his judgment?

Reading an anthology of his various writings, published posthumously by his niece Janet Sitchin, (2015, pp148-152), I came across what I think may be the answer. As a Jewish scholar, Sitchin pored over Biblical texts and other Hebrew writings as thoroughly as he did the earlier ancient texts of Mesopotamia. In his wide reading, he discovered a possible correlation between the Babylonian sar and a journey of 3,500 years (Sitchin 1996), mentioned in a conversation between a heretic and the Jewish savant Rabbi Gamliel:

"...tradition avers that the distance between heaven and earth would take a journey of 3,500 years to cover." (Lehrman 1961)

Sitchin then argued that this alluded to a journey to Nibiru, which in the Jewish text is referred to as a place called "Olam". The numerous biblical verses in which Olam appears indicate that it was deemed a physical place, not an abstraction (Sitchin 2015, p150). Other translations of Olam opt for terms like "everlasting" and "eternity", but Sitchin put forward a cogent argument that these later translations into English are misguided, and that Olam "is the term which stands for world" (ibid, p151).

In the discussion between the rabbi and the heretic, Earth and Olam are separated by seven heavens, each of 500 years. Hence, as Sitchin put it:

He then argued that this approximates to the orbital period of Nibiru, which he determines to be 3,600 years. Hence, Nibiru's orbit is confirmed by this ancient Hebrew text.

However, that is not exactly what is described here. The journey through the seven heavens is between one location and another that lasts 3,500 years, not the time it takes a world to periodically return. Indeed, one can quite imagine that the Jewish scribe of this text could have indicated such a return easily by using a phrase like "the return of the Kingdom of God", or something similar. Instead, what we have here is the description of a linear journey between Earth and another world which takes the traveler 3,500 years.

The implication of this is that incredibly long space journeys between Earth and Olam (corresponding, Sitchin argues, to the Sumerian Nibiru) must be made by the traveler seeking a destination beyond the seven heavens (which one might readily assume to be the other 7 planets in the solar system).

One might argue, then, that this makes reference to the frequency with which these journeys are made, and that those journeys are best made when Nibiru (Olam) is at its closest aspect. This would be during its perihelion passage, which Sitchin argued reaches as close to Earth as the asteroid belt. But, again, this is not actually what is said in this text, and given Sitchin's proclivity for taking these writings extremely literally, that must strike a note of caution.

How could this be placed in the context of the sensational announcement in early 2016 of the probable existence of Planet Nine (Batygin & Brown 2016)? Remember that various factors seem to link that proposed body with some of Sitchin's other orbital descriptions; long ellipse, orbital inclination, perihelion/aphelion positions

(Lloyd 2016a). Does this ancient Jewish description make more sense of what we are looking for?

Fig. 21-3 Red planet flare image

I would argue that this indicates that Olam is located way beyond the seven other solar system planets, and must be reached via a mind-boggling long journey through space, perhaps by some kind of celestial "ferry" of as yet unknown provenance (like a comet?). In other words, Planet X/9 lies well beyond the seven heavens, and is signposted by a second body moving in the space between them. This second body, perhaps a comet, is then the ferry Nibiru.

Jewish Dark Star Symbolism

As we have seen, Sitchin seems to have been strongly influenced by at least one Hebrew text when considering the likely orbital period of Nibiru/Marduk, a Planet X body which he argued for on the basis of his interpretation of various ancient Mesopotamian texts.

In the wake of his many books on the subject, there has been much discussion about the return date of Nibiru, in particular as a potentially catastrophic event. Such talk has been widespread for decades. It seems to feed into a feeling of dread of the "End of Days". This may be a peculiarity of our times, or may always been with us since ancient times. It may be part of human nature to think we live in special times. Either way, Sitchin's concept of Nibiru has become a magnet for all manner of claims and predictions of imminent catastrophe raining down from us from the heavens.

Perhaps it is not particularly surprising that such ideas are now circulating within fringe elements of Judaism, and are being openly discussed by various Rabbis (Berkowitz 2016).Whether this is truly significant or not, what are emerging from these discussions are Jewish textual fragments which contain fairly straightforward descriptions of a visitation of a complex celestial entity.

For instance, the appearance of a star at the End of Days is plainly spoken of in the Zohar, Judaism's primary mystical text, written by the 2nd century sage, Rabbi Shimon bar Yochai (Adler 2016). Included in the Zohar is this messianic prediction:

"A star will arise from the East side, flaming with all colors, and seven other stars will go around this star and make a war with it on all sides three times a day for 70 days, and all the people of the world will see."

There are a number of similar descriptions from Jewish texts available on the Internet (Travis 2016). This celestial entity is a complex, multi-colored affair with a retinue of seven other 'stars' in its wake, revolving around it. These could be seven moons orbiting a central Planet X body, and fits with other examples of *Dark Star* symbolism I have previously written about. Essentially, the complex nature of a **Dark Star** system is symbolized in ancient iconography and derivative works. It incorporates a central star, or large planet, with a number of accompanying planets or moons and symbolic wings. This configuration denotes the *Dark Star*, its planetary system, and the bow-shocked nebula encompassing it. An example would be a classic ancient Winged Disk.

Fig. 21-4 Ashur image

This large central planet/sub-brown dwarf is contained within a colorful aura, and accompanied by its own system of worlds. This seems to me to be a failed dwarf star system (if it was a 'real' star, you would not see the accompanying planets for the intensity of light). Sometimes, this imagery was associated with a Mesopotamian deity, like Ashur. In this case, it is associated with a returning Messiah. The passage was written during an historical period of time when there was a great deal of expectation about the imminent coming of a Messiah.

Initially, Sitchin proposed that the Jewish Count of Years, dating back to 3,760BC, marked a perihelion passage of Nibiru (he changed his mind in later years).The Jewish calendar correlates strongly with the calendar of the city of Nippur in ancient Sumer. Significantly, this date also related to an Akkadian description of the return of the god Anu to Earth at an early point in Sumerian history (Sitchin 1990, p26; Sitchin 1993, pp110-2). There is a strong cultural mix at play here. It seems likely that the Jewish traditions were informed by contemporary Mesopotamian ones. This runs in parallel with Sitchin's own interests as a researcher.

The appearance of Nibiru in 3,760BCE would then have placed the next return in 160BCE, during the Graeco-Roman period. As is well known, there was a significant Messianic expectation at that time among the Jewish people of the Levant. I have previously argued that this may have meant an expected return of a star in the skies two thousand years ago (2005).

In order to investigate the evidence from Judaism further, it is useful to review some of the unearthed data which slowly emerged from the Dead Sea Scrolls. Information sourced directly from the Jewish Qumran sect during the Graeco-Roman period should hold clues about cross-cultural mythical ties between Judaism and Mesopotamia. There was clearly an expectation of the Messiah in the Qumran community and, although there are no New Testament writings among the scrolls, the later Qumran sect was contemporaneous with early Jewish Christians in Jerusalem.

In the scrolls, the references to messianic activity seem to point to the coming of "three messianic figures: a prophet, star/scepter and priest" (Davies, Brooke & Callaway 2002). The Star is alluded to in the Oracle of Balaam:

"A star shall arise from Jacob and a scepter from Israel". (Numbers 24:15-17)

This plurality of messianic figures fits neatly with the Melchizedek scrolls of Cave 11 which portend a heavenly deliverer (not an earthly Messiah) in the form of a high priest. There is a conviction within the scrolls of being "in communion with the heavenly cult and the angels".

Then, in the Psalms Scroll from Cave 11, we find a familiar number:

> "...He composed 3,600 psalms and 364 songs for singing before the altar for the continual daily sacrifice, for every day of the year, and 52 songs for the Sabbath offerings, and 30 songs for the new moons, for festivals and for the Day of Atonement". (11QPs27)

Note that 364 is the number of days of the Jewish solar calendar used by the writers of the scrolls. The number 3,600 is given prominence by denoting the overall number of Psalms, and is suggestive of great significance. Yet, unlike the other numbers, it does not correlate with any known calendrical pattern. It does, however, suggest a Mesopotamian origin from within the sexagesimal system. Zecharia Sitchin claimed that this number denoted the orbital period of Nibiru.

The use of the number 3600 in this context demands an explanation as to what celestial cycle is being alluded to by the Jewish authors of these ancient scrolls. This argument is strengthened by the seriousness with which the Qumran community held the calendar as a whole. It appears from the scrolls that one of the main reasons that this sect withdrew from the mainstream Jewish religion was that they had a fundamental disagreement with the Temple priests about which calendar should be used. The Qumran sect favored a solar calendar of 364 days, where all months had 30 days and four days were inserted (quarterly).This could be contrasted with the official Jewish calendar of 12 lunar months, or 354 days.

An explanation for their abhorrence of the priestly use of the official Jewish calendar might stem from their wish to ensure that festival days did not fall upon Sabbaths. But the 3,600 figure may require us to look at this important sectarian conflict more closely. If Nibiru's expected return was indeed the basis for dating systems and calendars, like that of Nippur and the Jewish count of years, then the actual number of days that makes up a year is of absolutely crucial significance. Sitchin notes that the "intercalation" between the solar and lunar year requires an adjustment of "10 days, 21 hours, 6 minutes and about 45.5 seconds", and seems to assume that this was taken into account in ancient times (Sitchin 1993, p19).

Over the course of 3,600 years, the difference between the use of lunar and solar calendars makes a margin of error of about 100 years. So, if Jewish traditions imported significant chunks of the belief systems of ancient Mesopotamia, including the importance of their more significant sexagesimal numbers, then it was obviously important to the priest-class which calendar was to be used – particularly if the dating over extraordinarily long periods of time were to be adhered to.

The Dead Sea Scrolls may highlight the problems with Jewish dating methods, but they also clarify the significance of the 3,600 number. Combine this with the sect's expectation of the imminent return of a heavenly Messiah, and a powerful picture emerges that is in keeping with Sitchin's general hypothesis about Planet X.

Was Nibiru the Prophesied Star?

Let us assume Sitchin was right, and 3,600 denotes solar years, unfettered by any sectarian arguments over the technicalities of the ancient Jewish calendar. If we take 3,760BCE as an established benchmark for the appearance of Nibiru, as per Sitchin's original analysis, then it is not just the return date that is of interest. It would then have placed the next return in 160BCE, during the Graeco-Roman period (Alford 1996).

This takes us into the territory of the awaited Messianic Star. Given the way that Nibiru is otherwise connected to the beginning of calendars (that 3,760BCE date links in with the Jewish count of years and the Calendar of Nippur), then it is reasonable to suppose that the advent of Christ might also fit with the Return. In terms of starting calendars, the (expected) return of Nibiru – in the form of the appearance of the Messianic Star – would sit well with history.

Sitchin later pulled the date of the last return back to coincide with the destruction of the First Temple of Jerusalem by the Babylonians in the 6th Century BCE. He argued that Nibiru came early, setting off a wave of Messianic prophecy and political intrigue prior to the predicted time (Sitchin 2007 & 2010). Even Sitchin's most ardent devotees would admit that these Nibiru chronologies became highly muddled through the course of his **Earth Chronicles** series.

There does not appear to be an historical record of a return of this celestial body on either occasion, however. Perhaps that is because:

a. It simply does not exist;

b. The orbital period is much longer than Sitchin suggested;

c. It failed to appear as predicted, leading to a sense of loss and disappointment during a period of Messianic fervor in the Levant.

If the Sumerian Sar of 3,600 is derived from the orbital period of Nibiru, then the nature of this body needs re-considering. Sitchin's original timings incorporated the starting points of Near Eastern calendars around 3,760BCE, as well as a suggested Flood event projected back to about 11,000BCE. This date seems to offer us something more concrete in terms of evidence of a cosmic event.

End of the Ice Age

In the time that modern humans have lived on Earth, there is period of time which keeps cropping up involving possible catastrophism. That period occurred towards the end of the last glacial period. Zecharia Sitchin argued that an immense flood catastrophically ended the last Ice Age (1976, p402). At the time, he was intrigued by the possibility that the Biblical Flood had been caused by a catastrophic slippage of part, or all, of the Antarctic ice sheet. A resultant tsunami swept around the world, causing devastating flooding and burying stranded fauna, like the mastodons. This idea had originally been proposed by John Hollin (1972), a British graduate student at Princeton University. Further arguments for a sudden, catastrophic flood at that time involve unusual patterns of erosion of landscape features in Washington State (Sullivan 1995).

Fig. 21-5 Flood image

Sitchin considered it possible that a close passage of Nibiru might have triggered this sudden collapse of the Antarctic ice sheet. In his writings, he drew parallel between the narrative of the Biblical Flood with earlier texts of the Sumerian tales of the Deluge hero Ziusudra, as well as the Mesopotamian Flood hero Utnapishtim (who was featured in the Babylonian Epic of Gilgamesh). Although this mechanism of sudden collapse is almost certainly incorrect, there was a significant loss of the Antarctic ice shelf into the sea during the global warming which followed the Younger Dryas period (Andrews 2016).

This date of ~11,000 BCE remains of great interest, however, as it coincides neatly with a catastrophic event which is thought by many to have occurred at the Younger Dryas boundary. Indeed, the probable cause of this sudden cataclysmic melting of extensive parts of the northern ice sheet more closely insinuates a cosmic event than Hollin's Antarctic ice sheet slippage. This marked a transition into a particularly cold period across the northern hemisphere known as the Younger-Dryas, which lasted from about 12,900 to around 11,700 years ago.

The Younger Dryas Impact Hypothesis

In the thousands of years running up to 10,900 BCE, there had been a gradual warming of the earth. At this point, the northern hemisphere was suddenly plunged back into the freezer. Many have advocated a sudden, catastrophic event causing this unexpected change in the global climate. One of these hypotheses is the "Younger Dryas Impact hypothesis". This involves an airburst or impact of cometary fragments above or into the extensive northern ice cap.

Generally, scientists take a gradualist approach towards the formation of geophysical surface features. Yet, the majority of the 100,000 years or so during which modern humans have lived, toiled and built communities together took place under Ice Age conditions. Much of the North American and European continents were covered in ice caps, and conditions elsewhere were arid. Much of the Earth's water was tied up in engorged ice caps and glaciers, and the world's sea levels were significantly lower than they are now. As humans tend to favor coastal communities, then any vestiges of prehistoric Ice Age villages, towns – even cities – would inevitably have been entirely inundated by rising sea levels as glacial periods drew to a close. Any evidence of Ice Age stone structures or monuments would have been consumed by the seas, and would currently be found lying below hundreds of feet of water.

One of the central questions about this alteration in our climate, and the effect it had on modern humans, is how rapid the change was. Did this change take place over centuries, decades, years, months, or even just days? Opinions differ as to whether this was a gradual change, with the melting of ice caps occurring over months or years, triggering the inevitable migration of peoples as sea levels gradually rose; or whether there were geophysical events bringing about sudden disastrous Earth-changes.

British author Graham Hancock has passionately advocated that natural cataclysmic events had a significant impact upon Ice Age peoples and their lost civilizations (Hancock 1995, Hancock 2015). Hancock argues for an underlying reality behind the ancient Atlantis myth, as described by the great Greek philosopher Plato, whereby a sudden melting of the northern ice caps catastrophically inundated coastal Ice Age civilizations.

Citing the 2007 paper by Firestone, Kennett and West, which first proposed the 'Clovis Comet' (Firestone et al. 2007), Hancock draws upon a growing body of scientific papers presented in support of an extinction-level cosmic event around 10,900 BCE. Critics of the Younger Dryas impact hypothesis remain unconvinced (Holiday et al. 2014; 2016).

The timing of this contentious event coincides with Sitchin's original timing for the return of Nibiru, which he linked to sudden shifts in human development:

"...agriculture, circa 11,000 B.C., the Neolithic culture, circa 7,500 B.C., and the sudden civilization of 3,800 B.C. took place at intervals of 3,600 years" (Sitchin 1976, p415).

As previously noted, for Sitchin, the date of 3,760BCE was a central tenet of his chronology for the return of Nibiru. It was a significant calendar marker in the Near East, marking the beginning of the Nippur calendar and the Jewish count of days. More recently, it has emerged that the first settlements along the Nile were also established around 3,700 to 3,600 BCE (Morelle 2013).If Sitchin's chronology was correct, then this marker sets a previous return of Nibiru back to the Younger-Dryas transition, and a possible moment of shattering cataclysm.

Göbekli Tepe

The remarkably ancient site of Göbekli Tepe in Turkey demonstrates how far humanity had already come as the Ice Age drew to a close. The ancient site has set back the clock of ancient civilizations by thousands of years. It is thought to date to about 10,000BCE. Much of it remains underground, still, and is in the process of being slowly unearthed. As that process continues, evidence is emerging that the key to understanding this most ancient of sites lies in the sky. However, deciphering any archaeo-astronomical clues is made tricky by questions about the relative chronologies of various parts of the site.

Those interested in archaeo-astronomy have been quick to discern sky markers to constellations depicted at Göbekli Tepe. These may indicate that these ancient post-Ice Age folk were trying to tell us something about what had taken place during the Younger-Dryas transition.

A number of hypotheses have emerged to explain the unusual mix of monuments and symbols emerging from this well-preserved site. For instance, the Italian archaeo-astronomer Giulio Magli has speculated that the reappearance of the star Sirius, as the skies shifted over millennia, was the driving force behind the construction of this remarkable ancient observatory. He notes that many of the animals depicted upon the monumental stones at the site (in particular, the rather remarkable 'Pillar 43') symbolize star constellations.

Although this site is considerably older than other monuments in the Levant, it is not inconceivable that the region's astrological traditions are extremely old, and that these constellation identities are the same as those identified by the later ancient Mesopotamian civilizations. Significantly, there are other curious similarities between three bags depicted upon a particularly detailed Göbekli Tepe stela and the Babylonian Kudurru tradition of three sky houses (Magli 2016).

So, could the Turkish site be a very ancient precursor to the civilizations that emerged much later in the Levant? This seems to be the implication of the shared symbolism. Pillar 43 features a number of animalistic symbols, which together seem to depict a sizable star field. It seems to show the constellation of Scorpio, with an enigmatic central disk lying within the non-zodiacal constellation of Cygnus (Vahradyan & Vahradyan 2010). The central belt of Orion also seems to feature heavily in the monument's ancient alignments (Collins 2013).

Above the wing of the central vulture is a circular disk, which Magli argues is the star Sirius. This strange depiction seems to me to also share a passing similarity to the Winged Disk symbolism later employed by the Mesopotamians and the Egyptians. The broader astronomical context seems befitting, too. It seems possible that this symbolism is an early precursor to the Winged Disk as seen in later ancient Mesopotamian and Egyptian iconography. This entire site at Göbekli Tepe may be paying homage to an ancient event in the sky.

The Biblical Deluge has been connected with Mount Ararat, on the Armenian/Turkish border, which is not a million miles away from Göbekli Tepe. If a global tsunami was triggered by the Clovis Comet, then survivors of this great flood would have shifted ground away from the submerged coastal areas and river valleys up into the mountains. This traumatic period was marked by a desperate fight for survival and renewal which led in turn to an oral tradition that lasted through to Biblical times. It became the legend of Noah, as well as similar myths all over the globe.

Comet Swarms

To supplement the warnings in these oral traditions, the ancients constructed complex stone sites to depict the appearance of the Clovis Comet. This, I suggest, is what Göbekli Tepe was all about. In which case, could the Clovis Comet be connected to the return of Nibiru? Conceivably, this strange disk on pillar 43 held by an anthropomorphic vulture may be the earliest known depiction of Nibiru, shortly to be followed by a cataclysm which almost destroyed the ancient world.

Fig. 21-6 Clovis comet image

Was the Clovis comet part of a swarm? Researcher Barry Warmkessel et al. argue that in-coming comet swarms need not be accompanying debris of a cometary Planet X, like Sitchin's Nibiru, but rather a regular resonant function of a *Dark Star* (which they name Vulcan) lurking beyond the Kuiper belt (Blackburn, Kawamoto, & Warmkessel, 1997). This concept of comet swarms assumes that the Earth and other planets are periodically subjected to a blunderbuss-style bombardment. However, as we have seen, the cratering record suggests that any profound regularity to these events takes place over tens of millions of years (Rampino & Caldeira 2015), not just thousands of years. The scales are out of kilter with one another by several orders of magnitude. Therefore, it is difficult to see how comet swarms measured in the timescales of thousands of years can viably predict cometary impact events.

Nonetheless, it is worth noting something else connected with these time periods. Planet Nine's orbit is of the order of 25,000 years, and is thought to currently lie near to its furthest point from the Sun. Setting the clock back nearly 13,000 years would place it near to its closest point. In other words, if the projections for Planet Nine are correct, then its last perihelion was coincident with the Younger-Dryas transition.

Could it conceivably have approached close enough to Earth to have been a visible object during perihelion? Would that then explain connections with Orion, which I have argued is the planet's perihelion constellation? Such highly speculative thinking could be applied to any number of ancient sites, including the Pyramids at Giza.

Planet X/9 is probably further out in the solar system than Sitchin's famous number of 3,600 years implies. I think that Nibiru's perihelion passage through the inner solar system every 3,600 years or so, could only really work if Nibiru was simply a highly significant comet, or possibly a dwarf planet X object. Otherwise, the disruption to the inner solar system would be self-evident.

That said, the regular return of a great comet would provide a sensible marker for the ancients to have set their calendars to; or against which to record their long King Lists. This blazing comet may have become mythologized as the expected Messianic Star of the Graeco-Roman era (although this 'star' failed to make much of an impression in historical records from that time). If one of those perihelion passages was coincident with the impact of the Clovis Comet, then that implies a swarm effect. However, lunar cratering does not bear out such regularity in comet bombardment, which would surely be the implication of a repeated passage of a comet swarm every 3,600 years.

On its own, a comet does not help us to explain the *Dark Star* symbolism depicted in these ancient accounts. However spectacular it might be visually, a comet is not a planetary system. Neither does the comet notion sit well with Sitchin's assertion about a habitable Planet X body. For that to work, we still need a massive planet, or **Dark Star**.

The Dark Star Revisited

Having weighed up an immense amount of evidence through the course of this book, and considered a plethora of new concepts and theories, we seem to have come full circle. The evidence that has emerged in recent years about the existence of Planet Nine has rejuvenated the Planet X debate, and created a renewed hope that a new planet will soon be found in the outer solar system. However, that excitement is dampened down by a continuing lack of progress with finding what should be a sizable planet. This frustration reflects a century or more of Planet X lore. **Plus ça change, plus c'est la même chose.**

When I first read the Planet Nine paper in early 2016, I saw the potential for a fit with an even larger body, located still further away from the Sun. The sliding scale could extend quite a long way out, I thought. I had already recognized, way back when I wrote **Dark Star**, that the solution to this mystery lay closer to the Oort clouds than to the Kuiper belt. Even if Sitchin had somehow gotten on the right track with his highly adventurous theory, he had massively underestimated the distances and time periods involved with a viable Planet X orbit. I felt that, to a lesser extent, the Caltech team might also be underestimating the beast.

The problem for me was that a massive planet screams out for discovery in the infrared. Yet, WISE had not found it. This, too, seemed to place a different set of constraints upon the solution to this mystery. The slow realization that the conditions within interstellar space were different to those within the Sun's heliosphere enabled me to work on a solution to this problem. I was trying to seek a solution to a problem that did not seem to exist for other researchers and scientists. I was questioning safe assumptions about the variability of conditions in space. Set upon a critical quest, I found much to interest me.

Many scientists were working on similar problems, but as perceived from their own particular research angle. Scientists can be subject to something of a natural silo-mentality. None seemed to be wondering whether ejected planetessimals could continue to grow as they moved through interstellar space. Why would they?

Similarly, those interested in wide orbit planets (usually brown dwarfs or sub-brown dwarfs) did not seem particularly interested in the fact that many of these massive exoplanets spend a large proportion of their time in interstellar space, beyond the confines of their parent star's heliosphere. Perhaps that is because the heliosphere is generally considered to be a protective environment. However, the star at its center clears space of light debris, dust and gas through a combination of drag effects and the action of its stellar wind. As a result, the very materials which might help to build planets are systematically removed.

Fig. 22-1 Winged disk image

Things are different in interstellar space. The 'clearing' mechanism works more slowly. There is also plenty of material out there, including sticky, greasy aliphatic hydrocarbons which seem eminently suitable for sticking materials together. All one needs is a gravitational influence big enough to draw these materials together. What could be better than a sub-brown dwarf? Its failed stellar status means that it is incapable of blasting material away from itself in the way stars do. Yet, its massive gravitational power will entangle interstellar material as it cuts through space. Free-floating sub-brown dwarfs are like wandering trawlers netting shoals of fish, or like blue whales ingesting krill. Why wouldn't they continue to grow in size over millions, or billions, of years?

It seems common sense, but I could not find any scientific paper exploring such an idea. There may be one I have missed, or the concept may suffer a fatal flaw which has eluded me so far. If so, no one has pointed anything out in the two years since I first wrote about this in a scientific paper (2016b) and Online.

A Nemesis-like object out among the comet clouds is moving through interstellar space. Sometimes the space the Sun moves through contains just a little local fluff, as now. Occasionally, it is dense with the interstellar medium within gigantic molecular clouds, dusty lanes within the galactic spiral arms, or glorious nebulae. While scientists have speculated on the effect the Sun's movement through such a dense medium in space might have upon the Earth, little or no thought has been extended to planets stuck outside the heliosphere. Yet, we know that such planets exist, both in the form of free-floating objects, and wide binary objects. Sometimes, it is difficult to tell between them.

Wide Orbit sub-Brown Dwarf Revealed

A known sub-brown dwarf, previously thought to float on its own through interstellar space, has shown unexpected parental attachment (Deacon, Schlieder, & Murphy 2016). The two are separated by more than 4,500 AU, making this the widest orbit yet discovered between planet and star. The sub-brown dwarf 2MASS J2126-8140, which lies about 104 light years away, is thought to be between 11 – 15 Jupiter masses, and at somewhere between 10 and 45 million years old, is relatively young.

At this kind of age, an object of this size is still burning up some nuclear fuel, and gives out its own light. Ultra-cool dwarfs like 2MASS J2126-8140 eventually 'go out', becoming much darker planets – more like a smaller, smoldering equivalent of Jupiter. These older **Darker Stars** become extremely difficult to spot. Certainly, at over 4000AU distance, it would reflect back only the tiniest amount of light from its parent red dwarf star. Given how these objects emit their own light for only a tiny fraction of their full lifetimes (they easily outlive our relatively fast-burning Sun), we can extrapolate from this chance finding that there are potentially a huge number of *Darker Stars* in wide orbits awaiting discovery around stars.

Fig. 22-2 Binary star image

Astronomers remain puzzled by how such an orbit might have come about, and how it might be sustainable. Very wide orbit objects create headaches, particularly regarding their formation conditions (Vigan et al. 2017). These tend to be either due to direct gravitational collapse, or core accretion. This issue has always been an argument set out against a Nemesis/Tyche/massive Planet X object in our own solar system, as this concept does not fit well with currently understood models of solar system formation. But as more evidence comes forward that these types of exoplanets are actually reasonably common, as I believe them to be, then the old models of planetary system formation will require some adjustment. Finding a precedent such as this will establish that it is at least possible, no matter how unlikely it was considered to be theoretically. If it can happen there, it can also have happened here.

The Winged Disk

I have argued in this book that this rogue "system" is a nebulous and complex entity, taking the form physically and symbolically of the Winged Disk. The central dwarf star appears to be surrounded by some kind of aura, or halo, swept back by galactic winds to create the classic Winged Disk look.

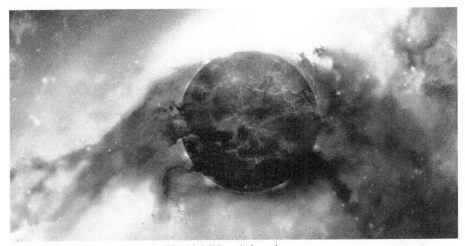

Fig. 22-3 Winged planet image

There is a scientific precedent for a planet taking on the appearance of a comet: A Neptune-size planet that lies about 33 light years away in the constellation Leo. It is known as GJ 436b (Ehrenreich et al. 2015). GJ 436b is a relatively small 'hot Jupiter' gas giant swinging around its parent red dwarf star in a very tight circle, whose radius is just 3 million kilometers. The cloud of gas emanating from this boiling world is truly awe-inspiring, even larger than the volume of the parent red dwarf star. The resultant tail swept back from the star extends about 15 million kilometers into the planetary system.

This is what a 'hot' planetary comet looks like! As the evaporating gases are not entirely swept away by the stellar winds, the red dwarf "allows the huge cloud to gather" around GJ 436b (Bhanoo 2015): A nebulous planet indeed.

Another planet which seems way too big for its small parent star has been found orbiting a young red dwarf star (designated HATS-6) some 500 light years away. Again, the planet lies extremely close to the parent M-dwarf star – a tenth of the distance of Mercury to the Sun (Hartman et al. 2015). It whizzes around the M-dwarf in just 3.3 days, making little old Mercury appear positively tortoise-like in comparison. The odd thing is that the planet has the mass of Saturn, but is the size of Jupiter. Like the previous example, it is "quite a puffed up planet" (ANU 2015). The discovery of this stretched out world calls into question standard models of planet formation, none of which is able to predict such a massive gaseous world in such close proximity to a young red dwarf star.

These planetary distortions take place in extreme proximity to red dwarf stars. Yet, their existence indicates that our notion of regular oblate spheroid worlds has simply been conditioned by what we already know. The universe is beginning to reveal its planetary secrets to us, and even among the tiny slice we have glimpsed at, the variety is incredible. By comparison, we know nothing of dark planets floating through interstellar space, and we can only assume that these, too, obey the rules. So far, the rules have been found wanting.

The Return of Nemesis

Let us say that massive free-floating planets do indeed envelop themselves in a nebula of collected materials from interstellar space, as I have been arguing. Once that world then penetrates into the heliosphere of a mainstream star, like the Sun, then that nebula would be subject to the fierce solar wind. Much of that accumulated material would be swept back from the Sun by the solar wind. This would create a highly visible Winged Disk effect which, like a comet, would significantly dwarf the planet and moons within.

In this way, then, any incursion would be a truly spectacular event. The cometary 'planet' would be far more easily observed than a regular planet. This follows the same argument as that for any comet approaching the Sun.

However, such an incident would be necessarily extremely rare. If this were to occur every few thousand years, then insufficient materials would have built up around the planet in interstellar space in the meantime. Each perihelion passage would rob the planet of its glorious mane, and, once de-cloaked, Planet X would slink off back into interstellar space like a sheared sheep. Instead of a few thousand years, it might take hundreds of thousands, possibly millions of years, to rejuvenate its cloak.

Such a hypothesis, then, applies better to Nemesis than Nibiru. Incursions of a sub-brown dwarf into the planetary zone would be disruptive, and visually spectacular. The cometary planet might even bring catastrophism in its glorious wake. This may yet account for an extinction cycle on Earth, not simply through bombarding the planets with rocks from afar, but by applying a far more personal touch during the reunion with its siblings. Between times, a highly distant, massive Planet X becomes enshrouded within a nebula of accumulated interstellar medium, preventing our astronomers from detecting it quite as easily as they think they should.

Nemesis and Nibiru offer two extremes of the Planet X spectrum. There may be a middle way, and that may take the form of a more massive Planet Nine, located farther out than Caltech has calculated. If such a massive world had an orbit which oscillated over time, then this could periodically bring it within the ~200AU distance advocated as the perihelion distance for Planet Nine. During these rare flyby perihelion passages, the gravitational effects that lead to the creation of the extended scattered disk would be enacted, even in the case of an object which generally maintains its distance. This may provide another reason why Planet Nine currently defies all attempts to find it: Its current distance is an underestimate, as, quite possibly, is its mass.

Just to be clear, I am suggesting that the extinction level events on Earth are indeed coincident with a flyby passage of a massive Planet X body, but that these events are rare examples of the object's closest approaches during its oscillating, or librating, orbital pattern.

Extinction Periodicity

You will recall that the Nemesis theory describes a distant, sub-stellar companion orbiting the Sun. This object, probably of several Jupiter masses, is thought to be located tens of thousands of astronomical units away, deep in the outer Oort cloud of comets. Its proposed existence (Davis, Hut & Muller 1984; Whitmire & Jackson 1984) was a response to a discernible pattern of extinction level events on the Earth (Raup & Sepkoski 1984).

Both theories proved equally contentious. Many doubted that the alleged periodicity of extinctions was accurate. Others argued that such a loosely-bound sub-stellar object in a highly elliptical orbit – which might extend out beyond a light year – is likely to itself be perturbed by external influences; and its own periodicity would therefore vary. Nemesis skeptics point to a lack of discovery of such an object by finely-tuned infrared sky surveys, like 2MASS and WISE (Prigg 2008). We have explored reasons to think that these searches are not as effective at finding such objects as is generally believed.

In 2010, a new study into a possible extinction cycle supported a marked periodicity of extinction level events in the fossil record. The periodicity of these events had been adjusted down to 27 million years, which had taken into account a generally accepted re-evaluation of geological time periods. Additionally, the researchers found that this periodicity extended over a much longer geological time period than previously thought – of up to 500 million years (Melott & Bambach 2010).

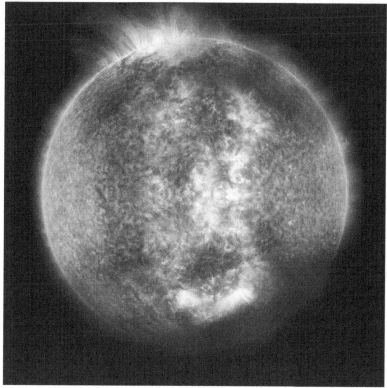

Fig. 22-4 Nemesis image

In the meantime, a complicating factor had been added to the mix. Nemesis theorist, Richard Muller had argued in 2002 that there had been a significant change in the massive planet's orbit. This had occurred, he thought, around 400 million years ago. This adjustment was proposed in response to a noted uptick in impacts on the lunar surface around that time. Muller suggested that the very wide orbit planet had shifted from a broadly circular orbit to a more elliptical one. This had induced a fresh pattern of periodic comet bombardment, aligned with the extinction cycle in the fossil record (Muller 2002).

More recently, scientists examining the link between Extinction Level Events (ELEs) and impact craters over the last 260 million years concluded that there is indeed a viable connection between them. Once again, the phasing of the periodicity is between ~26 million years for the cratering and ~27 million years for the ELEs (Rampino & Caldeira 2015). If one takes into account the level of uncertainty in these chronologies (cumulatively around 1.3 million years), then the periodicity can neatly coalesce into a ~26.5 million year cycle. In other words, we can meet in the middle between the fossil record and the cratering record in a viable way.

Of further concern, a statistical census of comets suggests that the extrapolated populations of comets in the outer solar system are insufficient to account for such regular extinction level events on Earth (Kaib & Quinn 2009). The protecting influence of Jupiter often prevents long-period comets from finding their way into the inner solar system. The cratering record does seem to speak for itself in this regard. The fact that there do not seem to be enough comets to have caused these routine events indicates that something else may be involved, bypassing the planetary zone's defenses.

This is where a modified version of the Nemesis theory could work well. If a Dark Star object was capable of brushing into the planetary zone every 26-27 million years or so, then this would readily explain the significant uptick in planetary bombardment. After all, this object would visit along with a significant retinue of materials, which could immerse itself into the Sun's compressed heliosphere. Rather than showering down comets from afar, the oscillation of the Dark Star's elliptical orbit could make it the direct cause of these extinction level events.

The Kozai Mechanism

Let me be clear: I am not suggesting that the Dark Star has an orbital period of around 27 million years. Far from it: Rather, its orbital period would be measured in the tens, possibly hundreds of thousands of years. However, because its orbit librates over time, this oscillating motion may periodically draws the *Dark Star* closer to the Sun than usual during perihelion. This cosmic concertina effect can be driven by the "eccentric Kozai-Lidov mechanism", which will generate ebb and flow in the evolving orbits of a binary star system object over time, particularly with respect to planetary eccentricity and orientation (Naoz 2016).

These oscillating effects upon one body within the system involve the influence of an externally located third party. There is the potential for this physical mechanism to apply in two distinctly different scenarios:

1. Let us say that we have a three party Planet X system. As well as the Sun and a distant sub-brown dwarf Dark Star located among the comets, we also have a Planet X body in motion between them. Instead of having a nice steady orbit over time, this loosely bound Planet X body is subject to the perturbing influence of the Dark Star beyond. As a result, its orbital eccentricity and inclination oscillate over time. At times, then, Planet X enjoys a wide, more circular orbit around the Sun. At others, its orbit becomes far more eccentric, drawing it closer to us during perihelion.

2. The loosely bound *Dark Star* companion is itself subject to the influence of the external forces, like the galactic tide. It has been shown that the galactic tide can routinely affect binary star systems through a process similar to that of the Kozai-Lidov mechanism (Correa-Otto, Calandra & Gil-Hutton 2017). Given how wide the Dark Star's orbit is, then the effect of external gravita-

tional influences will be felt that much more strongly. The same argument that applies to comets also works for a massive planet within their midst.

Assuming both a Dark Star and a 'regular' Planet X object exist concurrently, both of these distant planets in our system could have their orbits radically altered over time by external influences. The Dark Star is affected by the galactic tide as the Sun bobs its way around the Milky Way; and Planet X/9 is affected by both this and the Dark Star as well. These loosely bound systems oscillate over long time periods, with Planet X/9 and the Dark Star beyond possibly locked into some kind of resonance relationship between them. The implication of this is that the closest distances that either of these bodies could get to perihelia will vary over time – possibly by quite a margin.

If either of them infringe upon the planetary zone, they might bring with them catastrophe. These oscillations over time could then explain why both cratering and extinction patterns are regular, but spread out over periods of time which are multiples of the resonant orbital periods of these bodies. That is because, for the most part, the Kozai mechanism means that Planet X and/or the Dark Star keep their respective distances. It is only during those periods when the eccentricity increases that the extinction stakes are raised.

Habitability and Tidal Locking

The Dark Star system need not just be the harbinger of mass death. It may also be a repository of life. I have long argued that the existence of a distant sub-brown dwarf in our solar system would provide the potential for life to exist, even thrive out there (Lloyd 2005). In recent years, it has become more acceptable to speak of dwarf star systems as potentially being good places to find life. While my main interest is in super-cool sub-brown dwarfs, these objects are difficult to find, and even tougher to study. To give us a flavor of this potential, however, we can learn a lot from their bigger sisters, the red dwarfs.

Red dwarfs are a lot easier to find than brown dwarfs, being that much brighter. They are also more prevalent than yellow dwarfs, like our Sun. Their milder energy output means that habitable planets orbiting red dwarfs would need to be that much closer than the Earth is to the Sun. Practically, though, that need not present too much of a problem, given the sheer abundance and variety of exoplanets in the galaxy. What might be a problem, though, is whether water can survive on the surface of a planet tucked in close to a red dwarf. This has become something of a talking point among astronomers.

The closer a planet gets to its parent star, the more likely it will become tidally locked. This effect is demonstrated by the Moon as it orbits around the Earth. The Moon always shows the same side to us, rotating on its axis at the same rate as it orbits around the Earth. If the Earth was its source of light, rather than the Sun, then the Moon would have a permanently light side and a permanently dark side. This is the scenario for a planet in the habitable zone of a brown dwarf star.

However, if the Moon had its own atmosphere, this synchronizing of orbital period and rotational period would not be assured. Such is the case for Venus –for instance – whose dense atmosphere is a complicating factor. The Sun's tug upon this atmosphere helps to maintain Venus's rotation. This effect is applicable to quite thin planetary atmospheres, too (Leconte et al. 2015).

Things get more complicated if the parent body is a dwarf star. Let us assume that an exoplanet becomes tidally locked to its parent dwarf star. The dark side of such a planet might freeze out, while the volatiles on the light side might be driven off, bringing about extremes of climate that necessarily preclude the existence of liquid water, and thus life. Or there might be a runaway greenhouse effect. Either way, the planet might look rather different to those in our solar system
.

In my second novel **The Followers of Horus** (Lloyd 2010) I imagined a habitable moon of the **Dark Star** as the "Eye of Horus". This moon was in a tidally locked orbit around the **Dark Star**, with one side eternally immersed in its dim red light. The other side was permanently frozen out. Such planets have since been given the moniker "eyeball Earths" (Choi 2013), following an examination of the planetary environment of the exoplanet Gliese 581g (Pierrehumbert 2010). This planet, located some 20 light years away, is located in the "Goldilocks Zone" of the red dwarf Gliese 58.

Computer modeling by geophysicists suggests that buffering mechanisms involving ocean currents and weather patterns would buffer the planetary environment on a tidally locked planet in close proximity to a red dwarf star (Yang et al. 2014). For desiccation to occur, the planet would need to have less than half of the Earth's water to begin with, creating special conditions necessary for the feared 'water-trapping' scenario. This is good news, particularly as red dwarf systems have several other factors in their favor for nurturing life. Red and brown dwarfs burn their fuels much slower than the Sun, and last much longer as stars. They also emit less in the way of life-endangering radiation. There is a potential for a greater window of opportunity for life to evolve during the lifetimes of these planets.

Whether cooling brown dwarfs are sufficiently energetic to create the right long-term conditions for life within their systems is uncertain (Barnes & Heller 2012). Their initial bursts of ultraviolet light and heat may rid their planets of water, too, leaving them arid early on. This is the same issue troubling scientists modeling the water cycle in the inner solar system, and I do not think that this is an insurmountable problem. After all, the Earth has bountiful water irrespective of this theoretical issue.

In the case of a sub-brown dwarf, the moon would have to be very close to the primary body to sit within its habitable zone. Rather than being heated directly, the moon's warmth would be generated internally from tidal friction. It is not certain whether tidal locking would occur, though – there is great variation in the solar system regarding whether moons become tidally locked to their parent planets, and many factors are at play. While sub-surface oceans seem to be a routine feature of gas giant moons, it is less clear whether a sub-brown dwarf could support a room-temperature atmosphere around an orbiting terrestrial world. There is no precedent for this within the solar system. That does not mean it is impossible.

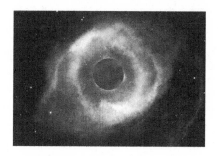

Conclusion

Many astronomers have sought out the elusive Planet X for over a century and more, without success. Many other astronomers consider this hunt for another planet to be entirely futile. Yet, even the skeptics would agree that the emerging picture of the outer solar system is puzzling.

The objects in the extended scattered disk beyond the Kuiper belt, like Sedna, have challenged theoretical models which had previously predicted order. Advances in computer technology have enabled complex, randomized simulations to be conducted in the hope of explaining how the apparently impossible orbits of these objects arose. Despite the general, ingrained cynicism among scientists towards Planet X, leading astrophysicists specializing in the dynamics of the outer solar system have proposed the existence of an undiscovered planet – a distant puppet-master pulling the strings of these anomalous cosmic bodies.

From the beginning of this century, there has been a growing movement of support for Planet X within the academic astrophysics community. It began with a few maverick papers regarding the non-random patterning of long period comets. Then it shifted ground inwards, focusing on the unpredictable bodies in the extended disk between the Kuiper belt and the Oort cloud. Careful computer modeling revealed orbital patterns which were suggestive of the presence of another planet.

The main advocates of this theory are not speculative researchers, or doomsday prophets, but rather leading theorists and astronomers in their specialized fields. However, aware of the negative connotations from decades of alarmist speculation about Planet X, the object they seek needed re-branding. Following the contentious demotion of Pluto to dwarf planet status, Planet X was renamed Planet Nine.

Since the well-publicized release of the theoretical evidence to support the existence of Planet Nine, attempts to find it have come up with nothing. This is despite widespread scientific interest in the subject, the use of ever-more powerful telescopes in infrared and visible light, and the greatest motivation of all – the hope of going down in history as the discoverer of a new planet in the solar system.

Further discoveries of scattered disk objects have either reinforced the notion of a perturbing body, or neutralized the observed effect – depending upon your perspective. It is curious, though, that while the indirect evidence for Planet X has strengthened considerably in recent years, along with the observational power brought to bear to find it – it remains as elusive as ever. Scientists may simply have come to the wrong conclusions about where it lies, and have concentrated their search in the wrong place. Or, it may be more difficult to find than had been expected.

The presence of a massive Planet X body is perplexing in many ways. Current models of planet formation cannot account for it having formed in its current location. Instead, astrophysicists consider whether it might have been captured by the Sun, or whether it formed among the regular planets and was displaced early on in the history of the solar system. Both possibilities are problematic, and represent low probability events. These considerations fuel skepticism about the existence of another sizable planet, but they also highlight other issues. The contentious presence of Planet X takes its place alongside other anomalies in the solar system which are equally puzzling.

The solar system may seem, on the face of it, an orderly construct which evolved steadily from the Sun's protoplanetary disk. However, as the old cliché goes, the devil is in the details. The outer solar system is particularly problematic. There is a substantial missing chunk of mass beyond Neptune – a planet that is inexplicably more massive than its inner sibling, Uranus. These anomalies suggest that the outskirts of the planetary zone have been shaped by an external influence (Pfalzner et al. 2018). On top of the glaring irregularities of the Kuiper belt and diffuse scattered disk beyond, the apparent order elsewhere within the planetary zone belies a muddled arrangement of interlacing comets and asteroids, and irregular moons. Complexity is the order of the day in our postmodern solar system.

The arrangement of deuterium-to-hydrogen isotope ratios of water found on solar system bodies is far from the linear pattern predicted. The Earth has more in common with asteroids than comets. This suggests either a large-scale shuffling of materials within the solar system, or a point of common origin between the Earth/Moon system and the asteroid belt.

Once considered dead rocks, or dirty snowballs, the minor bodies of the solar system are also more active than previously thought. Unexpected surface features on dwarf planets imply ongoing geophysical processes within these ancient objects. Sub-surface oceans seem ubiquitous, and even Mars is now known to have a liquid water lake below its surface (Orosei et al. 2018). Nothing is quite what it seems, and this all suggests that the solar system is a far more active environment than previously thought possible.

The presence of one or more additional planets is in keeping with this unexpected set of features. Planet X potentially provides a driving force behind periodic change; or, at the very least, an unsettling influence. It stands as an alternative to the ever more complex mechanisms that constitute the Nice model. Instead of four major planets migrating across the solar system in tandem, spilling minor bodies all over the place, a single incursion of a massive Planet X body through the solar system can, at a stroke, explain the wealth of anomalies present.

It matters little whether it was originally a dislodged planet within the solar system, or one that formed at the periphery of the solar system. It may have been a planet that entered our system as a free-floating entity, or was part of the planetary retinue of a star passing close to the Sun.

If such a body encountered the early Earth/Moon system some 4 billion years ago, then the catastrophic aftereffects of that encounter would have been written all over the solar system. This is indeed what we see. Such an encounter provides common ground between the Earth and the asteroids. It provides the trigger point for a long-standing period of chaos, with one terrestrial planet being shunted out of place, rather than four gaseous behemoths.

The weird thing here is that this scenario has been spoken of before. It is an important part of what is generally considered to be a pseudo-scientific theory, based upon one man's quirky interpretation of ancient Mesopotamian myths. In the world of grown-ups, it simply should not be possible that this turns out to be the correct solution to all of these anomalies. But, what if it is? What, if by some bizarre twist of fate, the fringe theory turned out to be right after all? That it has some kind of bizarrely accurate predictive power? Astrophysicists can re-brand Planet X, and dismiss the obvious correlations all they want. The presence of Planet X still potentially represents an uncomfortable truth.

Yet, this planet continues to elude astronomers. All things being equal, it should have been found by now. In this book, I have tried to square that difficult circle. I have argued that assumptions about how planets in interstellar space appear may be wide of the mark. Although astronomers are well aware of the power of the solar wind to clear out interplanetary space of dust and gas, they do not appear to have considered the planetary science outside of that domain. Why should they have? Their interest lies with detectable planets within star systems, and these are extremely difficult to find and study.

Dark planets located beyond the heliospheres of their parent suns are practically impossible to find, unless they are young brown dwarfs (and therefore at the lower end of the stellar spectrum). Besides, these objects behave differently, creating their own, short-lived stellar wind and heliosphere. In no time at all, they become dark, traveling loners, hiding in the shadows beyond the glare of the Milky Way's streetlights.

No one specializes in the study of interstellar planets. After all, there is practically nothing tangible to actually study yet. Our only hope of understanding free-floating planetary mass objects is to find one close to home, by sheer good fortune, or to study other visitors from interstellar space, like 1I/'Oumuamua. As we have seen, this bizarre object resets several compasses, and not just because of its rapid tumbling motion. It compels us to rethink what is floating around out there, and why this material might be very different from what we initially imagined.

In this book, I have set out an idea about how planets located beyond the heliopause can continue to accrete material from the interstellar medium, albeit intermittently. I have argued that, in the absence of the solar wind, a dynamic equilibrium is set up between in-falling materials, and collisional cascades of materials captured into orbit around the planet. This creates a complex local environment, mostly in the form of a nebula enshrouding the main planet. This nebula creates an indistinct target for astronomers seeking out the pinpoint light of a planet that they would expect to see, and masks the en-wrapped planet's heat signature. As a result, candidate objects may have been dismissed many times over, as they more closely resemble a fuzzy background object, like a distant galaxy, or dark nebula. I can only hope that some of the ideas in this book will help to make future searches more inclusive of fuzzy, nebulous objects.

What I have not been able to do is take this hypothesis to the next stage. I have not yet been able to get a science paper about this published. Worse, I have failed to interest research scientists in the potential of these ideas. Without their technical support, I am unable to conduct scientific experimentation to back up my theoretical claims. This is frustrating, but not entirely surprising. These are busy people, and I am not one of their graduate students.

I do note with some optimism, however, that sometimes solutions come out of unusual places. An example is the discovery of the mechanism for oxygen release from a cometary surface. The potential for creating atmospheric environments through particle bombardment of the surface is not lost on me. This could be readily applied to planets in interstellar space. The issue, as ever, is heat. Where comets undergoing such processes are being heated by the Sun during perihelion, planets beyond the heliosphere occupy an extremely frigid environment indeed.

This brings me back to the material I first covered on my websites, and in the book **Dark Star**. Since publication in 2005, it has become clear that sub-surface oceans exist in many bodies of the solar system. Some of these are moons that are internally warmed by their parent giant planets. Enceladus is a spectacular example, but by no means the only one. Even Pluto seems to be getting in on the act, and its partner is tiny Charon. So, internal heating processes seem to be more significant than previously thought.

Scientists scouring the data from WISE have ruled out a massive planet in the solar system. I am not sure that they have been looking for the right thing, and I think they have overestimated the strength of both visible light and infrared making it back to us. I still think, therefore, that there may be a very massive planet, a Dark Star, out there. It would have to be enshrouded in gas and dust, safely cocooned within its own celestial bubble. Within this local environment, the right conditions might exist for a sustainable atmosphere on a planet orbiting this sub-brown dwarf. There could also be exposed oceans, rather than the sub-surface ones detected on so many moons in the solar system.

In this way, I envisage habitable planets in interstellar space, driven by tidal heating, and protected by a nebula shroud. It is true that these would be relatively dark environments – at least, as seen from the outside: The powerful central planet might provide an aurora display exacerbated by the local mass of materials and objects in its midst. It is true that this faint polar light would be insufficiently intense to escape the nebula cloud in any meaningful way. But it might be sufficient to drive some biological processes locally, perhaps even photosynthesis. An enshrouded Dark Star might also intermittently flare up to provide sufficient life-supporting light and heat.

This is, of course, highly speculative. But it provides new ideas to consider, and to test. At the moment, we simply do not know what the conditions in interstellar space will be like around free-floating planets, or wide orbit planets located beyond heliospheres. It is a given that they will be the same as within heliospheres – that everything will be very familiar about these planets. However, that is rarely how things turn out in science.

You know, I don't think anyone's really ever thought about it. Why should they have? As Plato once noted, **necessity is the mother of invention**. These ideas have only occurred to me because I have been grappling with the multiple and often arcane problems of Planet X for so long.

Many people ask me two questions about Nibiru: Where is it? When it will return? I will be truthful when I say that I do not know the answer to the first question any better than anyone else. I have favored locations, like the Solar Apex or Sagittarius, but there is insufficient evidence to prove the case for either.

The second question is, I believe, misplaced. If the Universe is amusing itself with us, then some of what Zecharia Sitchin wrote may turn out to be accurate. But I think he made some incorrect assumptions, too. I think his proposal about the orbital period of this body being 3,600 years is wide of the mark – probably by an order, or even two, of magnitude. I believe that this number was just an educated guess on his part, based upon the name given to an important number in the complex sexagesimal system. It was always a little too neat, to my mind.

Perhaps because of its very simplicity, it became a central plank of Sitchin's theory. It offered his readers a predictive clarity that the notion of a **Sar** honestly did not merit. By comparison, the Planet Nine hypothesis offers a far more realistic proposition, with an orbital period measured in the tens of thousands of years. Moreover, the science behind it is rigorous.

Then there is the not insignificant problem of the ongoing disruptive effect a continually returning planet would have on the asteroid belt. The Lidov-Kozai mechanism could prove useful for understanding the oscillating adjustments to Planet X's orbit over time, but it also creates an issue within the planetary zone, too. If Nibiru returns frequently, then it would have affected objects in our midst, destabilizing the order we see along the ecliptic plane. So, if a substantial Planet X body does return, it must be an infrequent event, with sufficient time between perihelion passages for the solar system to settle back down again.

If I am right, then there is no doomsday planet to concern ourselves with. That is not to say that a rogue planet did not rock our world at some point in the past, though; certainly around four billion years ago, and likely every ~27 million years or so over the last 500 million years. Both of the planets which featured in the catastrophic encounter which triggered the Late, Heavy Bombardment migrated permanently away from the conflict zone: The Earth inwards, the rogue planet outwards. Only the rubble of the asteroid belt was left behind, providing a permanent reminder of the cataclysm.

This latter point is where Sitchin's work may be prophetic. His interpretation of the ancient Mesopotamian creation mythology may turn out to have some basis in truth. It accounts for many of the anomalies in the solar system, and represents a far more straightforward solution than the current Nice model of solar system formation and evolution. Sitchin's contentious version of Planet X reflects some of the ideas put forward to account for the presence of Planet Nine. It explains how Earth got its water, and why it's the same as the asteroids, not the comets. It is a viable hypothesis, no matter how unpalatable its origins.

That said, Planet Nine is a super-Earth, or mini-Neptune, according to the astrophysicists. If Planet Nine was the rogue planet which battered the solar system 4 billion years ago, then its mass is still an open question. At 1000AU, or so, it is clearly located far too close to be a sub-brown dwarf. A sub-brown dwarf at this distance would be causing much greater gravitational disturbances than those noted for the extended scattered disk. The location of a Dark Star would need to be much deeper out among the comets: Perhaps not as far as Tyche or Nemesis, but certainly further out than Planet Nine. Its orbit might be counted in hundreds of thousands of years. Its 'return' every 27 million years would represent a relatively close flyby, resulting from its oscillating orbit, which either disrupts comets or brings with it a quite different retinue of objects.

If my hypothesis about nebulae aggregating around interstellar planets has any truth to it, then it is possible that this massive world has been missed by the sky searches charged with locating these kinds of planets in the solar neighborhood. They could have overlooked it, or simply not recognized it for what it is if they did spot it.

This Dark Star might even be part of a three-body solution, encompassing a lesser intermediary. In that case, Nibiru is either a minor dwarf Planet X body, or a spectacular, ruddy comet. It is locked into a resonance orbit with the Dark Star and, as a "ferry", merely signposts us towards the much greater entity beyond.

That massive planet out among the comets could provide a habitat for life. It is not necessary to invoke spectacularly big planets to generate the conditions for life out there. A super-Earth might suffice, either within its own right, or upon one or more of its moons. The possibilities here are limited to sub-surface oceanic life, however. A mini-Neptune would be very cold, and a super-Earth frozen over. Binary planets can create mutually complex conditions, as with Pluto/Charon. But, these possibilities seem limited.

Planet Nine offers less potential for life than a more distant sub-brown dwarf. As a result, I still favor the sub-brown dwarf scenario. In the hunt for this object, the criminal profiling might prove to be correct; it is the culprit's strange appearance that is causing problems with its capture.

The Dark Star is capable of causing the same effects currently attributed to Planet Nine, but it is located much further out, probably within the inner Oort cloud. Indeed, the objects within the inner Oort cloud might be under its influence in the same way that Neptune shepherds the TNOs. We might then need to consider the denizens of the inner Oort clouds as TXOs.

I would argue that the very existence of the Hills cloud implies a massive body within its immediate vicinity, as well as being an occasional 'Doomsday Planet'. When the oscillations of its very wide orbit allow, it is also a crucible of life among the comets.

The **Dark Star** offers us a crucial link between myth and science.

References

A & G (2015). What shape is the heliosphere? Astronomy and Geophysics, 1 April 2015, 56(2): 2.8. doi: 10.1093/astrogeo/atv048

Acedo, L. (2016). On the effect of ocean tides and tesseral harmonics on spacecraft flybys of the Earth. Monthly Notices of the Royal Astronomical Society, 463(2): pp 2119-2124

Acedo, L. (2017). Anomalous accelerations in spacecraft flybys of the Earth. Astrophysics and Space Science, 362(12). Available at: https://www.researchgate.net/publication/320944078_Anomalous_accelerations_in_spacecraft_flybys_of_the_Earth

Acedo, L. (2017a). Correspondence from Luis Acedo, 11 December 2017

Acedo, L. (2017b). Correspondence from Luis Acedo, 12 December 2017

Acedo, L., Piqueras, P. & Moraño, J. (2018). A possible flyby anomaly for Juno at Jupiter. Advances in Space Research, 61(10): pp 2697-2706

Adler, R. (2016). Nibiru, Planet X, gaining traction among Rabbinic voices as harbinger of End of Days. Beit Shemesh. Breaking Israel News, 22 February 2016, Available at: http://www.breakingisraelnews.com/61965/nibiru-coming-bringing-messiah-wake-jewish-world/

Agle, D. & Brown, D. (2017). Jupiter's auroras present a powerful mystery. Pasadena, CA: NASA/JPL. 6 September 2017. Available at: https://www.missionjuno.swri.edu/news/jupiters-aurora-presents-a-powerful-mystery

Albarede, F. et al. (2013). Asteroidal impacts and the origin of terrestrial and lunar volatiles. Icarus, 22(1): pp 44-52

Alford, A. (1996). Gods of the New Millennium. Eridu books

Alford, A. (1998). The Phoenix Solution. London: Hodder & Stoughton

Altwegg, K. et al. (2015). 67P/Churyumov-Gerasimenko, a Jupiter family comet with a high D/H ratio. Science, 347(6220)

Amos, J. (2014). Mars Maven mission arrives in orbit. London: BBC. 22 September 2014. Available at: http://www.bbc.co.uk/news/science-environment-29253788

Ananthaswamy, A. (2015). Stellar intruder's daring fly-by of the solar system. London: New Scientist. 19 February 2015. Available at: https://www.newscientist.com/article/dn27004-stellar-intruders-daring-fly-by-of-the-solar-system/

Anderson, J. et al. (2001). Study of the anomalous acceleration of Pioneer 10 and 11. Physical Review D. 65 (8)

Andrews, D. et al. (2006). Ptolemy: An instrument aboard the Rosetta lander Philae, to unlock the secrets of the solar system. In: 37th Lunar and Planetary Science Conference, 13-17 March 2006. Available at: http://oro.open.ac.uk/8183/1/Ptolemy_-_An_Instrument_Aboard_the_Rosetta_Lander_Philae,_to_Unlock_the_Secrets_of_the_Solar_System.pdf

Andrews, R. (2016). Colossal Antarctic ice shelf collapsed at end of last Ice Age. 22 February 2016. Available at: http://www.iflscience.com/environment/colossal-antarctic-ice-shelf-collapsed-end-last-ice-age

Anglés-Alcázar, D. et al. (2017). The cosmic baryon cycle and galaxy mass assembly in the FIRE simulations. MNRAS, 470, (4): pp4698-4719

ANU (2015). New exoplanet too big for its star challenges ideas about how planets form. Canberra: Australian National University Press Release, 1 May 2015. Available at: http://www.anu.edu.au/news/all-news/new-exoplanet-too-big-for-its-star

ASU (2016). Ceres: The tiny world where volcanoes erupt ice. Tempe, AZ: Arizona State University Press Release, 1 September 2016. Available at: https://asunow.asu.edu/20160901-ceres-asu-tiny-world-where-volcanoes-erupt-ice

Bailer-Jones, C. (2015). Close encounters of the stellar kind. Astronomy & Astrophysics, 575: A35

Bailey, E., Batygin, K. & Brown, M. (2016). Solar obliquity induced by Planet Nine. The Astronomical Journal, 152(5)

Bandfield, J. et al. (2018). Widespread distribution of OH/H2O on the lunar surface inferred from spectral data. Nature Geoscience, 11: pp 173–177

Bannister, M. et al (2017). Col-OSSOS: Colors of the Interstellar Planetesimal 1I/'Oumuamua. The Astrophysical Journal Letters, 851(2). Available at: http://iopscience.iop.org/article/10.3847/2041-8213/aaa07c/meta

Barboni, M. et al. (2017). Early formation of the Moon 4.51 billion years ago. Science Advances, 3(1): e1602365

Barnes, R. & Heller, R. (2012). Habitable planets around white and brown dwarfs: The perils of a cooling primary. Astrobiology, 13(3): pp 279-291

Barr, A. & Collins, G. (2014). Tectonic activity on Pluto after the Charon-forming impact. Icarus, 246: pp 146-155

Barucci, M. A. et al. (2008). Composition and surface properties of Transneptunian Objects and Centaurs. In: The Solar System Beyond Neptune. University of Arizona space science series. Tucson, AZ: University of Arizona Press , pp143-160. Available at: http://www.lpi.usra.edu/books/ssbn2008/7005.pdf

Batygin, K. (2012). A primordial origin for misalignments between stellar spin axes and planetary orbits. Nature, 491: pp418-20

Batygin, K. & Brown, M. (2016). Evidence for a distant giant planet in the solar system. The Astronomical Journal, 20 January 2016, 151(2). Available at: http://iopscience.iop.org/article/10.3847/0004-6256/151/2/22

Batygin, K. & Brown, M. (2016a). Generation of highly inclined Trans-Neptunian Objects by Planet Nine. Pasadena: Online. 17 October 2016. Available at: https://arxiv.org/pdf/1610.04992v1.pdf

Batygin, K. (2017). The search for Planet Nine: Status update (part 1). Pasadena: Online. 30 June 2017. Available at: http://www.findplanetnine.com/2017/06/status-update-part-1.html

Batygin, K. (2017a). The search for Planet Nine: Status update (part 2). Pasadena: Online. 2 July 2017. Available at: http://www.findplanetnine.com/2017/07/status-update-part-2.html

Batygin, K. & Morbidelli, A. (2017). Dynamical Evolution Induced by Planet Nine. The Astronomical Journal, 154(6). Available at: https://arxiv.org/pdf/1710.01804.pdf

Beatty, K. (2017). Astronomers spot first known interstellar "comet". 25th October 2017 Available at: http://www.skyandtelescope.com/astronomy-news/astronomers-spot-first-known-interstellar-comet/

Becker, J. et al. (2018). Discovery and dynamical analysis of an Extreme Trans-Neptunian Object with a high orbital inclination. Submitted to the Astrophysical Journal, 16 May 2018. Available at: https://arxiv.org/pdf/1805.05355.pdf

Begelman, M. & Rees, M. (1976). Can cosmic clouds cause climatic catastrophes? Nature, 26: pp298 - 299

Bell, E. et al. (2015). Potentially biogenic carbon preserved in a 4.1 billion-year-old zircon. Proceedings of the National Academy of Sciences, 112(47): pp14518-21

Belyaev, M. & Rafikov, R. (2010). The dynamics of dust grains in the outer solar system. The Astrophysical Journal, 723: pp1718–1735

Bennett, J. (2017). New Horizons spacecraft on its way to mysterious Kuiper Belt Object. 7 July 2017. Available at: http://www.popularmechanics.com/space/deep-space/a27218/new-horizons-spacecraft-mysterious-kuiper-belt-object/

Berkowitz, A. (2016). Super-comet Nibiru could threaten entire planet, warns prominent Rabbi. Beit Shemesh. Breaking Israel News, 4 March 2016. Available at: http://www.breakingisraelnews.com/62890/prominent-rabbi-confirms-nibiru-possible-jewish-world/

Best, W. et al. (2017). The young L Dwarf 2MASS J11193254−1137466 Is a planetary-mass binary. The Astrophysical Journal Letters, 843(1): L4

Bhanoo, S. (2015). A planet with a tail nine million miles long. New York: The New York Times, 25 June 2015. Available at: https://www.nytimes.com/interactive/projects/cp/summer-of-science-2015/latest/exoplanet-tail

Bieler, A. et al. (2015). Abundant molecular oxygen in the coma of Comet 67P/Churyumov-Gerasimenko. Nature, 526: pp 678–681

Billings, L. (2015). Astronomers skeptical over "Planet X" claims. Springer Nature: Scientific American, 10 December 2015. Available at: http://www.scientificamerican.com/article/astronomers-skeptical-over-planet-x-claims/

Billings, L. (2018). Looking for Planet Nine, astronomers gaze into the abyss. Springer Nature: Scientific American. 22 March 2018. Available at: https://www.scientificamerican.com/article/looking-for-planet-nine-astronomers-gaze-into-the-abyss/

Bond, A. & Hempsell, M. (2008). A Sumerian observation of the Kofels' Impact Event. Kidlington: Writersprintshop Limited

Borenstein, S. (2010). Pluto turning red due to solar flares, Southwest Research Institute in Boulder Colorado determines. New York, NY: Huffington Post. 6 April 2010. Available at: http://www.huffingtonpost.com/2010/02/04/pluto-turning-red-due-to-_n_450100.html

Bottke, W. et al. (2015). Dating the Moon-forming impact event with asteroidal meteorites. Science, 348(6232): pp 321-323

Bottke, W. & Andrews-Hanna, J. (2017). A post-accretionary lull in large impact on early Mars. Nature Geoscience, 10(5)

Bouquet, A. et al (2015). Possible evidence for a methane source in Enceladus' ocean. Geophysical Research Letters. 42: pp1334-1339. Available at: https://agupubs.onlinelibrary.wiley.com/doi/epdf/10.1002/2014GL063013

Boyle, A. (2010). The case for Pluto: how a little planet made a big difference. Hoboken: John Wiley & Sons

Boyle, R. (2016). Planet Nine may have tilted entire solar system except the Sun. London: New Scientist, 19 July 2016. Available at: https://www.newscientist.com/article/2098029-planet-nine-may-have-tilted-entire-solar-system-except-the-sun/

Breakthrough Initiatives (2017) Breakthrough listen to observe interstellar object 'Oumuamua. 11 December 2017. Available at: https://breakthroughinitiatives.org/news/14

Brennan, P. (2017). The super-Earth that came home for dinner. Pasadena, CA: NASA. 4 October. Available at: 2017 https://www.jpl.nasa.gov/news/news.php?release=2017-259

Brown, D. et al. (2015). NASA confirms evidence that liquid water flows on today's Mars. Pasadena, CA: NASA/JPL. 28 November 2015. Available at: https://www.nasa.gov/press-release/nasa-confirms-evidence-that-liquid-water-flows-on-today-s-mars

Brown, D. et al. (2015a). NASA mission reveals speed of solar wind stripping Martian atmosphere. Pasadena, CA: NASA/JPL. 5 November 2015. Available at: https://www.nasa.gov/press-release/nasa-mission-reveals-speed-of-solar-wind-stripping-martian-atmosphere

Brown, M. (2008). The dwarf planets. Pasadena: California Institute of Technology. Available at: http://web.gps.caltech.edu/~mbrown/dwarfplanets/

Brown, M. (2010). Why I killed Pluto and why it had it coming. New York: Spiegel & Grau

Brown, M. (2016). Where is Planet Nine? Pasadena: Online, 25 January 2016. Available at: http://www.findplanetnine.com/p/blog-page.html

Brown, M. (2017). Observational bias and the clustering of distant eccentric Kuiper Belt Objects. The Astronomical Journal, 154(2)

Browne, M. (1993). Evidence for Planet X evaporates in spotlight of new research. New York: The New York Times, 1 June 1993. Available at: http://www.nytimes.com/1993/06/01/science/evidence-for-planet-x-evaporates-in-spotlight-of-new-research.html

Brunini, A. & Melita, M. (2002).The existence of a planet beyond 50AU and the orbital distribution of the classical Edgeworth-Kuiper Belt Objects. Icarus, 1, 160: pp32-43

Bryant, T. (2016). Failed star creates its own spotlight in the universe. Newark, DE: University of Delaware Press Release. 13th June 2016. Available at: http://www.udel.edu/udaily/2016/june/failed-star/

Buie, M. & Folkner, W. (2015). Astrometry of Pluto from 1930-1951 observations: the Lampland Plate 17 collection. Astronomical Journal, 149,(1): p13

Burrows, A., Sudarsky, D. & Hubeny, I. (2006). L and T Dwarf models and the L to T Transition. The Astrophysical Journal, 2006, 640(2). Available at: http://iopscience.iop.org/article/10.1086/500293/pdf

Campbell, B. & Morgan, G. (2018). Fine scale layering of mars polar deposits and signatures of ice content in nonpolar material from multiband SHARAD data processing. Geophysical Research Letters, 45(4): pp 1759-1766

Canup, R. (2005). A giant impact origin of Pluto-Charon. Science, 307(5709): pp 546-550.

Capaccioni, F. et al. (2015). The organic-rich surface of comet 67P/Churyumov-Gerasimenko as seen by VIRTIS/Rosetta. Science, 347(6220)

Carlisle, C. (2017). Milky Way may be made with swapped gas. New York: Sky and Telescope Magazine. 27 July 2017. Available at: http://www.skyandtelescope.com/astronomy-news/milky-way-may-be-made-with-swapped-gas-2707201723/

Carnegie Institution for Science (2017). Hunting for giant planet analogs in our own backyard. Washington D.C.: CIS. 1 March 2017. Available at: www.sciencedaily.com/releases/2017/03/170301130527.htm

Carnegie Institution for Science (2018). When do aging brown dwarfs sweep the clouds away? Washington D.C.: CIS. 26 February 2018. Available at: https://carnegiescience.edu/news/when-do-aging-brown-dwarfs-sweep-clouds-away

Carrozzo, F. et al. (2018). Nature, formation, and distribution of carbonates on Ceres. Science Advances, 4(3): e1701645

Cesare, C. (2015). Volcanic rock hints at source of Earth's water. Nature. 12 November 2015. Available at: https://www.nature.com/news/volcanic-rock-hints-at-source-of-earth-s-water-1.18779

Choi, C. (2011). Milky Way owes its shape to crashes with dwarf galaxy. New York: Space.com. 14 September 2011. Available at: https://www.space.com/12952-milkyway-galaxy-shape-galactic-crash.html

Choi, C. (2013). Eyeball Earths. Washington DC: Astrobiology Magazine. 25 April 2013. Available at: https://www.astrobio.net/meteoritescomets-and-asteroids/eyeball-earths/

Choi, C. (2017). Did Pluto's weird red spots result from crash that spawned Charon? New York: space.com. 30 January 2017. Available at: https://www.space.com/35501-pluto-red-spots-charon-moon-collision.html

Chu, J. (2014). 'Rogue' asteroids may be the norm: A new map of the solar system's asteroids shows more diversity than previously thought. Cambridge, MA: Massachusetts Institute of Technology News Release. 29 January 2014 Available at: http://news.mit.edu/2014/rogue-asteroids-may-be-the-norm-0129

CIT press release (2004). Planetoid found in Kuiper belt, maybe the biggest yet. Pasadena: California Institute of Technology, 20 February 2004. Available at: https://spaceflightnow.com/news/n0402/20kuiper

Clark, S. (2005). Far-out worlds, just waiting to be found. New Scientist, 20 July 2005, 2509. Available at: https://www.newscientist.com/article/mg18725091-500-far-out-worlds-just-waiting-to-be-found/

Clarke, A. C. (1973) Rendezvous with Rama. London: Gollancz 1973

Clavin, W. (2012). WISE finds few brown dwarfs close to home. Pasadena: NASA/JPL. 8 June 2012, Available at: https://www.nasa.gov/mission_pages/WISE/news/wise20120608.html

Clavin W. (2014). Stormy stars? NASA's Spitzer probes weather on brown dwarfs. Pasadena: NASA/JPL. 7 January 2014. Available at: https://www.nasa.gov/jpl/spitzer/brown-dwarf-20140107

Clavin, W. (2017). Caltech chemical engineer explains oxygen mystery on comets. Pasadena, CA: Caltech. 8 May 2017. Available at: http://www.caltech.edu/news/caltech-chemical-engineer-explains-oxygen-mystery-comets-61080

Clavin, W. & Harrington, J. (2014). NASA's WISE survey finds thousands of new stars, but no 'Planet X'. Pasadena: JPL/CIT, 7 March 2014. Available at: http://www.jpl.nasa.gov/news/news.php?release=2014-075

Cleeves, L.I. et al. (2014). The ancient heritage of water ice in the solar system. Science, 345(6204): pp 1590-1593

Collins, A. (2013). Göbekli Tepe: Its cosmic blueprint revealed. Available at: http://www.andrewcollins.com/page/articles/Gobekli.htm

Cooper, J. et al. (2003). Proton irradiation of Centaur, Kuiper Belt, and Oort cloud objects at plasma to cosmic ray energy. Earth, Moon, and Planets, 92, 1-4: pp261-277

Connelly, J., et al. (2012). The absolute chronology and thermal processing of solids in the solar protoplanetary disk. Science, 338(6107): pp651–655

Connor, S. (2015). Philae lander data show comets could have brought 'building blocks of life' to Earth. The Independent: Online. Available at: 30 July 2015 http://www.independent.co.uk/news/science/philae-lander-data-show-comets-could-have-brought-building-blocks-of-life-to-earth-10427773.html

Cornette, A. (2014). Correspondence received from Al Cornette, 14 October 2014

Cornette, A. (2017). Re: The solar system just had an interstellar visitor. Now it's gone. 26 October 2017. Available at: https://groups.google.com/forum/#!topic/dark-star-planet-x/1sGhPILl8Zk

Correa-Otto, J., Calandra, M. & Gil-Hutton, R. (2017). A new insight into the Galactic potential: A simple secular model for the evolution of binary systems in the solar neighbourhood. Astronomy & Astrophysics, 600, A59

Couper, H. & Henbest, N. (2002). The hunt for Planet X. New Scientist, 14th December 2002, pp30-4

Crane, L. (2017). NASA fires Voyager 1's engines for the first time in 37 years. London: New Scientist, 5 December 2017. Available at: https://www.newscientist.com/article/2155460-nasa-fires-voyager-1s-engines-for-the-first-time-in-37-years/

Crockett, C. (2014). Shadow planet: Strange orbits in the Kuiper belt revive talk of a Planet X in the solar system. Science News, 29 November 2014, 186(11): p18

Crockett, C. (2015). Quest to trace origin of Earth's water is 'a complete mess'. 5 August 2015. Available at:

https://www.sciencenews.org/article/quest-trace-origin-earth%E2%80%99s-water-%E2%80%98-complete-mess%E2%80%99

Ćuk, M. & Stewart, S. (2011). The puzzle of lunar inclination. EPSC-DPS Joint Meeting 2-7 October 2011, Nantes, France. p 580

Ćuk, M. (2018) 1I/'Oumuamua as a tidal disruption fragment from a binary star system. Astrophysical Journal Letters, 852(1). Available at: http://iopscience.iop.org/article/10.3847/2041-8213/aaa3db/meta

Currie, T. et al. (2012). Direct Imaging Confirmation and Characterization of a Dust-Enshrouded Candidate Exoplanet Orbiting Fomalhaut. The Astrophysical Journal Letters, 760(2)

Davey, M. (2014). Massive 'ocean' lies beneath the Earth's Surface. London: The Guardian. 14 June 2014. Available at: http://www.theguardian.com/science/2014/jun/13/earth-may-have-underground-ocean-three-times-that-on-surface

Davidson, K. (2005). Ice ages linked to galactic position: Study finds Earth may be cooled by movement through Milky Way's stellar clouds. San Francisco: San Francisco Chronicle, 25 June 2005. Available at: http://www.sfgate.com/news/article/Ice-ages-linked-to-galactic-position-Study-2620619.php

Davies, P., Brooke, G. & Callaway, P. (2002). The complete world of the dead sea scrolls. London: Thames & Hudson

Davis, M., Hut, P. & Muller, R. (1984). Extinction of species by periodic comet showers. Nature, 308(5961): pp715–717

Davis, N, (2017). Mysterious object seen speeding past sun could be 'visitor from another star system'. London: The Guardian. 27 October 2017. Available at: https://www.theguardian.com/science/2017/oct/27/mysterious-object-detected-speeding-past-the-sun-could-be-from-another-solar-system-a2017-u1

Deacon, N., Schlieder, J. & Murphy, S. (2016). A nearby young M dwarf with a wide, possibly planetary-mass companion. Monthly Notices of the Royal Astronomical Society, 457(3): pp3191–3199

Deienno, R. et al. (2016). Is the Grand Tack model compatible with the orbital distribution of main belt asteroids? Icarus, 272: pp 114-124

de la Fuente Marcos, C. & de la Fuente Marcos, R. (2014). Extreme trans-Neptunian objects and the Kozai mechanism: signalling the presence of trans-Plutonian planets. Monthly Notices of the Royal Astronomical Society Letters, 443(1): L59-L63
de la Fuente Marcos, C. & de la Fuente Marcos, R. (2016). Finding Planet Nine: a Monte Carlo approach.

Monthly Notices of the Royal Astronomical Society, 459(1): L66-L70

de la Fuente Marcos, C. & de la Fuente Marcos, R. (2016a). Finding Planet Nine: apsidal anti-alignment Monte Carlo results. Monthly Notices of the Royal Astronomical Society, 462(2): p. 1972-1977. Available at: https://arxiv.org/pdf/1607.05633.pdf

de la Fuente Marcos, C., de la Fuente Marcos, R. & Aarseth, S. (2016). Dynamical impact of the Planet Nine scenario: N-body experiments. Monthly Notices of the Royal Astronomical Society: Letters, 460(1): L123-L127

de le Fuente Marcos, C. (2017). Correspondence received from Carlos de le Fuente Marcos, 8 August 2017

de la Fuente Marcos, C. & de la Fuente Marcos, R. (2017a). Evidence for a possible bimodal distribution of the nodal distances of the extreme trans-Neptunian objects: avoiding a trans-Plutonian planet or just plain bias? 21 Jun 2017, MNRAS: Letters 471: L61-L65

de le Fuente Marcos, C. (2018). Correspondence received from Carlos de le Fuente Marcos, 8 March 2018

de la Fuente Marcos, D., de la Fuente Marcos, R. & Aarseth, S. (2018). Where the Solar system meets the solar neighbourhood: Patterns in the distribution of radiants of observed hyperbolic minor bodies. MNRAS, accepted for publication 2 February 2018.

Demontis, P., LeSar, R. & Klein, M. (1988). New high-pressure phases of ice. Phys. Rev. Lett. 60: pp2284–2287. Available at: https://pdfs.semanticscholar.org/a7bc/9cd13abc414339973276a00cb67423f5ec21.pdf

Devitt, T. (2014). Oldest bit of crust firms up idea of cool early Earth. Madison, WI: University of Wisconsin Press Release. 23 February 2014. Available at: https://news.wisc.edu/oldest-bit-of-crust-firms-up-idea-of-a-cool-early-earth/

Devlin, H. (2015). Nasa's Curiosity rover finds water below surface of Mars. London: The Guardian. 13 April 2015. Available at: http://www.theguardian.com/science/2015/apr/13/nasas-curiosity-rover-finds-water-below-surface-of-mars

Dickson, J. et al. (2015). Recent climate cycles on Mars: Stratigraphic relationships between multiple generations of gullies and the latitude dependent mantle. Icarus, 252: pp83-94

Dickson, J. et al. (2015a). Formation of gullies on Mars by water at high obliquity: Quantitative integration of global climate models and gully distribution. 46th Lunar and Planetary Science Conference. Available at: http://www.hou.usra.edu/meetings/lpsc2015/pdf/1035.pdf

Donaldson, J. et al (2016). New parallaxes and a convergence analysis for the TW Hya Association. The Astrophysical Journal, 833(1). Available at: http://iopscience.iop.org/article/10.3847/1538-4357/833/1/95

Doressoundiram, A. et al. (2008). Color properties and trends of the Transneptunian Objects. In Barucci, A. et al. (Eds) (2008). The Solar System Beyond Neptune. Tucson: University of Arizona Press

Dorminey, B. (2013). Main Asteroid Belt no remnant of exploded planet. Forbes: Online. 31 Jan 2013. Available at: http://www.forbes.com/sites/brucedorminey/2013/01/31/why-our-main-asteroid-belt-is-hardly-the-remnant-of-an-exploded-planet/

Draine, B. (2009). Perspectives on interstellar dust inside and outside of the Heliosphere. Space Science Reviews, 143(1): pp 333-345

Drake, N. (2015). Floating mountains on Pluto—You can't make this stuff up. Washington D.C.: National geographic. 9 November 2015. Available at: https://news.nationalgeographic.com/2015/11/151109-astronomy-pluto-nasa-new-horizons-volcano-moons-science/

Duhamel, J. (2013). Ice Ages and Glacial Epochs – What's the difference? Oro Valley, AZ: Arizona Daily Independent, 24 November 2013. Available at: https://arizonadailyindependent.com/2013/11/24/ice-ages-and-glacial-epochs-whats-the-difference/

Dundas, C. et al. (2017). Granular flows at recurring slope lineae on Mars indicate a limited role for liquid water. Nature Geoscience, 10: pp 903–907

Dvorak, R., Loibnegger, B. & Maindl. T. (2015). On the probability of the collision of a Mars-sized planet with the Earth to form the Moon. June 2015.

Dyches, P. (2016). Saturn spacecraft not affected by hypothetical Planet 9. Pasadena, CA: NASA/JPL. 8 April 2016. Available at: https://www.jpl.nasa.gov/news/news.php?feature=6200

Edgeworth, K. (1943). The evolution of our planetary system. Journal of the British Astronomical Association, 53, pp181–186

Ehrenfreund, P. & Cami, J. (2010). Cosmic carbon chemistry: From the interstellar medium to the early Earth. Cold Spring Harbor Perspectives in Biology: Online. Available at: http://www.geology.wisc.edu/~astrobio/docs/Ehrenfreund_Cami_2010_Orig_Life.pdf

Ehrenreich, D. et al. (2015). A giant comet-like cloud of hydrogen escaping the warm Neptune mass exoplanet GJ 436b. Nature, 522: pp 459–461

Emspak, J. (2017). Brown dwarfs have strong magnetic fields just like real stars. London: New Scientist. Available at: https://www.newscientist.com/article/2147636-brown-dwarfs-have-strong-magnetic-fields-just-like-real-stars/

ESA (2011). Did Earth's oceans come from the comets? Paris: European Space Agency. 5 October 2011. Available at: http://www.esa.int/Our_Activities/Space_Science/Herschel/Did_Earth_s_oceans_come_from_comets

ESA (2014). First measurements of comet's water ratio. Paris: European Space Agency. 10 December 2014. Available at: http://sci.esa.int/rosetta/55117-first-measurements-of-comets-water-ratio/

ESA (2015). Hot water activity on icy moon's seafloor. Paris: European Space Agency press release. 11 March 2015. Available at: http://www.esa.int/Our_Activities/Space_Science/Cassini-Huygens/Hot_water_activity_on_icy_moon_s_seafloor

ESO (2018). ESO's VLT Sees ʻOumuamua getting a boost: New results indicate interstellar nomad ʻOumuamua is a comet. Munich: European Southern Observatory press release 1820. 27 June 2018. Available at: http://www.eso.org/public/news/eso1820/

Espley, J. et al. (2017). A comet engulfs Mars: MAVEN observations of comet Siding Spring's influence on the Martian magnetosphere. Geophysical Research Letters, 42(21): pp 8810-8818. Available at: https://agupubs.onlinelibrary.wiley.com/doi/full/10.1002/2015GL066300

Farihi, J., Gänsicke, B. & Koester, D. (2013). Evidence for water in the rocky debris of a disrupted extrasolar minor planet. Science, 342(6155): pp 218-220

Fernández, J., Gallardo T. & Brunini A. (2004). The scattered disc population as a source of Oort Cloud comets: evaluation of its current and past role in populating the Oort Cloud. Icarus, 172 (2): 372–381. Available at: http://www.sciencedirect.com/science/article/pii/S0019103504002210

Fernandez, J. (2010). On the existence of a distant solar companion and its possible effects on the Oort cloud and the observed comet population. The Astrophysical Journal, 726(1). Available at: http://iopscience.iop.org/article/10.1088/0004-637X/726/1/33;j

Fesenmaier, K. (2014). Getting to know super-Earths. Pasadena: Caltech, 15 October 2014. Available at: http://www.caltech.edu/news/getting-know-super-earths-44005

Fienga, A. et al. (2016). Constraints on the location of a possible 9th planet derived from the Cassini data. Astronomy and Astrophysics, March 2016, 587,L8

Filacchione, G. et al. (2016). Exposed water ice on the nucleus of comet 67P/Churyumov–Gerasimenko. Nature, 529: pp 368–372

Firestone, R. et al. (2007). Evidence for an extraterrestrial impact 12,900 years ago that contributed to the megafaunal extinctions and the Younger Dryas cooling. Proceedings of the National Academy of Sciences, 104(41): pp16016–16021.

Forbes, G. (1880). On comets. Proceedings of the Royal Society, Edinburgh, 10: p426, and Additional note on an ultra-Neptunian planet. Proc. Roy. Soc. Edinburgh, 11: p89

Fox, K. (2010). Kuiper Belt of many colors. Greenbelt, MD: GSFC. 27 October 2010. Available at: https://www.nasa.gov/topics/solarsystem/sunearthsystem/main/kuiper-colors.html

Fraser, W. et al. (2018). The tumbling rotational state of 1I/'Oumuamua. Nature Astronomy, 9 February 2018, 2: pp383–386

Frisch, P. et al. (2002). Galactic environment of the Sun and stars: Interstellar and interplanetary material. STScI Astrophysics of Life conference. Available at: http://cds.cern.ch/record/578504/files/0208556.pdf

Fu, R. et al. (2017). The interior structure of Ceres as revealed by surface topography. Earth and Planetary Science Letters, 476: pp 153-164

Galeazzi, M. et al. (2014). The origin of the local 1/4-keV X-ray flux in both charge exchange and a hot bubble. Nature, 512: pp171-3

Gagné, J. et al. (2015). BANYAN. VII. A new population of young substellar candidate members of nearby moving groups from the bass survey. The Astrophysical Journal Supplement Series, 219:33. Available at: http://iopscience.iop.org/article/10.1088/0067-0049/219/2/33/pdf

Gagné, J. et al. (2017). BANYAN. IX. The initial mass function and planetary-mass objects space density of the TW HYA association. The Astrophysical Journal Supplement Series, 228:18. Available at: http://iopscience.iop.org/article/10.3847/1538-4365/228/2/18/pdf

Gagne', J. et al. (2017a). SIMP J013656.5+093347 is likely a planetary-mass object in the Carina-Near moving group. The Astrophysical Journal Letters, 841(1). Available at: https://arxiv.org/pdf/1705.01625.pdf

Gagné, J. et al (2018). 2MASS J13243553+6358281 Is an Early T-type Planetary-mass Object in the AB Doradus Moving Group. Astrophysical Journal Letters, 854(2)

Gahm, G et al. (2103). Mass and motion of globulettes in the Rosette Nebula. Astronomy and Astrophysics, 555: A57

Galiazzo, M. et al. (2016). Possible origin(s) of Comet 67P/Churyumov-Gerasimenko. 47th Lunar and Planetary Science Conference. LPI Contribution No. 1903, p.1021

Gibbs, W. W. (2017). Is there a giant planet lurking beyond Pluto? New York: IEEE Spectrum, August 2017. Available at: http://spectrum.ieee.org/aerospace/satellites/is-there-a-giant-planet-lurking-beyond-pluto

Gies, D. & Helsel, J. (2005). Ice Age Epochs and the Sun's path through the galaxy. The Astrophysical Journal, 626(2): pp844-848. Available at: https://arxiv.org/PS_cache/astro-ph/pdf/0503/0503306v1.pdf

Gizis, J. et al. (2017). K2 Ultracool Dwarfs Survey. II. The white light flare rate of young brown dwarfs. The Astrophysical Journal, 845(1): 33

Gladman, B. et al. (2002). Evidence for an Extended Scattered Disk, Icarus, 157, pp269-279

Gomes, R. et al. (2005). Origin of the cataclysmic Late Heavy Bombardment period of the terrestrial planets. Nature, 435: pp466-469

Gomes, R. et al. (2006). A distant planetary-mass solar companion may have produced distant detached objects, Icarus, 184(2): pp589–601

Gomes, R. & Soares, J. (2012). Signatures of a putative planetary mass solar companion on the orbital distribution of TNOs and Centaurs. AAS/Division of Dynamical Astronomy Meeting, vol. 43 of AAS/Division of Dynamical Astronomy Meeting, May 2012, p 05.01

Gomes, R., Deienno, R. & Morbidelli, A. (2016). The inclination of the planetary system relative to the solar equator may be explained by the presence of Planet 9. The Astronomical Journal, 153(1)

Goudge, T. et al. (2015). Assessing the mineralogy of the watershed and fan deposits of the Jezero crater paleolake system, Mars. Journal of Geophysical Research, 120(4): pp775-808. Available at: https://agupubs.onlinelibrary.wiley.com/doi/full/10.1002/2014JE004782

Grazier, K. (2015). Jupiter: Cosmic Jekyll and Hyde. Astrobiology, 16(1): p23

Griffin, A. (2017) Huge object flying past Earth may be alien spacecraft, scientists say. London: The Independent. 12 December 2017. Available at: http://www.independent.co.uk/life-style/gadgets-and-tech/news/oumuamua-alien-spacecraft-proof-life-seti-breakthrough-latest-asteroid-ship-a8104771.html

Greffos, V. (2003). PLANETS – But How Many Are There In Our Solar System? Paris, France: Science & Vie. February 2003

Grosser, M. (1964). The search for a planet beyond Neptune, Isis, 55(2): pp163-183

        Timeless Voyager Press

Grossman, L. (2012). Why do we think Curiosity found an old Mars riverbed? London: New Scientist. 28 September 2012. Available at: http://www.newscientist.com/article/dn22319-why-do-we-think-curiosity-found-an-old-mars-riverbed.html

Grossman, L. (2013). Moon water came from young wet Earth. London: New Scientist. 9 May 2013. Available at: http://www.newscientist.com/article/dn23515-moon-water-came-from-young-wet-earth.html

Grzedzielski, S. et al. (2010). A possible generation mechanism for the IBEX Ribbon from outside the Heliosphere. The Astrophysical Journal Letters, 715(2)

Günay B. et al. (2018). Aliphatic hydrocarbon content of interstellar dust. Monthly Notices of the Royal Astronomical Society, sty1582, 18 June 2018. Available at: https://doi.org/10.1093/mnras/sty1582

Gunn, G. (1958). The Strange World of Planet X. Distributors Corporation of America

Gurumath, S., Hiremath, K. & Ramasubramanian, V. (2016). Missing planetary mass in the vicinity of Sun: Clues from host stars and exoplanetary data. 19th National Space Science Symposium, Thiruvananthapuram, February 2016

Hall, S. (2016). We are closing in on possible whereabouts of Planet Nine. London: New Scientist, 20 April 2016. Available at: https://www.newscientist.com/article/2084924-we-are-closing-in-on-possible-whereabouts-of-planet-nine

Hall, S. (2016a). Planet Nine might be an exoplanet stolen by the Sun. London: New Scientist, April 2016. Available at: https://www.newscientist.com/article/2082970-planet-nine-might-be-an-exoplanet-stolen-by-the-sun/

Hallis, L. et al. (2015). Evidence for primordial water in Earth's deep mantle. Science, 350: p795

Hancock, G. (1995). Fingerprints of the Gods. London: William Heinemann

Hancock, G. (2015). Magicians of the Gods. London: Coronet

Hand, E. (2015). Lopsided ice on the moon points to past shift in poles. Science: Online. 19 March 2015. Available at: http://news.sciencemag.org/space/2015/03/lopsided-ice-moon-points-past-shift-poles

Hand, E. (2016). Astronomers say a Neptune-sized planet lurks beyond Pluto. Science: Online. 20 June 2016. Available at: http://www.sciencemag.org/news/2016/01/feature-astronomers-say-neptune-sized-planet-lurks-unseen-solar-system

Hansen, B. & Zuckerman, B. (2017). Ejection of material — "Jurads" — from post main sequence planetary systems. Research Note submitted to AAS journals 19 Dec 2017. Available at: https://arxiv.org/abs/1712.07247

Harrington, J. & Villard, R. (2013). NASA's Hubble reveals rogue planetary orbit for Fomalhaut B. Washington D.C.: Hubble Space Telescope Press Release. 8 January 2013. Available at: https://www.nasa.gov/mission_pages/hubble/science/rogue-fomalhaut.html

Harrington, R. & Van Flandern, T. (1979). The satellites of Neptune and the origin of Pluto. Icarus, 39(1): pp131-136

Hartman, J. et al. (2015). HATS-6b: A warm Saturn transiting an early M Dwarf star, and a set of empirical relations for characterizing K and M Dwarf planet hosts. The Astronomical Journal, 149(5): p166

Hauri, E. et al. (2015). Water in the Moon's interior: Truth and consequences. Earth and Planetary Science Letters, 409: pp 252-264

Head, J. et al. (2003). Recent Ice Ages on Mars. Nature, 426(18/25): pp 797-802. Available at: http://planetary.brown.edu/planetary/documents/2957.pdf

Hecht, J. (2008). Distant object found orbiting Sun backwards. New Scientist. 5 September 2008. Available at: https://www.newscientist.com/article/dn14669-distant-object-found-orbiting-sun-backwards/

Hills, J. (1984). Dynamical constraints on the mass and perihelion distance of Nemesis and the stability of its orbit. Nature, 311(5987): pp636–638

Hills, J. (1985). The passage of a 'Nemesis'-like object through the planetary system. The Astronomical Journal. 90: pp1876-1882. Available at: http://adsabs.harvard.edu/full/1985AJ.....90.1876H

Hao, K. (2017). Let's take a moment to look at the mesmerizing images from Voyager 1. 7 December 2017. Available at: https://qz.com/1149254/lets-take-a-moment-to-look-at-some-of-the-best-images-from-voyager-1/

Holliday, V. et al. (2014). The Younger Dryas impact hypothesis: a cosmic catastrophe. Journal of Quaternary Science, 29(6) pp515–530. Available at: http://www.uwyo.edu/surovell/pdfs/jqs%202014.pdf

Holliday, V. et al. (2016). A blind test of the Younger Dryas impact hypothesis. PLoS One, 8 July 2016, 11(7): e0155470. Available at: https://www.ncbi.nlm.nih.gov/pmc/articles/PMC4938604/

Hollin, J. (1972). Interglacial climates and Antarctic ice surges. Quaternary Research 2: pp401-408.

Holman, M. (2016). Correspondence received from Matthew Holman, 18 August 2016

Holman, M. & Payne, M. (2016). Observational constraints on Planet Nine: Astrometry of Pluto and other Trans-Neptunian Objects. The Astronomical Journal, 9th September 2016, 152(4). Available at: https://arxiv.org/pdf/1603.09008v1.pdf

Holman, M. & Payne, M. (2016a). Observational constraints on Planet Nine: Cassini range observations. The Astronomical Journal, 152(4). Available at: http://arxiv.org/pdf/1604.03180.pdf

Hook, R. (2015). Where did all the stars go? Dark cloud obscures hundreds of background stars. European Southern Observatory press release, eso1501, 7 January 2015. Available at: http://www.eso.org/public/unitedkingdom/news/eso1501/

Hopkins, M., Harrison, T.M. & Manning, C. (2008). Low heat flow inferred from >4 Gyr zircons suggests Hadean plate boundary interactions. Nature, 456: pp493–496

Horányi, M. et al. (2015). A permanent, asymmetric dust cloud around the Moon. Nature, 522: pp324–326

Howell, E. (2014). Sedna: Possible Dwarf Planet Far From the Sun. Online: Space.com. 30th April 2014. Available at: http://www.space.com/25695-sedna-dwarf-planet.html

Howells, E. (2014a). UPDATE: NASA Senior Review Declines WISE Spacecraft Data Usage Idea. Online: Universe Today. 16 May 2014. Available at: https://www.universetoday.com/111946/future-uncertain-for-wise-spacecraft-after-review-declines-extension/

Howell, E. (2015). Could water have carved channels on Mars half a million years ago? Pasadena, CA: Astrobiology Magazine. 19 March 2015. Available at: https://www.astrobio.net/mars/could-water-have-carved-channels-on-mars-half-a-million-years-ago/

Howell, E. (2018). Earth resides in oddball solar system, alien worlds show. Online: Space.com. 16 January 2018. Available at: https://www.space.com/39390-alien-planets-reveal-our-strange-solar-system.html

Hsu, H-W. et al. (2015). Ongoing hydrothermal activities within Enceladus. Nature, 519: pp207–210

Hui, H. et al. (2013). Water in lunar anorthosites and evidence for a wet early Moon. Nature Geoscience, 6: pp 177–180

Hussmann, H. & Sohl, F. (2007). Subsurface oceans and deep interiors of large Trans-Neptunian Objects. In: 2nd European Planetary Science Congress (2007), EPSC2007-A-00480. European Planetary Science Congress (EPSC), 24 August 2007, Potsdam, Germany.

IAU Press Release 0603 (2006) IAU 2006 General Assembly: Result of the IAU resolution votes. Paris: International Astronomical Union. Available at: https://www.iau.org/news/pressreleases/detail/iau0603/

International Max Planck Research School (2018). Open PhD Project: Dynamos in giant planets and brown dwarfs. Munich: IMPRS. Available at: https://www.mps.mpg.de/phd/astrophysics-dynamo-giantplanets-browndwarfs

Iorio, L. (2011). On the anomalous secular increase of the eccentricity of the orbit of the Moon. Monthly Notices of the Royal Astronomical Society, 415(2): pp1266-1275

Iorio, L. (2012). Constraints on the location of a putative distant massive body in the Solar System from recent planetary data. Celestial Mechanics and Dynamical Astronomy, 112(2): pp 117–130

Iorio, L. (2014). Planet X revamped after the discovery of the Sedna-like object 2012 VP113? Monthly Notices of the Royal Astronomical Society Letters, 444(1): L78-L79

Iorio, L. (2017). Preliminary constraints on the location of the recently hypothesized new planet of the Solar System from planetary orbital dynamics. Astrophysics and Space Science, January 2017, 362: 11. Available at: https://arxiv.org/pdf/1512.05288.pdf

Iorio, L. (2017a). Correspondence received from Lorenzo Iorio, 9 August 2017

Jackson, A. et al. (2018). Ejection of rocky and icy material from binary star systems: Implications for the origin and composition of 1I/'Oumuamua. Monthly Notices of the Royal Astronomical Society: Letters, 478(1): L49–L53

Janson, M. (2012). Infrared non-detection of Fomalhaut b - Implications for the planet interpretation. The Astrophysical Journal, 747(2): 166

Jenkins, A. & Villard, R. (2018). Hubble telescope discovers substellar objects in the Orion Nebula. NASA: Space Telescope Science Institute, 12 January 2018. Available at: https://www.nasa.gov/feature/goddard/2018/hubble-finds-substellar-objects-in-the-orion-nebula

Jenner, N. (2014). New dwarf planet hints at giant world far beyond Pluto. New Scientist, 26 March 2014. Available at: http://www.newscientist.com/article/dn25301-new-dwarf-planet-hints-at-giant-world-far-beyond-pluto.html

Jewitt, D. & Luu, J. (1993). Discovery of the candidate Kuiper Belt Object 1992 QB1. Nature, 362(6422): pp730–732

Jewitt, D. (2002). From Kuiper Belt Object to cometary nucleus: The missing ultrared matter. The Astronomical Journal. 123(2): pp1039-1049. Available at: https://pdfs.semanticscholar.org/64d7/7c6d3b97a64bf-6653557c030f8d919bc5f4a.pdf

Jewitt, D. et al. (2017). Interstellar interloper 1I/2017 U1: Observations from the NOT and WIYN Telescopes. The Astrophysical Journal Letters, 850:L36

Jewitt, D. (2017). Correspondence received from Dave Jewitt, 9 August 2017

Jilkova, L. et al. (2015). How Sedna and family were captured in a close encounter with a solar sibling. Monthly Notices of the Royal Astronomical Society, 453(3): pp3157-3162

Joergens, V. et al. (2013). OTS 44: Disk and accretion at the planetary border. Astronomy and Astrophysics Letters, 558: L7

Johnson, B. et al. (2015). Impact jetting as the origin of chondrules. Nature, 517: pp339–341

Johnsson, A. et al. (2014). Evidence for very recent melt-water and debris flow activity in gullies in a young mid-latitude crater on Mars. Icarus, 235: pp37-54

Kaib, N. & Quinn, T. (2008). The formation of the Oort cloud in open cluster environments. Icarus, 197: 221

Kaib, N. & Quinn, T. (2009). Reassessing the source of long-period comets. Science, 325: pp1234-1236. Available at: http://citeseerx.ist.psu.edu/viewdoc/download?-doi=10.1.1.840.9973&rep=rep1&type=pdf

Kaib, N. et al. (2013). Planetary system disruption by Galactic perturbations to wide binary stars. Nature, 493: pp381–384

Kalas, P. et al. (2008). Optical Images of an exosolar planet 25 light-years from Earth. Science, 322(5906): pp1345–8

Kaplan, S. (2015). Scientists claimed they found elusive 'Planet X.' Doubting astronomers are in an uproar. The Washington Post, 11 December 2015. Available at: https://www.washingtonpost.com/news/morning-mix/wp/2015/12/11/scientists-claimed-they-found-the-elusive-planet-x-now-astronomers-are-in-an-uproar/

Karlsson, N. et al. (2015). Volume of Martian mid-latitude glaciers from radar observations and ice-flow modelling. Geophysical Research Letters, 42(8): pp2627-2633

Kaufman, M. (2016). The case strengthens for "Planet 9". Many Worlds, 1 September 2016. Available at: http://www.manyworlds.space/index.php/2016/09/01/the-case-strengthens-for-planet-9/

Keane, J. et al. (2016). Reorientation and faulting of Pluto due to volatile loading within Sputnik Planitia. Nature, 540: pp 90–93

Keeter, W. (Ed.) (2018). New Horizons captures record-breaking images in the Kuiper belt. Pasadena, CA: NASA. 8th February 2018. Available at: https://www.nasa.gov/feature/new-horizons-captures-record-breaking-images-in-the-kuiper-belt

Kegerreis, J. et al. (2018). Consequences of giant impacts on early uranus for rotation, internal structure, debris, and atmospheric erosion. The Astrophysical Journal, 861(1)

Kennedy, G. & Wyatt, M. (2010). Collisional evolution of irregular satellite swarms: detectable dust around Solar system and extrasolar planets. Monthly Notices of the Royal Astronomical Society, 412(4): pp 2137–2153

Kenney, J., Abramson, A. & Bravo-Alfaro, H. (2015). HST and HI imaging of strong ram pressure stripping in the Coma Spiral NGC 4921: Dense cloud decoupling and evidence for magnetic binding in the ISM. The Astronomical Journal, 150(2): 59

Kenyon, S. et al. (2014). Fomalhaut b as a cloud of dust: Testing aspects of planet formation theory. The Astrophysical Journal, 786(1),

Kenyon, S. & Bromley, B. (2004). Stellar encounters as the origin of distant Solar System objects in highly eccentric orbits. Nature, 432(7017): pp598–602

Kenyon, S. & Bromley, B. (2015). Formation of super-Earth mass planets at 125–250 au from a solar-type star. The Astrophysical Journal, 806(1): 42. Available at: http://iopscience.iop.org/article/10.1088/0004-637X/806/1/42/pdf

Kenyon, S., Najita, J. & Bromley, B. (2016). Rocky planet formation: Quick and neat. The Astrophysical Journal, 831(1)

Khain, T., Batygin, K. & Brown, M. (2018). The Generation Of The Distant Kuiper Belt By Planet Nine From An Initially Broad Perihelion Distribution. 1 May 2018. Accepted for publication in the Astronomical Journal. Available at: https://arxiv.org/pdf/1804.11281.pdf

King, L. (Translator) (1902). Enuma Elish: The epic of creation. Available at: http://www.sacred-texts.com/ane/enuma.htm

Kirkpatrick, D. et al. (2011). The first hundred brown dwarfs discovered by the Wide-field Infrared Survey Explorer (WISE). Astrophysical Journal Supplement Series, 22 November 2011, 197(2)

Klačka, J. (2013). Comparison of the solar/stellar wind and the Poynting-Robertson effect in secular orbital evolution of dust particles. Monthly Notices of the Royal Astronomical Society, 436(3): pp2785-2792

Kohler, S. (2016). Could Ultracool Dwarfs Have Sun-Like Activity? Washington D.C.: American Astronomical Society. 9 November 2016. Available at: http://aasnova.org/2016/11/09/could-ultracool-dwarfs-have-sun-like-activity/

Kohler, S. (2017). Discovery of a Free-Floating Double Planet? Washington D.C.: American Astronomical Society. 5 July 2017. Available at: http://aasnova.org/2017/07/05/discovery-of-a-free-floating-double-planet/

Kramer, M. (2017). MAVEN: NASA's orbiter mission to Mars — Mission details. Online: Space.com. 30 August 2017. Available at: https://www.space.com/23617-nasa-maven-mars-mission.html

Kroupa, P. (2011). The universality hypothesis: binary and stellar populations in star clusters and galaxies. Computational Star Formation Proceedings IAU Symposium No. 270. Available at: https://www.cambridge.org/core/services/aop-cambridge-core/content/view/A003A085F0153148F10E215967988B1D/S1743921311000305a.pdf

Kuchner, M. (2017). The Colors of Cold Brown Dwarfs. Available at: https://blog.backyardworlds.org/2017/03/22/the-colors-of-cold-brown-dwarfs/

Kuchner, et al. (2017). The First Brown Dwarf Discovered by the Backyard Worlds: Planet 9 Citizen Science Project. Astrophysical Journal Letters, 841(2)

Kuiper, G. (1951). On the Origin of the Solar System. Astrophysics: a topical symposium (J. Hynek, Ed.) New York: Mc Graw-Hill, pp357-424

Landau, E. (2015). Gullies on Vesta suggest past water-mobilized flows. Pasadena: NASA/JPL, 20 January 2015. Available at: https://www.nasa.gov/jpl/dawn/gullies-on-vesta-suggest-past-water-mobilized-flows

Laughlin, G. & Batygin, K. (2017). On the consequences of the detection of an interstellar asteroid. Research Notes of the AAS 1(1). Available at: http://iopscience.iop.org/article/10.3847/2515-5172/aaa02b/meta

Laskar, J. & Robutel, P. (1993). The chaotic obliquity of the planets. Nature, 361(6413): pp 608-612

Lawler, S. et al. (2016). Observational signatures of a massive distant planet on the scattering disk. The Astronomical Journal, 153(1): 33

Leconte, J. et al. (2015). Asynchronous rotation of Earth-mass planets in the habitable zone of lower-mass stars. Science, 347(6222): pp. 632-635

Leggett, S. et al. (2017). The Y-Type Brown Dwarfs: Estimates of mass and age from new astrometry, homogenized photometry and near-infrared spectroscopy. The Astrophysical Journal, 842(2). Available at: http://iopscience.iop.org/article/10.3847/1538-4357/aa6fb5/pdf

Lehrman, S. (1961). The world of the midrash. New Jersey: Thomas Yoseloff

Leshin L. et al. (2013). Volatile, isotope and organic analysis of Martian fines with the Mars Curiosity Rover. Science, 341: 6153. Available at: http://mars.jpl.nasa.gov/files/msl/Science-2013-Leshin-.pdf

Levison, H. et al. (2007). Origin of the Structure of the Kuiper Belt during a dynamical instability in the orbits of Uranus and Neptune. Icarus, 196(1): pp 258-273

Levison, H. et al. (2011). Late Orbital Instabilities in the Outer Planets Induced by Interaction with a Self-gravitating Planetesimal Disk. The Astronomical Journal, 142(5): 152

Li, G. & Adams, F. (2016). Interaction cross sections and survival rates for proposed solar system member Planet Nine. Accepted for publication in ApJL, 26 April 2016. Available at: http://arxiv.org/abs/1602.08496

Lin, Y.,et al. (2017). Evidence for an early wet Moon from experimental crystallization of the lunar magma ocean. Nature Geoscience, 10: pp 14–18

Linder, E. & Mordasini, C. (2016). Evolution and magnitudes of candidate Planet Nine. Astronomy and Astrophysics, 589, April 2016

Liseau, R. et al. (2015). A new submm source within a few arcseconds of α Centauri: ALMA discovers the most distant object of the solar system. 8 December 2015. Available at: https://arxiv.org/abs/1512.02652

Lloyd A. (1999). Ancient to Modern. Leeds: UFO Magazine, September 1999, p76

Lloyd A. (2001). Synopsis of the Dark Star Theory. Leeds: UFO Magazine, Aug 2001, pp50-5

Lloyd, A. (2002). The Great Water Conundrum. Gloucester: Online. 2 April 2002. Available at: http://www.darkstar1.co.uk/water.html

Lloyd, A. (2004). Alessandro Morbidelli and the Origin of Sedna. Gloucester: Online. 15 November 2004. Available at: http://www.darkstar1.co.uk/Morbidelli.html

Lloyd, A. (2005). Dark Star: The Planet X Evidence. Santa Barbara: Timeless Voyager Press

Lloyd, A. (2005a). Sedna: A Clue to Nibiru. Gloucester: Online. 15 March-21 April 2005. Available at: http://www.darkstar1.co.uk/sedna.htm

Lloyd, A. (2009). Ezekiel One. Santa Barbara: Timeless Voyager Press

Lloyd, A. (2010). The Followers of Horus. Santa Barbara: Timeless Voyager Press

Lloyd, A. (2014). New Object in Outer Solar System Hints at Planet X. Gloucester: Online. 27 March 2014. Available at: http://www.andylloyd.org/darkstarblog13.htm

Lloyd, A. (2016). Massive Planet X Now Urgently Sought by Top Planet-Hunters. Gloucester: Online. 20 January 2016. Available at: http://www.andylloyd.org/darkstarblog34.htm

Lloyd, A. (2016a). Planet Nine constellations predicted by Sitchin, and IRAS. Gloucester: Online. 26 January 2016. Available at: http://www.andylloyd.org/darkstarblog34.htm

Lloyd, A. (2016b). The cumulative effect of intermittent interstellar medium inundation upon objects in the outer solar system. Academia.net, 7 February 2016. Available at: https://www.academia.edu/21700220/The_Cumulative_Effect_of_Intermittent_Interstellar_Medium_Inundation_Upon_Objects_In_The_Outer_Solar_System

Lowe, D., Byerly, G. & Kyte, F. (2014). Recently discovered 3.42–3.23 Ga impact layers, Barberton Belt, South Africa: 3.8 Ga detrital zircons, Archean impact history, and tectonic implications. Geology, 42(9): pp 747-750.

Luhman, K. et al. (2011). Discovery of a candidate for the coolest known brown dwarf. The Astrophysical Journal Letters, Volume 730, Number 1, 25th February 2011

Luhman, K. (2013). Discovery of a binary brown dwarf at 2 pc from the Sun. The Astrophysical Journal Letters, 20 March 2013, 767:L1. Available at: http://iopscience.iop.org/article/10.1088/2041-8205/767/1/L1/pdf

Luhman, K. (2013a). A search for a distant companion to the Sun with the Wide-Field Infrared Survey Explorer. The Astrophysical Journal, 24December 2013, 781(1)

Luhman, K. (2014). Discovery of a ~250K brown dwarf at 2pc from the Sun. Astrophysical Journal Letters, 21 April 2014. 786(2). Available at: http://iopscience.iop.org/article/10.1088/2041-8205/786/2/L18/pdf

Lunning, N. et al. (2015). Olivine and pyroxene from the mantle of asteroid 4 Vesta. Earth and Planetary Science Letters, 418: pp 126-135

Lykawka, P. and Mukai, T. (2008). An Outer Planet Beyond Pluto and the Origin of the Trans-Neptunian Belt Architecture. The Astronomical Journal, 135(4): p1161

Lynds. B. (1962). Catalogue of Dark Nebulae. Astrophysical Journal Supplement, 7: p1

Magli, G. (2016). Sirius and the project of the megalithic enclosures at Gobekli Tepe. Basel: Nexus Network Journal, 17: pp 1-11. Available at: https://arxiv.org/ftp/arxiv/papers/1307/1307.8397.pdf

Malhotra, R., Volk, K. & Wang, X. (2016). Corralling a distant planet with extreme resonant Kuiper Belt Objects. The Astrophysical Journal Letters, 824(2): L22-L28. Available at: http://arxiv.org/pdf/1603.02196v2.pdf

Malhotra, R. (2017). Correspondence received from Renu Malhotra, 4 July 2017

Malin, M. & Edgett, K. (2000). Evidence for recent groundwater seepage and surface runoff on Mars. Science, 288: pp2330–2335.

Mann, A. (2016). New icy world with 20,000-year orbit could point to Planet Nine. Science. Washinton: AAAS, 17 October 2016. Available at: http://www.sciencemag.org/news/2016/10/new-dwarf-planet-points-planet-nine

Mamajek, E. et al. (2015). The closest known flyby of a star to the Solar System. The Astrophysical Journal Letters, 800(1): L17

Mandelbaum, R. (2018). Scientists Observe Incredible New Kind of Ice Thought to Exist in Uranus' Center. 6 February 2018. Available at: https://gizmodo.com/scientists-observe-incredible-new-kind-of-ice-thought-t-1822761008

Mandt, K. et al. (2014). Protosolar ammonia as the unique source of Titan's nitrogen. The Astrophysical Journal Letters, 788: L24. Available at: http://iopscience.iop.org/article/10.1088/2041-8205/788/2/L24/pdf

Mangold, N. et al. (2010). Sinuous gullies on Mars: Frequency, distribution, and implications for flow properties. Journal of Geophysical Research, 115: E11001, Available at: http://www.ipgp.fr/~mangeney/Mangoldetal_JGRsinuousgullies.pdf

Marcus, R. et al. (2011). Identifying collisional families in the Kuiper belt. The Astrophysical Journal, 733(1). Available at: https://www.cfa.harvard.edu/~rmarcus/preprints/RMarcus_ApJ_submitted.pdf

Marshall, M. (2014). Earth's early life endured long asteroid bombardment. London: New Scientist. 15 August 2014. Available at: https://www.newscientist.com/article/dn26055-earths-early-life-endured-long-asteroid-bombardment

Martin, E. (2002). The birth and evolution of brown dwarfs. Available at: http://www2.ifa.hawaii.edu/CSPF/presentations/bdtutorial/bdtutorial.pdf

Martin-Torres, F.J. et al. (2015). Transient liquid water and water activity at Gale Crater on Mars. Nature Geoscience, 13 April 2015, 8: pp357–361

Masetti, M. (2018). Webb vs Hubble Telescope. NASA: Goddard Space Flight Center. Available at: https://jwst.nasa.gov/comparison_about.html

Matese, J., Whitman, P. & Whitmire, D. (1999). Cometary evidence of a massive body in the outer Oort Clouds. Icarus, 141(2): pp354-366

Matese, J. et al. (2006). A wide binary solar companion as a possible origin of Sedna-like objects. Earth, Moon, and Planets, 97(3–4): pp459–470

Matese et al. (1995). Periodic modulation of the Oort cloud comet flux by the adiabatically changing galactic tide. Icarus, 116(2): pp255-268

Matese, J. & Whitmire, D. (2011). Persistent evidence of a Jovian mass solar companion in the Oort cloud. Icarus, 211(2): pp 926-938. Available at: https://arxiv.org/PS_cache/arxiv/pdf/1004/1004.4584v1.pdf

Matese, J. & Whitmire, D. (2011a). Searching the WISE preliminary catalog for massive planets in the solar system. EPSC-DPS Joint Meeting 2011, Nantes, France, 2-7 Oct 2011. Available at: http://www.ucs.louisiana.edu/~jjm9638/Nantes.pdf

Matthews, R. (2002). Mysterious force holds back Nasa probe in deep space. London: The Telegraph, 10 February 2002. Available at: http://www.telegraph.co.uk/news/science/space/1384420/Mysterious-force-holds-back-Nasa-probe-in-deep-space.html

Mauk, B et al. (2017). Discrete and broadband electron acceleration in Jupiter's powerful aurora. Nature, 549: pp 66–69

Max Planck Institute for Astronomy (2018). Disks and extrasolar planets. Heidelberg: MPIA. Available at: http://www.mpia.de/en/PSF/research/disks-exoplanets

McCord, T. et al. (2012). Dark material on Vesta from the infall of carbonaceous volatile-rich material. Nature, 491(7422): pp 83-86

McCrea, W. (1975). Ice ages and the Galaxy. Nature, 255: pp 607 -609

Medvedev, M. & Melott, A. (2006). Do extragalactic cosmic rays induce cycles in fossil diversity? The Astrophysical Journal, 664(2). Available at: http://iopscience.iop.org/article/10.1086/518757

Meech, K. (2015). Origins of water in the solar system leading to habitable worlds. International Astronomical Union 29th General Assembly, Honolulu. 4 August 2015. Available at: http://www.scifac.hku.hk/event/FM15/abstracts/Meech.pdf

Meech, K. et al. (2017). A brief visit from a red and extremely elongated interstellar asteroid. Nature, 552, pp378–381. Available at: https://www.nature.com/articles/nature25020

Meisner, A. et al. (2017). A 3pi Search for Planet Nine at 3.4 microns with WISE and NEOWISE. MNRAS preprint. Available at: https://arxiv.org/pdf/1712.04950.pdf

Melott, A. & Bambach, R. (2010). Nemesis Reconsidered. Monthly Notices of the Royal Astronomical Society Letters, 407: L99–L102

Messenger S. et al. (2014). Dust in the solar system: Properties and origins. Conference: The Life Cycle of Dust in the Universe: Observations, Theory, and Laboratory Experiments, 1/2014, Taipei, Taiwan, Volume: PoS(LCDU 2013)016

Meyer, D. et al. (2017). On the existence of accretion-driven bursts in massive star formation. Monthly Notices of the Royal Astronomical Society: Letters, 464(1): L90-L94

Micheli, M. et al. (2018). Non-gravitational acceleration in the trajectory of 1I/2017 U1 ('Oumuamua). Nature, 27 June 2018

Milliken, R. & Li, S. (2017). Remote detection of widespread indigenous water in lunar pyroclastic deposits. Nature Geoscience, 10: pp 561–565

Millot, M. et al (2018). Experimental evidence for superionic water ice using shock compression. Nature Physics, 14: pp297–302

Minor Planet Center (2017). Comet C/2017 U1 (PAN-STARRS). 25 October 2017 Cambridge, MA: MPC. Available at: http://www.minorplanetcenter.net/mpec/K17/K17U11.html

MIT Technology Review (2011). Where Did Triton Come From? Cambridge, MA: Massachusetts Institute of Technology. 6 June 2011. Available at: https://www.technologyreview.com/s/424200/where-did-triton-come-from/

Mollière, P. & Mordasini, C. (2012). Deuterium burning in objects forming via the core accretion scenario. Astronomy & Astrophysics, 547: A105

Moore, J. et al. (2016). The geology of Pluto and Charon through the eyes of New Horizons. Science. 351(6279): pp1284-93

Morbidelli, A. (2010). Review: A coherent and comprehensive model of the evolution of the outer solar system. 28 October 2010. Available at: https://hal.archives-ouvertes.fr/hal-00530284/document

Morbidelli, A. & Levison, H. (2004). Scenarios for the origin of the orbits of the Trans-Neptunian Objects 2000 CR105 and 2003 VB12 (Sedna). The Astronomical Journal, 128(5): pp2564–2576.

Morelle, R. (2013). New timeline for origin of ancient Egypt. London: BBC, 4 September 2013. Available at: http://www.bbc.co.uk/news/science-environment-23947820

Morley, C. (2012). Swirling, patchy clouds on a teenage brown dwarf. 28 February 2012. Available at: https://astrobites.org/2012/02/28/swirling-patchy-clouds-on-a-teenage-brown-dwarf/

Muller, R. (2002). Measurement of the lunar impact record for the past 3.5 billion years, and implications for the Nemesis theory. Geological Society of America, Special Paper 356, pp 659–665. Available at: http://muller.lbl.gov/papers/Lunar_impacts_Nemesis.pdf

Müller, A. (2018). High winds prevented discovery of Planet Nine yet. Online blog. 27 February 2018. Available at: https://www.grenzwissenschaft-aktuell.de/high-winds-prevented-discovery-of-planet-nine-yet20180227/

Murphy, E. (1999). Prosaic explanation for the anomalous accelerations seen in distant spacecraft. Physical Review Letters, 83: 1890

Murray, J. (1999). Arguments for the presence of a distant large undiscovered solar system planet. Monthly Notices of the Royal Astronomical Society, 309(1): pp31–34.

Mustill, A., Raymond, S. & Davies, M. (2016). Is there an exoplanet in the Solar system? Monthly Notices of the Royal Astronomical Society Letters, 460(1): L109-L113. Available at: https://academic.oup.com/mnrasl/article/460/1/L109/2589689/Is-there-an-exoplanet-in-the-Solar-system

Najita, J. (2017). A familiar-looking messenger from another solar system. National Optical Astronomy Observatory, NOAO Release 17-06. Available at: https://www.noao.edu/news/2017/pr1706.php

Naoz, S. (2016). The eccentric Kozai-Lidov effect and its applications. Annual Review of Astronomy and Astrophysics, 54(1): pp. 441 - 489

Napier, W. (2007). Pollination of exoplanets by nebulae. International Journal of Astrobiology, 6(3): pp. 223-228

NASA/JPL Press Release (2011). Discovered: Stars as cool as the human body, Pasadena: NASA, 24 August 2011. Available at: https://science.nasa.gov/science-news/science-at-nasa/2011/23aug_coldeststars/

NASA/JPL (2013). NASA's GRAIL Mission solves mystery of moon's surface gravity. Pasadena, CA:

NASA. 30 May 2013. Available at: http://www.nasa.gov/mission_pages/grail/news/grail20130530.html

NASA (2017). Small asteroid or comet 'visits' from beyond the Solar System. Pasadena: JPL. 26 October 2017. Available at: https://www.jpl.nasa.gov/news/news.php?feature=6983

NASA's Juno Mission (2017). 16 November 2017. Available at: https://twitter.com/NASAJuno

NASA Spitzer (2017). Brown Dwarf Weather Animated Chart. YouTube: NASA Spitzer. Posted 17 August 2017. Available at: https://www.youtube.com/watch?v=m-8VoMt11eb0

Nellis, W. (2015). Unusual magnetic fields of Uranus and Neptune: Metallic Fluid Hydrogen. Modern Physics Letters B (Condensed Matter), 29(1). Available at: https://arxiv.org/ftp/arxiv/papers/1503/1503.01042.pdf

Nesvorný, D (2011). Young solar system's fifth giant planet? The Astrophysical Journal Letters, 742(2).

Nesvorný, D. (2015). Jumping Neptune can explain the Kuiper belt kernel. The Astronomical Journal, 150: 68

Neveu, M. & Rhoden, A. (2017). The origin and evolution of a differentiated Mimas. Icarus, 296: pp 183-196

Nimmo, F. et al. (2016). Reorientation of Sputnik Planitia implies a subsurface ocean on Pluto. Nature, 540: pp 94–96

Nimura, T. et al. (2016). End-cretaceous cooling and mass extinction driven by a dark cloud encounter. Gondwana Research, 37: pp301–307

Núñez, J. et al. (2016). New insights into gully formation on Mars: Constraints from composition as seen by MRO/CRISM. Geophysical Research Letters. 43(17): pp 8893-8902

O'Connor, J. & Robertson E. (2000). Babylonian numerals. Available at: http://www-groups.dcs.st-and.ac.uk/~history/HistTopics/Babylonian_numerals.html

Opher, M. et al. (2009). A strong, highly-tilted interstellar magnetic field near the Solar System. Nature, 462, 24/31. Available at: http://www-personal.umich.edu/~tamas/TIGpapers/2009/2009_Opher_nature.pdf

Opher, M. et al. (2015). Magnetized jets driven by the Sun: The structure of the heliopause revisited. Astrophysical Journal Letters, 19 February 2015, 800(2). Available at: http://iopscience.iop.org/article/10.1088/2041-8205/800/2/L28

Orosei, R. et al. (2018). Radar evidence of subglacial liquid water on Mars. Science, 25 July 2018: eaar7268

Orrman-Rossiter, K. & Gorman, A. (2015). Finding Pluto—the hunt for Planet X. London: The Conversation, 16 July 2015. Available at: http://theconversation.com/finding-pluto-the-hunt-for-planet-x-44508

O'Toole, T. (1983). Possibly as large as Jupiter. Washington D.C.: The Washington Post, 30 December 1983. Available at: https://www.washingtonpost.com/archive/politics/1983/12/30/possibly-as-large-as-jupiter/1075b265-120a-4d40-9493-a8c523b76927/

Pahlevan, K. & Morbidelli, A. (2015). Collisionless encounters and the origin of the lunar inclination. Nature, 527, pp 492–494

Pavlov, A. et al. (2005). Passing through a giant molecular cloud: "Snowball" glaciations produced by interstellar dust. Geophysical Research Letters, 82: p 3705

Pearson, D. et al. (2014). Hydrous mantle transition zone indicated by ringwoodite included within diamond. Nature, 507: pp 221–224

Perkins, S. (2016). How Pluto got its heart of ice. AAAS: Science. 19 September 2016. Available at: http://www.sciencemag.org/news/2016/09/how-pluto-got-its-heart-ice

Pfalzner, S. et al. (2018). Outer solar system possibly shaped by a stellar fly-by. Accepted by The Astrophysical Journal, 9 July 2018.

Phillips, T. (2001). The Solar Wind at Mars. Pasadena, CA: NASA/JPL. 31 January 2001. Available at: http://science.nasa.gov/science-news/science-at-nasa/2001/ast31jan_1/

Phillips, T. (2009). Voyager makes an interstellar discovery. Pasadena: NASA. Available at: https://science.nasa.gov/science-news/science-at-nasa/2009/23dec_voyager

PhysOrg (2017) No alien 'signals' from cigar-shaped asteroid: researchers. PhysOrg: Online. 14 December 2017. Available at: https://phys.org/news/2017-12-alien-cigar-shaped-asteroid.html thanks to John

Pierrehumbert, R. (2010). A palette of climates for Gliese 581g. The Astrophysical Journal Letters, 726 (1)

Pieters, C. et al. (2009). Character and spatial distribution of OH/H2O on the surface of the Moon seen by M3 on Chandrayaan-1. Science, 326(5952): pp. 568-572

Plait, P. (2008). The Planet X saga: Science. Online: Bad Astronomy. Available at: http://www.badastronomy.com/bad/misc/planetx/science.html

Plait, P. (2011). No, there's no proof of a giant planet in the outer solar system. Waukesha, WI: Discovery Magazine, 14 February 2011. Available at: http://blogs.discovermagazine.com/badastronomy/2011/02/14/no-theres-no-proof-of-a-giant-planet-in-the-outer-solar-system/

Press Association (2014). Saturn's moon Enceladus has huge ocean deep under icy surface. New York, NY: Huffington Post. 3 April 2014. Available at: http://www.huffingtonpost.co.uk/2014/04/03/saturns-moon-enceladus-has-huge-ocean_n_5086837.html

Prettyman, T. et al. (2016). Extensive water ice within Ceres' aqueously altered regolith: Evidence from nuclear spectroscopy. Science, aah6765

Prigg, M. (2008). The hunt for Planet X: Nasa fails to find mysterious giant body believed to have caused mass extinctions on Earth (and they now say it may not exist at all). London: Daily Mail. 7 March 2014. Available at: http://www.dailymail.co.uk/sciencetech/article-2576009/The-hunt-Planet-X-Nasa-fails-mysterious-giant-body-caused-mass-extinctions-Earth-say-not-exist-all.html

Rabinowitz, D. et al. (2008). The youthful appearance of the 2003 EL61 collisional family. The Astronomical Journal. 136(4): 1502–1509.

Radigan, J. et al. (2012). Large amplitude variations of an L/T transition brown dwarf: Multi-wavelength observations of patchy, high-contrast cloud features. The Astrophysical Journal, 750(2). Available at: http://iopscience.iop.org/article/10.1088/0004-637X/750/2/105/pdf

Radigan, J. (2015). Weather on substellar worlds: Brown dwarfs. Online blog. 1 June 2015. Available at: https://blogs.stsci.edu/universe/category/brown-dwarfs/

Rampino, M. & Caldeira, K. (2015). Periodic impact cratering and extinction events over the last 260 million years. Monthly Notices of the Royal Astronomical Society, 454(4): pp 3480-3484

Ramsay, L. (2018). NASA's Webb Telescope to investigate mysterious brown dwarfs. Pasadena, CA: NASA. 4 January 2018. Available at: https://www.nasa.gov/feature/goddard/2018/nasa-s-webb-telescope-to-investigate-mysterious-brown-dwarfs

RAS-WEB (2017). Juno isn't exactly where it's supposed to be. the flyby anomaly is back, but why does it happen? The Renfrewshire Astronomical Society: Online blog. Available at: http://renfrewshireastro.co.uk/juno-isnt-exactly-where-its-supposed-to-be-the-flyby-anomaly-is-back-but-why-does-it-happen

Raup, D. & Sepkoski, J. (1984). Periodicity of extinctions in the geologic past. Proceedings of the National Academy of Sciences, 81(3): pp801–805. Available at: http://www.pnas.org/content/pnas/81/3/801.full.pdf

Raymond, S. & Izidoro, A. (2017). The empty primordial asteroid belt. Science Advances, 3, e1701138. Available at: https://arxiv.org/pdf/1709.04242.pdf

Raymond, S. et al (2018). Implications of the interstellar object 1I/'Oumuamua for planetary dynamics and planetesimal formation, MNRAS, sty468, 26 February 2018

Redd, N. (2014). Could 'Planet X' still be lurking out there? Discovery News, 28 March 2014. Available at: http://www.seeker.com/could-planet-x-still-be-lurking-out-there-1768430462.html

Redfern, M & Henbest, N (1983). Has IRAS found a tenth planet? New Scientist, 10 November 1983

Rickman, H., Valsecchi, G. & Froeschlé, C. (2001). From the Oort cloud to observable short-period comets - I. The initial stage of cometary capture. Monthly Notices of the Royal Astronomical Society, 325(4): pp. 1303-1311

Rievers, B. & Lämmerzahl, C. (2011). High precision thermal modelling of complex systems with application to the flyby and Pioneer anomaly. Annalen der Physik. 523 (6): 439.

Rincon, P. (2006). Pluto vote 'hijacked' in revolt. London: BBC, 25 August 2006. Available at: http://news.bbc.co.uk/1/hi/sci/tech/5283956.stm

Rincon, P. (2016). Planet Nine's profile fleshed out. London: BBC, 10 April 2016. Available at: http://www.bbc.co.uk/news/science-environment-35996813

Route, M. (2016). The discovery of solar-like activity cycles beyond the end of the main sequence? Astrophysical Journal Letters, 830: L27

Russell, D. (1979). The enigma of the extinction of the dinosaurs. Annual Review of Earth and Planetary Sciences, 7: p163

Saal, A. et al. (2008). Volatile content of lunar volcanic glasses and the presence of water in the Moon's interior. Nature, 454(7201): pp 192-5

Saal, A. et al. (2013). Hydrogen isotopes in lunar volcanic glasses and melt inclusions reveal a carbonaceous chondrite heritage. Science, 340(6138): pp 1317-1320

Sadavoy, S. & Stahler, S. (2017). Embedded binaries and their dense cores. Monthly Notices of the Royal Astronomical Society, 469(4): pp3881-3900

Sagan, C. (1963). Direct contact among galactic civilizations by relativistic interstellar spaceflight. Planetary and Space Science, 11: pp485-498, Available at: https://ntrs.nasa.gov/archive/nasa/casi.ntrs.nasa.gov/19630011050.pdf

Sample, I. (2014). Curiosity Rover's discovery of methane 'spikes' fuels speculation of life on Mars. London: The Guardian. 16 December 2014. Available at: http://www.theguardian.com/science/2014/dec/16/methane-spikes-mars-fuel-speculation-life-nasa-curiosity

Sample, I. (2014a). Methane on Mars: does it mean the Curiosity rover has found life? London: The Guardian. 17 December 2014. Available at: http://www.theguardian.com/science/2014/dec/17/methane-mars-curiosity-rover-life

Sanders, R. (2017). New evidence that all stars are born in pairs. Berkeley, Ca: U.C. Berkeley. 13 June 2017. Available at: http://news.berkeley.edu/2017/06/13/new-evidence-that-all-stars-are-born-in-pairs/

Sarafian, A. et al (2014). Early accretion of water in the inner solar system from a carbonaceous chondrite-like source. Science, 346(6209): pp623-6

Schlaufman, K. (2018). Evidence of an upper bound on the masses of planets and its implications for giant planet formation. The Astrophysical Journal, 853(1)

Schlyter, P. (1997). Hypothetical planets. 3 September 1997. Available at: https://www.physics.upenn.edu/nineplanets/hypo.html

Schmandt, B. et al. (2014). Dehydration melting at the top of the lower mantle. Science, 344(6189): pp1265-1268

Schmedemann, N. et al. (2016). Timing of optical maturation of recently exposed material on Ceres. Geophysical Research Letters, 43(23): pp 11,987-11,993

Scholz, A. et al. (2012). Substellar objects in nearby young clusters (SONYC). VI. The Planetary-Mass Domain of NGC 1333. The Astrophysical Journal, 756(1): 9 August 2012. Available at: http://iopscience.iop.org/article/10.1088/0004-637X/756/1/24/meta

Schon, S., Head, J. & Milliken, R. (2009). A recent ice age on Mars: Evidence for climate oscillations from regional layering in mid-latitude mantling deposits. Geophysical Research Letters, 36(15). Available at: https://agupubs.onlinelibrary.wiley.com/doi/full/10.1029/2009GL038554

Sci News (2017). Astronomers hunt for free-floating planetary-mass objects in solar neighbourhood, 1 March 2017. Available at: http://www.sci-news.com/astronomy/free-floating-planetary-mass-objects-solar-neighborhood-04662.html

Scherer, K. (2000). Drag forces on interplanetary dust grains induced by the interstellar neutral gas. Journal of Geophysical Research, 105(A5): pp 10329-10342

Schneider, A. et al (2017). A 2MASS/AllWISE Search for extremely red L Dwarfs -- The discovery of several likely L type members of Beta Pic, AB Dor, Tuc-Hor, Argus, and the Hyades. The Astronomical Journal, 153(4)

Schulz, R. et al. (2015). Comet 67P/Churyumov-Gerasimenko sheds dust coat accumulated over the past four years. Nature (letter), 518(7538): pp216–218

Schwamb, M. (2007). Searching for Sedna's sisters: Exploring the inner Oort cloud. Pasadena: CalTech, 18 September 2007. Available from: https://web.archive.org/web/20130512221422/http://www.astro.caltech.edu/~george/option/candex07/schwamb_report.pdf

Shankman, C. et al. (2017) OSSOS VI. Striking biases in the detection of large semimajor axis Trans-Neptunian Objects. Accepted for publication by The Astronomical Journal, 19 June 2017. Available at: https://arxiv.org/abs/1706.05348

Shaviv, N. (2003). The spiral structure of the Milky Way, cosmic rays, and Ice Age Epochs on Earth. New Astronomy, 8(1): pp39-77. Available at: http://www.phys.huji.ac.il/~shaviv/articles/long-ice.pdf

Shipway, A. & Shipway, S. (2008). Orbital period of a planet: Based on Kepler's third law. Online. Available at: http://www.calctool.org/CALC/phys/astronomy/planet_orbit

Siegler, M. et al. (2011). Effects of orbital evolution on lunar ice stability. Journal of Geophysical Research: Planets. 116(E3)

Sitchin, J. (Ed) (2015). The Anunnaki Chronicles. Rochester: Bear & Co

Sitchin, Z. (1976). The twelfth planet. New York: Avon

Sitchin Z. (1990). The lost realms. New York: Avon

Sitchin, Z. (1990a). Genesis revisited. New York: Avon

Sitchin, Z. (1993). When time began. New York: Avon

Sitchin, Z. (1996). Divine encounters. New York: Avon

Sitchin, Z. (1998). The cosmic code. New York: Avon
Sitchin, Z. (2004). The case of the French [sic] astronomer. New York: Online. Available at: http://www.sitchin.com/frenchastron.htm

Sitchin, Z. (2007). The end of days. New York: William Morrow & Company

Sitchin, Z. (2010). There were giants upon the Earth. Rochester: Bear & Company

Sleep, N. & Lowe, D. (2014). Physics of crustal fracturing and chert dike formation triggered by asteroid impact, ~3.26 Ga, Barberton greenstone belt, South Africa. Geochemistry, Geophysics, Geosystems, 15(4): pp

1054-1070. Available at: https://agupubs.onlinelibrary.wiley.com/doi/epdf/10.1002/2014GC005229

Sneiderman, P. (2018). Johns Hopkins scientist proposes new definition of a planet. John Hopkins University Press release, 22 January 2018. Available at: http://releases.jhu.edu/2018/01/22/johns-hopkins-scientist-proposes-new-definition-of-a-planet/

Soderblom, J. et al. (2015). The fractured Moon: Production and saturation of porosity in the lunar highlands from impact cratering. Geophysical Research Letters, 42(17): pp6939-6944

Sokol, J. (2017). New haul of distant worlds casts doubt on Planet Nine. Washington: American Association for the Advancement of Science, 21 June 2017. Available at: http://www.sciencemag.org/news/2017/06/new-haul-distant-worlds-casts-doubt-planet-nine

Sonett, C. et al. (1987). Interstellar shock waves and 10Be from ice cores. Nature, 330: pp458 - 460

Spanish Foundation For Science And Technology (2016). Extreme Trans-Neptunian Objects lead the way to Planet Nine. (FECYT Press Release). Astronomy Now, 13 June 2016. Available at: https://astronomynow.com/2016/06/13/extreme-trans-neptunian-objects-lead-the-way-to-planet-nine/

Spiegel, D. et al. (2011). The deuterium-burning mass limit for brown dwarfs and giant planets. The Astrophysical Journal, 727(1). Available at: http://iopscience.iop.org/article/10.1088/0004-637X/727/1/57/meta

Standish, E. (1993). Planet X: No dynamical evidence in the optical observations. The Astronomical Journal, 105(5): pp2000-2006. Available at: http://adsabs.harvard.edu/full/1993AJ....105.2000S

Stephant, A. & Robert, F. (2014). The negligible chondritic contribution in the lunar soils water. Proceedings of the National Academy of Sciences, 111(42): pp 15007-15012

Stephens, T. (2016). New analysis adds support for a subsurface ocean on Pluto. Santa Cruz, CA: UC Santa Cruz. 16 November 2016. Available at: https://news.ucsc.edu/2016/11/pluto-ocean.html

Starr, M. (2016). Is this what Planet Nine looks like? San Francisco: CNET, 8 April 2016. Available at: http://www.cnet.com/uk/news/is-this-what-planet-nine-looks-like/

Stern, S. (1990). ISM-induced erosion and gas-dynamical drag in the Oort Cloud. Icarus, 84: pp447-466

Stiles, L. (2008). Gamma-Ray Evidence Suggests Ancient Mars Had Oceans. Tucson, AZ: University of Arizona Press Release. 17 November 2008. Available at:

https://uanews.arizona.edu/story/gamma-ray-evidence-suggests-ancient-mars-had-oceans

Stolte, D. (2017). UA scientists and the curious case of the warped Kuiper Belt. University of Arizona Press Release, 20 June 2017. Available at: https://uanews.arizona.edu/story/ua-scientists-and-curious-case-warped-kuiper-belt

Strain, D. (2018). Collective gravity, not Planet Nine, may explain the orbits of 'detached objects'. CU Boulder Today. 4 June 2018. Available at: https://www.colorado.edu/today/2018/06/04/collective-gravity

Strampelli, G. et al. (2018). A HST/WFC3 Search for substellar companions in the Orion Nebula Cluster. 231st AAS Meeting, abstract # 414.07

Strobel, N. (2010). Kepler's laws of planetary motion. Bakersfield, CA: Astronomy Notes. 6 April 2010. Available at: http://www.astronomynotes.com/history/s7.htm

Stuurman, C. et al. (2016). SHARAD detection and characterization of subsurface water ice deposits in Utopia Planitia, Mars. Geophysical Research Letters, 43(18): pp9484-9491

Sullivan, W. (1995). New theory on ice sheet catastrophe Is the direst one yet. New York: New York Times, 2 May 1995 Available at: http://www.nytimes.com/1995/05/02/science/new-theory-on-ice-sheet-catastrophe-is-the-direst-one-yet.html

Sumi, T. et al. (2011). Unbound or distant planetary mass population detected by gravitational microlensing. Nature, 473: 349

Takahashi, F. et al. (2014). Reorientation of the early lunar pole. Nature Geoscience, 7: pp 409–412

Talbert, T. (Ed.) (2015). The rich color variations of pluto. Pasadena, CA: NASA. 14 July 2015. Available at: https://www.nasa.gov/image-feature/the-rich-color-variations-of-pluto

Talbert, T. (Ed.) (2015a). Stunning nightside image reveals pluto's hazy skies. Pasadena, CA: NASA. 24 July 2015. Available at: http://www.nasa.gov/feature/stunning-nightside-image-reveals-pluto-s-hazy-skies

Tamayo, D. & Burns, J. (2013). Circumplanetary debris disks and consequences of an eccentric Fomalhaut B. American Astronomical Society, DPS meeting #45, October 2013

Tamblyn, T. (2015). Saturn's moon Enceladus has a global subsurface ocean. New York, NY: Huffington Post. 16 September 2015. Available at: http://www.huffingtonpost.co.uk/2015/09/16/saturns-moon-enceladus-giant-subsurface-ocean_n_8144346.html

Tennyson, A. (1846). The golden year. In Poems. 4th Edn.

Thackeray, A. & Wesselink, A. (1965). A photometric and spectroscopic study of the cluster IC 2944. Monthly Notices of the Royal Astronomical Society, 131: 121

Thaddeus, P. (1986). Molecular clouds and periodic events in the geologic past. The Galaxy and the Solar System, Tucson, AZ: University of Arizona Press, pp. 61-68

The Sky Live (2018). 1I/2017 U1 (Oumuamua). Available at: https://theskylive.com/oumuamua-tracker (Accessed: 9th March 2018)

The Space Academy (2017). NASA's $1 billion Jupiter probe just sent back stunning new photos of Jupiter. 7 November 2017. Available at: http://www.thespaceacademy.org/2017/11/nasas-1-billion-jupiter-probe-just-sent.html

Thies, I. et al. (2010). Tidally induced brown dwarf and planet formation in circumstellar disks. The Astrophysical Journal, July 2010, 717(1): pp 577-585. Available at: http://adsabs.harvard.edu/abs/2010ApJ...717..577T

Thomas, P. et al. (2005). Differentiation of the asteroid Ceres as revealed by its shape. Nature, 437(7056): pp 224-226

Thomas, P. et al. (2016). Enceladus's measured physical libration requires a global subsurface ocean. Icarus, 264: pp 37-47

Toth, I. (2005). Connections between asteroids and cometary nuclei. Proceedings on the International Astronomical Union. 1: S229, pp 67-96

Touma, J. & Wisdom, J. (1993). The chaotic obliquity of Mars. Science, 259 (5099): pp 1294–1297

Travis, D. (2016). The Star of Mashiach. 5 February 2016. Available at: http://nibiruiscoming.blogspot.co.uk/2016/02/the-star-of-mashiach.html

Trujillo, C. & Sheppard, S. (2014). A Sedna-like body with a perihelion of 80 astronomical units. Nature, 507: pp471–474

Tschauner, O. et al. (2018). Ice-VII inclusions in diamonds: Evidence for aqueous fluid in Earth's deep mantle, Science, 359(6380): pp. 1136-1139

Tsehmeystrenko, V. (2013). The FU Orionis phenomenon. Variable Stars Observer Bulletin, Sept-Oct 2013. Available at: http://www.vs-compas.belastro.net/bulletin-pdf/article/2-3.pdf

Tsiganis, K. et al. (2005). Origin of the orbital architecture of the giant planets of the Solar System. Nature, 435(7581): pp459-461

Tsiganis, K. (2015). Planetary science: How the solar system didn't form. Nature, 528: pp 202–204

Ulmer, E. (1951). The man from Planet X. Beverly Hills: United Artists

Vahradyan, V. & Vahradyan, M. (2010). About the astronomical role of "Qarahunge" monument. Available at: http://www.anunner.com/vachagan.vahradyan/About_the_Astronomical_Role_of__%E2%80%9CQarahunge%E2%80%9D_Monument_by_Vachagan_Vahradyan,_Marine_Vahradyan

Valley, J. et al. (2014). Hadean age for a post-magma-ocean zircon confirmed by atom-probe tomography. Nature Geoscience, 7: pp 219-223.

Vigan, et al. (2017). The VLT/NaCo large program to probe the occurrence of exoplanets and brown dwarfs at wide orbits. IV. Gravitational instability rarely forms wide, giant planets. Astronomy and Astrophysics, 603

Villanueva, G. et al. (2015). Strong water isotopic anomalies in the Martian atmosphere: Probing current and ancient reservoirs. Science, 348(6231): pp. 218-221

Vitense, C. et al. (2012). An improved model of the Edgeworth-Kuiper debris disk. Astronomy and Astrophysics, 540(A30)

Vitense, C. et al. (2014). Will New Horizons see dust clumps in the Edgeworth-Kuiper belt? The Astronomical Journal, Vol. 147(6)

Vlemmings, W. et al. (2015). The serendipitous discovery of a possible new solar system object with ALMA. 8 December 2015, subsequently 'withdrawn until further detections are obtained'.

Volk, K. & Malhotra, R. (2017). The curiously warped mean plane of the Kuiper belt. Accepted for publication in The Astronomical Journal, 19 June 2017. Available at: https://www.lpl.arizona.edu/~renu/malhotra_preprints/kb-plane-2.pdf

Wallis, M. & Wickramasinghe, N. (2004). Interstellar transfer of planetary microbiota. Monthly Notices of the Royal Astronomical Society. 348(1): pp52-61.

Walsh, K. et al. (2011). A low mass for Mars from Jupiter's early gas-driven migration. Nature, 475: pp206–209

Blackburn, R., Kawamoto, S. & Warmkessel, B. (1997). Vulcan, comets and the impending catastrophe. Available at: http://barry.warmkessel.com/1997Paper.html

Webster, G. (2012). Spacecraft monitoring martian dust storm. Pasadena, CA: NASA/JPL. 21 November 2012. Available at: http://www.nasa.gov/mission_pages/mars/news/mars20121121.html

Webster, G., Brown D. & Cantillo, L. (2016). Second cycle of Martian seasons completing for Curiosity Rover. Pasadena, CA: NASA/JPL. 11 May 2016. Available at: https://www.nasa.gov/feature/jpl/second-cycle-of-martian-seasons-completing-for-curiosity-rover

Weissman, P. & Levison, H. (1997). Origin and evolution of the unusual object 1996 PW: Asteroids from the Oort cloud? The Astrophysical Journal, 488: L133–L136

Wenz, J. (2017). NASA eyes Neptune and Uranus for missions in the 2030s. New Scientist, 16 June 2017. Available at: https://www.newscientist.com/article/2137606-nasa-eyes-neptune-and-uranus-for-missions-in-the-2030s/

Whitmire, D. & Jackson, A. (1984). Are periodic mass extinctions driven by a distant solar companion? Nature, 308 (5961): pp713–715

Wickramasinghe, D. & Allen, D. (1980). The 3.4-micron interstellar absorption feature. Nature, 287: pp 518-519.

Wilford, J. N. (1987). Looking for Planet X: Old clues, new theory. New York: The New York Times, 1 July 1987. Available at: http://www.nytimes.com/1987/07/01/us/looking-for-planet-x-old-clues-new-theory.html

Witze, A. (2016). Astronomers spot distant world in Solar System's far reaches. Nature, 18th October 2016, doi:10.1038/nature.2016.20831. Available at: http://www.nature.com/news/astronomers-spot-distant-world-in-solar-system-s-far-reaches-1.20831

Wood, E. (1959). Plan 9 from outer space. Distributors Corporation of America: Valiant Pictures

Yang, J. et al. (2014). Water trapping on tidally locked terrestrial planets requires special conditions. The Astrophysics Journal, 796, L22

Yao, Y. & Giapis, K. (2017). Dynamic molecular oxygen production in cometary comae. Nature Communications, 8: 15298. Available at: https://authors.library.caltech.edu/77245/1/ncomms15298.pdf

Yates, J. et al. (2017). Atmospheric habitable zones in Y Dwarf atmospheres. Accepted to The Astrophysical Journal, 17 February 2017. Available at: http://www.jackyates.co.uk/publications/ahz.pdf

Ying-Tung Chen et al. (2016). Discovery of a new retrograde Trans-Neptunian Object: Hint of a common orbital plane for low semi-major axis, high inclination TNOs and Centaurs. The Astrophysical Journal Letters, 827(2): L24

Yuhas, A. (2016). A possible ninth planet may be the reason for a tilt in our solar system. London: The

Guardian, 19 October 2016. Available at: https://www.theguardian.com/science/2016/oct/19/planet-nine-solar-system-tilt-astronomers

Yeghikyan, A. & Fahr, H. (2004). Effects induced by the passage of the Sun through dense molecular clouds: I. Flow outside of the compressed heliosphere. Astronomy & Astrophysics, 415: pp763-770

Yun, J. & Clemens, D. (1990). Star formation in small globules – Bart Bok was correct. Astrophysical Journal, Part 2 – Letters 365: L73

Zank, G. & Frisch, P. (1999). Consequences of a change in the galactic environment of the Sun. The Astrophysical Journal, 518: pp965-973

Zooniverse (2017). Discover the ninth planet in our Solar System with the Zooniverse and BBC Stargazing Live. Oxford: Zooniverse. Available at: https://www.zooniverse.org/projects/skymap/planet-9

Zygutis, D. (2017). The Sagan conspiracy. Pompton Plains, NJ: New Page Books

Image Files: Sources, Attributions, and Notes

Front Cover: Digital montage by Andy Lloyd, using Corel Paint Shop Pro Photo x2.

Fig 1-1; p14. "Solar Sys". Harman Smith and Laura Generosa (nee Berwin): JPL/NASA: Image in the Public Domain.

Fig 1-2; p16. "Kuiper Belt and Oort Cloud". NASA: Image in the Public Domain.

Fig 1-3; p17. "Voyagers in the Heliosheath". NASA/JPL – Caltech : Image in the Public Domain.

Fig 1-4; p22. "Solar System Plus One". Composite of NASA Images in the Public Domain, courtesy of Comfreak, under CC0 Creative Commons license; digitally altered by Andy Lloyd.

Fig 2-1; p28. "NASA's Swift Mission Observes Mega Flares from a Mini Star". NASA 's Goddard Space Flight Center/S. Wiessinger: Image in the Public Domain.

Fig 2-2; p31. "Marine Genus Biodiversity: Extinction Intensity". Wikimedia Commons, under GNU Free Documentation Licence 1.2. https://en.wikipedia.org/wiki/File:Extinction_intensity.svg

Fig 2-3; p33. "Oort cloud Sedna orbit". NASA/JPL-Caltech/R. Hurt: Image in the Public Domain.

Fig 3-1; p38. "WISE artists concept (as of 2006)". NASA: Image in the Public Domain.

Fig 3-2; p41. "Brown Dwarfs Comparison 01". Attribution: MPIA/V. Joergens in "Joergens, Viki, 50 Years of Brown Dwarfs - From Prediction to Discovery to Forefront of Research, Astrophysics and Space Science Library 401, Springer, ISBN 978-3-319-01162-2". Under the Creative Commons Attribution 3.0 Unported license.

Fig 3-3; p43. "2MASSJ22282889-431026 (Artist's Conception)" NASA/JPL-Caltech: Image in the Public

Domain. https://exoplanets.nasa.gov/resources/114/forecast-for-exotic-weather/

Fig 3-4; p45. "Star Clusters Rosette Nebula". WikiImages; under CC0 Creative Commons license, via Pixabay.

Fig 4-1; p49. "KBOs and Resonances". ["Distribution of Kuiper belt objects' orbits. Objects in resonance are plotted in red (Neptune trojans 1:1, plutinos 2:3, twotinos 1:2,...). Classical objects are plotted in blue. Scattered disk objects (not members of the Kuiper Belt) are shown in grey, Sednoids in yellow"]. JPL/Renerpho. under the Creative Commons Attribution-Share Alike 4.0 International license. https://commons.wikimedia.org/wiki/File:KBOsAndResonances.gif

Fig 4-2; p52. "M104 ngc4594 Sombrero Galaxy" NASA/ESA and the Hubble Heritage Team (STScI/AURA); Image in the Public Domain; digitally altered by Andy Lloyd.

Fig 4-3; p53. "Shattered Comet". NASA/JPL-Caltech: Image in the Public Domain.

Fig 4.4; p54. "Argument of Periapsis in Elliptical Orbit". ["Letters in the image denote: (A) Minor, orbiting body; (B) Major body being orbited by A; (C) Reference plane, e.g. the ecliptic; (D) Orbital plane of A; (E) Ascending node; (F) Periapsis; (ω) Argument of the periapsis. The red line is the line of apsides; going through the periapsis (F) and apoapsis (H); this line coincides with the major axis in the elliptical shape of the orbit. The green line is the node line; going through the ascending (G) and descending node (E); this is where the reference plane (C) intersects the orbital plane (D)"]. Under the terms of the GNU Free Documentation License. https://commons.wikimedia.org/wiki/File:Argument_of_Periapsis_in_Elliptical_Orbit.png

Fig 4-5; p57. "Planet Nine related clustering of small objects detected". Prokaryotes; Image in the Public Domain

Timeless Voyager Press

Fig 5-1; p61. "Ancient Constellations over ALMA". ESO/B. Tafreshi; under the Creative Commons Attribution 4.0 International license.

Fig 5-2; p62. "Plan 9 from Outer Space (Film Poster)". Distributors Corporation of America; Image in the Public Domain. https://commons.wikimedia.org/wiki/File:Plan_9_Alternative_poster.jpg

Fig 5-3; p65. "2003 UB313 compared to Pluto, 2005 FY9, 2003 EL61, Sedna, Quaoar, and Earth". NASA, ESA, and A. Feild (STScI); Image in the Public Domain

Fig 5-4; p67. "Mauna Kea Observatory". Alan L.; under the Creative Commons Attribution 2.0 Generic license.

Fig 5-5; p69. "Kuiper Belt Object (Artist's Concept)". NASA, ESA, and G. Bacon (STScI); Image in the Public Domain

Fig 5-6; p71. "Orbit of Planet X". Digital montage by Andy Lloyd, using Corel Paint Shop Pro Photo x2

Fig 5-7; p72. "Planet Nine". Comfreak; under CC0 Creative Commons license, via Pixabay; image digitally rendered by Andy Lloyd

Fig 6-1; p76. "Planet Nine Orbit". MagentaGreen/Prokaryotes; Image in the Public Domain.

Fig 6-2; p77. "Double sun halo above the Shuchinsk lake". TEHb K0CM0CA; under the Creative Commons Attribution-Share Alike 4.0 International license.

Fig 6-3; p80. "ALMA observatory equipped with its first antenna". ALMA (ESO/NAOJ/NRAO); under the Creative Commons Attribution 4.0 International Licence.

Fig 6-4; p84. "Radar Dishes". WikiImages: under CC0 Creative Commons license, via Pixabay.

Fig 7-1; p89. "Extreme trans-Neptunian objects, 9 objects plus 4 new ones in orange". Tomruen; under the Creative Commons Attribution-Share Alike 4.0 International license.

Fig 7-2; p93. "The argument of the periapsis for a "minor" object in an elliptic orbit around a larger object". ["Letters in the image denote: (A) Minor, orbiting body; (B) Major body being orbited by A; (C) Reference plane, e.g. the ecliptic; (D) Orbital plane of A; (i) – Inclination. The green line is the node line; going through the ascending and descending node; this is where the reference plane (C) intersects the orbital plane (D)"]. Under the terms of the GNU Free Documentation License. https://commons.wikimedia.org/wiki/File:Inclination_in_Elliptical_Orbit.png

Fig 7-3; p96. "Planet Nine 15 eTNO objects orbit relative to hypothetical Planet Nine". ["Anti-aligned orbits red, others blue"].Tomruen; under the Creative Commons Attribution-Share Alike 4.0 International license. Digitally altered by Bruce Stephen Holms.

Fig 7-4; p98. "WISE Sky Image". NASA/JPL-Caltech/UCLA; Image in the Public Domain.

Fig 8-1; p99. "Cold and Close Celestial Orb (Artist's Concept)". NASA/JPL-Caltech/Penn State University; Image in the Public Domain.

Fig 8-2; p101. "Monte Carlo Casino in Monaco". Preslitsky; under the Creative Commons Attribution-Share Alike 4.0 International license.

Fig 8-3; p102. "Planet Nine Comparison". PlanetUser; under the Creative Commons Attribution-Share Alike 4.0 International license.

Fig 8-4; p102. "Artist's impression of an ultra-short-period planet". ESA/Hubble/A. Schaller; under the Creative Commons Attribution 4.0 International licence.

Fig 8-5; p105. "K/T extinction event theory: An artist's depiction". NASA: Image in the Public Domain.

Fig 8-6; p106. "Protoplanetary Disk: An Artist's Concept". NASA Goddard; Image in the Public Domain.

Fig 8-7; p107. "Observatoire de la Côte d'Azur, Nice". Photograph by Andy Lloyd.

Fig 9-1; p112. "Pluto's size compared with the largest moons". Eurocommuter; Images by NASA in the Public Domain. Digitally altered by Andy Lloyd.

Fig 9-2; p114. "Large Kuiper belt objects". Eurocommuter; under the Creative Commons Attribution-Share Alike 3.0 Unported license.

Fig 9-3; p118. "Illustration of relative sizes, colours and albedos of the large trans-Neptunian objects". Eurocommuter~commonswiki; under the Creative Commons Attribution-Share Alike 3.0 Unported license.

Fig 9-4; p122. "Neptune's Fountain, Cheltenham". Photograph by Andy Lloyd.

Fig 9-5; p124. "Planet and Asteroids". UKT2; under CC0 Creative Commons license, via Pixabay.

Fig 10-1; p129. "Early Planetary System". GamOI; under CC0 Creative Commons license, via Pixabay.

Fig 10-2; p130. "Uranus Interior". Uranus-intern-de.png: FrancescoA/derivative work: WolfmanSF; Image in the Public Domain.

Fig 10-3; p132. "M51 CHANDRA2". NASA: Image in the Public Domain.

Fig 10-4; p134. "The Wreath Nebula". NASA/JPL-Caltech/WISE Team: Image in the Public Domain.

Fig 10-5; p140. "Dark Star ellipses". Digital montage by Andy Lloyd, using Corel Paint Shop Pro Photo x2.

Fig 10-6; p142. "Planet X and the Inner Oort Cloud". Digital montage by Andy Lloyd, using Corel Paint Shop Pro Photo x2.

Fig 11-1; p144. "The spectacular star-forming Carina Nebula imaged by the VLT Survey Telescope". ESO; under the Creative Commons Attribution 4.0 International licence. https://www.eso.org/public/images/eso1250a/

Fig 11-2; p145. "Rotating Disk". Buddy Nath; under CC0 Creative Commons license, via Pixabay.

Fig 11-3; p147. "Protoplanetary Disk". Pat Rawlings/ NASA: Image in the Public Domain.

Fig 11-4; p150. "Zodiacal Glow Lightens Paranal Sky". ESO/Y. Beletsky; under the Creative Commons Attribution 4.0 International license. http://www.eso.org/public/images/potw1348a/

Fig 11-5; p152. "Pioneer and Voyager Spacecraft, the Heliosphere and Solar System". NASA: Image in the Public Domain.

Fig 11-6; p157. "An artist's rendering of the minor planet 10199 Chariklo, with rings".ESO/L. Calçada/M. Kornmesser/Nick Risinger (skysurvey.org); under the Creative Commons Attribution 3.0 Unported license. http://www.eso.org/public/images/eso1410b/

Fig 12-1; p161. "The planetary disc surrounding the star RX J1615". ESO, J. de Boer et al.; under the Creative Commons Attribution 4.0 International License. https://www.eso.org/public/images/eso1640b/
Fig 12-2; p162. "Extrasolar Planet Fomalhaut B in Visible Light". NASA/ESA/T. Curie/U. Toronto: Image in the Public Domain.

Fig 12-3; p162. "Apkallu, Winged Genius with a Bird's Head". BurgererSF: Image in the Public Domain.

Fig 12-4; p164. "Furniture Plaque Carved in Relief with a Male Figure Grasping a Tree; Winged Sun Disc Above". Rogers Fund, 1959: Image in the Public Domain.

Fig 12-5; p166. "The Dark Nebula LDN 483". ESO; under the Creative Commons Attribution 4.0 International License. http://www.eso.org/public/images/eso1501a/

Fig 12-6; p167. "Annual parallax". P.wormer; under the Creative Commons Attribution-Share Alike 3.0 Unported licence.

Fig 12:7; p168. "Going Deep on Thackerays Globules". Dylan O'Donnell: Image in the Public Domain.

Fig 12-8; p169. "Artist's conception of circumstellar disk around AA Tauri". NASA: Image in the Public Domain.

Fig 12-9; p172. "NGC 1333-IRAS 4B (Artist's Concept)". NASA/JPL-Caltech: Image in the Public Domain.

Fig 12-10; p176. "A nebula containing gas, dust, and asteroids (Artist's Concept)". NASA Goddard: Image in the Public Domain.

Fig 13-1; p180. "Tau Geminorum and brown dwarf". Tyrogthekreeper at en.wikipedia; under the Creative Commons Attribution-Share Alike 3.0 Unported license.

Fig 13-2; p182. "Red Dwarf Planet". NASA, ESA and G. Bacon (STScI): Image in the Public Domain.

Fig 13-3; p184. "Young Brown Dwarf in TW Hydrae Family of Stars (Artist Concept)". NASA/IPAC/JPL-Caltech: Image in the Public Domain.

Fig 13.4; p186. "A Star Cluster in the Wake of Carina". ESO; under the Creative Commons Attribution 4.0 International License. http://www.eso.org/public/news/eso1416/

Fig 13-5; p189. "Orion Nebula - Hubble 2006 mosaic 18000". NASA, ESA, M. Robberto (Space Telescope Science Institute/ESA) and the Hubble Space Telescope Orion Treasury Project Team: Image in the Public Domain.

Fig 13-6; p191. "Cosmic Sparklers". NASA/CXC/JPL-Caltech/NOAO/DSS: Image in the Public Domain.

Fig 14-1; p196. "Oumuamua orbit at perihelion". Nagualdesign; Tomruen - Made with trajectory data from JPL Horizons; under the Creative Commons Attribution-Share Alike 4.0 International license.

Fig 14-2; p197. "Oumuamua Skypath-Sun". Tomruen; under the Creative Commons Attribution-Share Alike 4.0 International license.

Fig 14-3; "Eso1737f: Oumuamua Brightness Variation". ESO/K. Meech et al.; under the Creative Commons Attribution 4.0 International licence. https://www.eso.org/public/images/eso1737f

Fig 14-4; p202. "Oumuamua (Artist's impression)". Original: ESO/M. Kornmesser Derivative: Nagualdesign - Derivative of http://www.eso.org/public/images/eso1737a/, [shortened (65%) and reddened and darkened]; under the Creative Commons Attribution-Share Alike 4.0 International license.

Fig 14.5; p205. "Oumuamua Outgassing (Artist's impression)". ESA/Hubble, NASA, ESO, M. Kornmesser; under the Creative Commons Attribution-Share Alike 4.0 International license. https://www.eso.org/public/images/eso1820a/

Darker Stars – Andy Lloyd

Fig 15-1; p215. Digital montage by Andy Lloyd, using Corel Paint Shop Pro Photo x2.

Fig 15-2; p217. "An Artist's Impression of a Pioneer Spacecraft on its way to Interstellar Space". NASA Ames: Image in the Public Domain.

Fig 15-3; p220. "PIA21972 Jupiter Blues". NASA/JPL-Caltech/SwRI/MSSS/Gerald Eichstadt/Sean Doran: Image in the Public Domain.

Fig 15-4; p222. "Juno's View of Jupiter's Southern Lights". NASA/JPL-Caltech/SwRI/ASI/INAF/JIRAM: Image in the Public Domain.

Fig 15-5; p223. "Anatomy of Brown Dwarf's Atmosphere". NASA/JPL-Caltech: Image in the Public Domain.

Fig 15-6; p227. "Fountains of Enceladus PIA07758". NASA/JPL/Space Science Institute: Image in the Public Domain.

Fig 16-1; p231. "Pluto-01 Stern 03 Pluto Color TXT". [NASA's New Horizons spacecraft captured this high-resolution enhanced color view of Pluto on July 14, 2015]. NASA/JHUAPL/SwRI: Image in the Public Domain.

Fig 16-2; p233. "Blue hazes over backlit Pluto". [A near-true-color image of Pluto's bluish hazes taken by New Horizons].NASA/JHUAPL/SwRI: Image in the Public Domain.

Fig 16-3; p234. "Pluto Charon Moon Earth Comparison". NASA: Image in the Public Domain.

Fig 16-4; p237. "Mountainous Shoreline of Sputnik Planum (PIA20198)". NASA: Image in the Public Domain.
Fig 17-1; p242. "Mars Gullies". Malin Space Science Systems, MGS, JPL, NASA: Image in the Public Domain.

Fig 17-2; p244. "Mars Ice Age Orbit Variations". [The changing of Mars' axial tilt (obliquity) and its orbital eccentricity over the past 10 million years, along with the changing of the amount of sunlight (insolation) reaching the Martian surface]. NASA/Ames/J. Laskar: Image in the Public Domain.

Fig 17-3; p245. "Mars with Ocean - Looking at Hellas Basin" NASA/Greg Shirah: Image in the Public Domain.

Fig 17-4; p247. "Mars Curiosity Rover-Methane Source". ["This illustration portrays possible ways that methane might be added to Mars' atmosphere (sources) and removed from the atmosphere (sinks). NASA's Curiosity Mars rover has detected fluctuations in methane concentration in the atmosphere, implying both types of activity occur in the modern environment of Mars"]. NASA/JPL-Caltech: Image in the Public Domain.

Fig 17-5; p250. "Hydrogen Isotopes". Bruce Blaus; under Creative Commons Attribution 3.0 Unported license.
Fig 17-6; p253. "Chemical Alteration by Water, Jezero Crater Delta". NASA/JPL-Caltech/MSSS/JHU-APL: Image in the Public Domain.

Fig 17-7; p255. "Artist concept of MAVEN spacecraft". NASA's Goddard Space Flight Center: Image in the Public Domain.

Fig 18-1; p258. "The Earth seen from Apollo 17". NASA: Image in the Public Domain.

Fig 18-2; p259. "Chaos Monster and Sun God". ["Black and white crop of full engraving plate scan - from Plate 5 of the work "A second series of the monuments of Nineveh: including bas-reliefs from the Palace of Sennacherib and bronzes from the ruins of Nimrud ; from drawings made on the spot, during a second expedition to Assyria" (WH Layard)"] Editor Austen Henry Layard, drawing by L. Gruner: Image in the Public Domain.

Fig 18-3; p262. "Earth Crust Cutaway". Jeremy Kemp at en: Wikipedia: Image in the Public Domain.

Fig 18-4; p263. "The Galilean Satellites Europa, Ganymede, and Callisto". NASA/JPL/DLR: Image in the Public Domain.

Fig 18-5; p265. "Vesta from Dawn, 17 July 2011". NASA/JPL-Caltech/UCLA/MPS/DLR/IDA: Image in the Public Domain.

Fig 18-6; p270. "Cutaway view of asteroid 1". Ceres. NASA, ESA, and A. Feild (STScI): Image in the Public Domain. Further digital adaptation by Andy Lloyd.

Fig 19-1; p274. "LCROSS separated". NASA: Image in the Public Domain.

Fig 19-2; p276. "Moon Craters (Including Daedalus)". NASA: Image in the Public Domain.

Fig 19-3; p279. "The Gravity Field at the Mare Serenitatis (Sea of Tranquility)." ["The upper segment represents the topography - a fairly flat low region - and the lower segment shows the corresponding strong gravity field. This is a mascon."]. NASA: Image in the Public Domain.

Fig 19-4; p282. "The Inner Solar System, from the Sun to Jupiter". Mdf at English Wikipedia: Image in the Public Domain.

Fig 19-5; p284. "Moon Rock (Natural History Museum, Vienna)". Photograph by Andy Lloyd.

Fig 20-1; p268. "Comet Hartley 2". JPL/NASA: Image in the Public Domain.

Fig 20-2; p289. "Comet 67P" ESA/Rosetta/NAVCAM; under the Creative Commons Attribution-ShareAlike 3.0 IGO License. http://www.esa.int/var/esa/storage/images/esa_multimedia/images/2014/09/comet_on_19_september_2014_navcam/14832396-1-eng-GB/Comet_on_19_September_2014_NavCam.jpg

Fig 20-3; p292. "Champollion Comet Lander" ["(1997) --- A massive, dirty snowball the size of a mountain plows through the Solar System, venting water vapor as it begins its inbound loop around the Sun. The proposed Champollion spacecraft lander anchors to the irregular surface and collects both surface and core samples."]. NASA/JPL/Pat Rawlings: Image in the Public Domain.

Fig 20-4; p294. "Comet 67P on 24 September 2014 NavCam mosaic". ESA/Rosetta/NAVCAM; under the Creative Commons Attribution-ShareAlike 3.0 IGO Licence. http://www.esa.int/spaceimages/Images/2014/09/Comet_on_24_September_NavCam

Fig 20-5; p299. "Water Molecule "Hops" on Ceres". ["This graphic shows a theoretical path of a water molecule on Ceres. Some water molecules fall into cold, dark craters at high latitudes called "cold traps," where very little of the ice turns into vapor, even over the course of a billion years. Other water molecules that do not land in cold traps are lost to space as they hop around the dwarf planet"]. NASA/JPL-Caltech/UCLA/MPS/DLR/IDA: Image in the Public Domain.

Fig 21-1; p303. "Zecharia Sitchin" Lapavaestacaliente: Image in the Public Domain. https://commons.wikimedia.org/wiki/File:ZECHARIA_SITCHIN.jpg

Fig 21-2; p306. "Mesopotamian cylinder seal impression". ["1st Millennium seal showing a worshipper and a fish-garbed sage before a stylized tree with a crescent moon & winged disk set above it. Behind this group is another plant-form with a radiant star and the Star-Cluster (Pleiades cluster) above. In the background is the dragon of Marduk with Marduk's spear and Nabu's standard upon its back"]. Image in the Public Domain. https://commons.wikimedia.org/wiki/File:Mesopotamian_cylinder_seal_impression.jpg

Fig 21-3; p309. "Red Planet Flare". WikiImages: under CC0 Creative Commons license, via Pixabay. Further digital adaptation by Andy Lloyd.

Fig 21-4; p310. "Ashur". Image in the Public Domain. https://commons.wikimedia.org/wiki/File:Ashur_god.jpg

Fig 21-5; p314. "The Pannemaker" by Gustave Doré (1832-1883). Image in the Public Domain. https://commons.wikimedia.org/wiki/File:Gustave_Dor%C3%A9_-_The_Holy_Bible_-_Plate_I,_The_Deluge.jpg

Fig 21-6; p318. "Clovis Comet". Digital montage by Andy Lloyd, using Corel Paint Shop Pro Photo x2.

Fig 22-1; p322. "Veil Nebula - NGC6960". Ken Crawford: under the Creative Commons Attribution-Share Alike 3.0 Unported licence. Further digital adaptation by Andy Lloyd.

Fig 22-2; p323. "Binary Star". Digital montage by Andy Lloyd, using Corel Paint Shop Pro Photo x2.

Fig 22-3; p324. "Winged Planet". Digital montage by Andy Lloyd, using Corel Paint Shop Pro Photo x2.

Fig 22-4; p327. "Nemesis". Digital montage by Andy Lloyd, using Corel Paint Shop Pro Photo x2.

Back Cover: "Andy Lloyd". Photograph by Fiona Lloyd.

Timeless Voyager Press

Index

 Timeless Voyager Press

Timeless Voyager Press

Timeless Voyager Press

Timeless Voyager Press

Made in the USA
Coppell, TX
11 August 2020

32926023R00218